Musical Spaces

Musical Spaces

Place, Performance, and Power

edited by

James Williams

Samuel Horlor

Jenny Stanford Publishing

Published by

Jenny Stanford Publishing Pte. Ltd.
Level 34, Centennial Tower
3 Temasek Avenue
Singapore 039190

Email: editorial@jennystanford.com
Web: www.jennystanford.com

British Library Cataloguing-in-Publication Data
A catalogue record for this book is available from the British Library.

Musical Spaces: Place, Performance, and Power

Cover image courtesy: Polly Barnes

ISBN 978-981-4877-85-5 (Hardcover)
ISBN 978-1-003-18041-8 (eBook)

Contents

Preface

This volume's history extends back to a one-day conference held at Durham University, UK, in January 2017 with the title 'Geography, Music, Space'. Co-organized by Samuel Horlor (Music), with Alice Cree and Sarah M. Hughes (both Geography), the event's call for contributions attracted a small deluge of responses from scholars across the globe working in disciplines well beyond our own areas of experience. We soon realized that the intersection of music and space was a topic of far wider and deeper interest than we had originally imagined.

The event saw rich dialogue between scholars of music, geography, and various other disciplines, prompting the *Journal of Music, Health, and Wellbeing* (JMHW) (previously *Musicology Research*) to collate two issues of articles under the original conference title. These issues, which remain openly accessible, are the foundations for the chapters in this volume, precursors to many of the interweaving narratives on geography, music, and space found within the book. The revised and updated studies presented here deepen engagements with musical spaces, particularly as connected with issues of place, performance, and power.

Enormous thanks are due to Alice and Sarah for their inspiration and energy in creating and realizing the original event and to the Institute of Musical Research and Durham University for funding it. We also owe much gratitude to the contributing authors of *JMHW* for their original manuscripts and to those who developed their work further for this book. Many thanks also go to the visual artist Polly Barnes for her cartographic line artwork upon which the book's cover is based. She explains:

> The cover art for this volume evolved as part of an exploration of my social landscape and sense of self in relation to others. The locus of each contour and branch represents a creative, professional, or educational setting through which I have developed connections and relationships with others throughout my life.

We would like to extend special recognition to the team at Jenny Stanford Publishing, and especially to Jenny Rompas and Shivani Sharma, for the expertise and unfailing patience in guiding this project to publication.

James Williams
Samuel Horlor
Summer 2021

Notes on Contributors

James Edward Armstrong is an associate lecturer, research assistant, and year leader in Music Composition & Technology at the University for the Creative Arts, Farnham, UK. James' research combines music performance studies and environmental psychology to better understand person–environment transactions in musical settings. Additional research interests include music subcultures, student development in higher education music subjects, ambient and experimental music, and developing music and sound techniques for theatre in the dark application.

Diego Astorga de Ita earned a BSc in Environmental Science and an MSc in Biology from the Institute for Ecosystems and Sustainability Research (IIES) at the National Autonomous University of Mexico (UNAM). He is currently in the concluding stages of his PhD at the Department of Geography, Durham University, UK. His research looks at *son Jarocho* — the traditional music of the region of Sotavento in South-eastern Mexico — at how it connects to Sotaventine landscapes, and at the (de)construction of culture and nature through folk music in these spaces. His research interests include the geographies of music, the ontologies of culture/nature, and questions regarding the phenomenology and poetics of space.

Benjamin Davis studied English and related literature at the University of York, UK, and Theatre Directing at Goldsmiths' College, London, before becoming a staff director at Welsh National Opera from 2001 to 2011. He has since worked as a freelance opera director for national opera companies and festivals in the UK, the Netherlands, Denmark, France, Germany, Austria, Spain, Switzerland, Italy, Canada, North America, and China. His doctoral thesis was a practice-based study on performing realism in contemporary opera productions, sponsored by the Centre for Interdisciplinary Research in Opera and Drama (CIROD), Cardiff University, UK.

Oscar Galeev is a graduate of Leiden University, the Netherlands, and Peking University, China. He worked at Leiden Institute for Area Studies, and he is currently a Yenching Scholar and a researcher at

Peking University. His area studies research is focused on the politics and culture of the modern Middle East and Asia, in particular on the cultural consequences of urbanization and its impact on political participation in the Global South.

Jelena Gligorijević is a popular music scholar with a strong international academic record (spanning Serbia, Britain, and Finland), and with multiple research interests (from music festivals to queer karaoke). Her primary area of expertise is in issues of identity, power, and politics in Balkan popular music across the former Yugoslav region. This also applies to her doctoral dissertation *Contemporary Music Festivals as Micronational Spaces: Articulations of National Identity in Serbia's Exit and Guča Trumpet Festivals* (2019), which was on national identity in post-Milošević Serbia using the country's two major music festivals as case studies. Jelena is currently affiliated to University of Fribourg, Switzerland, where she is doing smaller-scale ethnographic research on hipster cultural practices surrounding St. Gallen's alternative music venue 'Palace'.

Shabnam Goli holds a PhD in ethnomusicology from the University of Florida, USA. She earned a Bachelor of Arts in English language and literature from Khayyam University of Mashhad, Iran, in 2008 and a Master of Music in Ethnomusicology from the University of Florida in 2015. Her interdisciplinary research draws on popular culture and music studies, sociology, diaspora studies, and philosophy to investigate the multifaceted role of music in expression of identities and in shaping communities. Her doctoral dissertation focuses on Persian popular music scenes in California to illuminate processes of identity negotiation and ethnic boundary making among Iranian migrants in the United States. Shabnam is also active as a drummer in Northern California.

Jonathon Grasse is a professor of music at California State University, Dominguez Hills, USA, and his research focuses on regional identity in Minas Gerais, Brazil. His articles and reviews have appeared in various journals, and in 2020, Bloomsbury Academic published his study of the 1972 LP *Clube da Esquina* (Corner Club) by Milton Nascimento, Lô Borges, and friends. His current book, *Hearing Brazil: Music and Histories in Minas Gerais*, is forthcoming from University Press of Mississippi, 2021.

Thomas Graves holds a bachelor's degree in Popular Music from the University of Kent, UK, and a master's degree in Ethnomusicology from SOAS, University of London. He is currently a PhD candidate at Durham University, UK, studying musical emotion and *qawwali.* His primary research interests lie in musical emotion, particularly in relation to lyrics, social context, and religion. His approach is multidisciplinary, with the aim of using ethnographic knowledge to inform psychological research in musical emotion, as well as adapting quantitative methods from music psychology to the ethnographic field. His area interests lie in South Asia, particularly *qawwali.* He is also interested in the expression of political ideology in music, English folk music, and applied ethnomusicology.

Samuel Horlor is a postdoctoral fellow at the Center for Ethnomusicology, Yunnan University, China. He specialises in research on street performance, Chinese pop, and music in urban life. Samuel is the author of *Chinese Street Music: Complicating Musical Community* (Cambridge University Press, 2021) and articles in journals including *Ethnomusicology Forum* and *Asian Music.* He has taught at Durham University, UK, and been an early career fellow at the Institute of Musical Research.

Artemis Ignatidou is an early career cultural historian, specializing in the cultural afterlife of Western art music since the nineteenth century. She holds a PhD in Modern European History (2018), and she is also a musician and performer. Her freelance research portfolio includes reports on human rights and anti-discrimination, musicology, art history, and political history.

Alex Jeffery's main writing and research interests are around popular music and narrativity, having conducted doctoral work on Gorillaz' third album and transmedia project *Plastic Beach* at City University, London. His first book, a series of essays on Donna Summer's Cinderella-themed concept album *Once Upon A Time,* was published in 2021 via Bloomsbury, and is followed by a major monograph, *Popular Music and Narrativity,* for the same publishers in 2022. He is currently based in Berlin.

Daithí Kearney — ethnomusicologist, geographer, and performer — is a lecturer in Music at Dundalk Institute of Technology, Ireland, and co-director of the Creative Arts Research Centre. His research is primarily focused on Irish traditional music but extends to include performance studies, community music, and music education. Daithí performs regularly as a musician, singer, and dancer. His albums include *Midleton Rare* (2012) with John Cronin, which is related to a wider research project on the music and musicians of the Sliabh Luachra region. Most recently he released an album of new compositions with collaborator Adèle Commins entitled *A Louth Lilt* (2017). A former chair of ICTM Ireland, Kearney was awarded the DkIT President's Prize for early career researcher 2015. His publications include contributions to the *Companion to Irish Traditional Music* (ed. Vallely, 2012), the *Encyclopaedia of Music in Ireland* (ed. White and Boydell, 2013), and *New Crops, Old Fields* (ed. Caldwell and Byers, 2016).

Alican Koc is a doctoral candidate at the Department of Art History & Communication Studies, McGill University, Canada. His doctoral research uses the appropriation of the term 'woke' following 2013's Black Lives Matter protests as a departure point into the aestheticization of critical consciousness in North American popular media. His broader interests largely pertain to aesthetics, affect theory, new media, and popular culture. Alican holds bachelor's and master's degrees in Socio-Cultural Anthropology from the University of Toronto, Canada.

David Leahy — referred to as the 'dancing bass meister' by Barre Philipps — has dedicated the last twenty years to navigating the area between music and dance. Originally from New Zealand, David works as an improvising double bass player, contact-improvising dancer, composer, educator, and researcher. He is a prominent member of the London free improvised music scene and is a long-standing member of the London Improvisers Orchestra as well as various other improvisation ensembles across Europe. David's PhD, *Musicians in Space* (2020), explores what happens when the performance of free improvisation expands outward into space, so as to remove the physical separation between the performer and the audience. David works as both a lecturer and dance musician in the dance faculty of Trinity Laban Conservatoire of Music and Dance, London. For further information, please go to www.dafmusic.com.

Anne Macgregor studied piano at the Royal Conservatoire of Scotland and worked as a freelance performer before undertaking her doctoral studies at the University of Nottingham, UK. It was through her specialism in vocal accompaniment that she first encountered

Ture Rangström's songs, and her research of them as expressions of identity was the first English-language scholarship to deal with his music. Anne is a teaching associate in Music Performance at the University of Sheffield, UK.

Joanne Mills is currently a PhD candidate at the University of Wolverhampton, UK. Her doctoral research unites two so far discrete areas within minimalism: 'music' and the 'visual arts' — to contribute new knowledge on immersive forms of practice, and in doing so, the research identifies composers La Monte Young and Terry Riley as pioneers within this contemporary, inter-disciplinary art form. Previously, she studied Media, Photography, and Fine Arts at the Universities of Sunderland and Wolverhampton, UK. An artist and researcher, Joanne's practice and research focuses on the liminal space between the real and the unreal and considers the relationship between audience, artwork, and environment.

Joseph Nelson is a PhD candidate in Musicology, with a doctoral minor in Comparative Studies in Discourse and Society, at the University of Minnesota, USA. His research focuses on seventeenth-century madsongs and the politics of noise, with special attention paid to Bethlem Hospital. He has presented his work at national and international conferences, including that of the American Musicological Society. He has also had work accepted by the Renaissance Society of America, Early Modern Soundscapes, and the Graduate Interdisciplinary Conference at the Arthur F. Kinney Center for Renaissance Studies.

Kiara Wickremasinghe is a PhD candidate in Anthropology at SOAS, University of London. She holds a Bloomsbury PhD studentship and is conducting ethnographic research as part of a wider UKRI ESRC funded study on mental health care in the United Kingdom. This study follows an NHS randomized controlled trial named 'Open Dialogue', a person-centred and social network approach to psychiatric crisis care. Kiara also holds a bachelor's degree in Geography from the University of Cambridge, UK, and a master's degree in Music in Development from SOAS, University of London. Her previous research has examined musical responses to disaster following the 2004 Tsunami, and the impact of collaborative music workshops for mental health service users undergoing recovery from psychosis.

Stephen Wilford is a music scholar based at the University of Cambridge, UK. His work focuses upon North African musics, particularly those of Algeria, and spans a range of traditional and contemporary styles. He is particularly interested in the role of technologies in the production and circulation of music among North African composers, performers, and listeners, and the connections between past and present musical practices. He is currently part of the ERC-funded project 'Past and Present Musical Encounters across the Strait of Gibraltar'. He has previously taught at City, University of London (where he completed his AHRC-funded PhD), Goldsmiths, University of London, and the University of Southampton, UK. He is a member of both the national committee of the British Forum for Ethnomusicology and the ethnomusicology-ethnochoreology committee of the Royal Anthropological Institute of Great Britain and Ireland, and he was an early career research fellow of the Institute of Musical Research (2016–2017). He is a fellow of Wolfson College, Cambridge, UK.

James Williams is an ethnomusicologist and senior lecturer at the University of Derby, UK, where he also runs the Integrated Master of Arts and Health Practice programme. He previously lectured in music composition at the University of Hertfordshire, UK (2012–2015), while completing his PhD at the University of Wolverhampton, UK (2016), on the collaborative and creative interactions between professional musicians. James' research focuses on anthropologies of music-making, investigating behavioural, socio-cultural, and creative processes in a range of contexts and subject areas, notably wellbeing and education.

Sunmin Yoon is an ethnomusicologist whose research focuses on Mongolian folk-song genres in relation to ecology and environment, place, rural/urban dialogues, and the interface of ideology and cultural heritage. Her work has appeared in journals such as *Asian Music*, *MUSICultures*, *Mongolian Studies*, and *Smithsonian Folkways Magazine*. She holds a PhD from the University of Maryland at College Park, USA, and currently teaches courses on Music and Asian Studies at the University of Delaware, USA.

Introduction: Musical Spaces

Samuel Horlor
Center for Ethnomusicology, Yunnan University,
No. 2 Cuihu Road, Kunming, Yunnan Province,
P. R. China 650091
samhorlor@protonmail.com

Introduction

The phrase 'musical spaces' encapsulates the two primary levels of interest shared by this volume's chapters as they examine geographical factors in musical practices and ideas. On the one level are the place-based identities that are meaningful in musical activity, those involving terrains and imaginaries from neighbourhoods to nations and expanding to the various kinds of translocality of the modern world. The often in-flux and contested natures of these identities make power relationships a key ingredient of meanings afforded at this scale of musical spaces. On the other level are musical environments in a microscopic sense — the stages, rooms, and outdoor venues where interactions of physical bodies and material contribute to the activity and representation of music. How people experience sites for music-making is part of the individual meanings they make, and their positioning and mobility in the immediate physical situation can afford different qualities of collective engagement.

Significantly, these two levels intersect; relationships forged in intimate musical environments feed and are fed by how people make sense of the broader territories of their lives. As Thomas Solomon puts it, 'musical performance is a social practice for place-making' (2000: 258). Imagining a local punk scene as an example, an ethos of belligerent political resistance might develop not only from its participants' frustrations about life in their town in general, but also from the energy of people sharing proximity and negotiating tight physical situations together at gigs (see, for instance, Tatro, 2014). With research from contemporary and historical musical contexts around the globe, this volume considers musical spaces with a close eye on place-based identities and power relations, intimate-scale performance practices, and intersections between these two levels.

Place

If the idea of musical spaces is to encompass place-based identities, it is important to clarify theoretical nuances that the terms 'place' and 'space' conjure, and how the ideas may overlap and diverge. Whiteley, Bennett, and Hawkins' edited volume *Music, Space and Place* (2005) is just one example of the notions being paired prominently in the framing of music scholarship on the topic, and the relationship between the two notions is debated extensively across cultural disciplines. Anthropologist Eric Hirsch summarizes a widespread idea that place is about 'a specific (subject) vantage point', while space has been 'divorced as much as possible from a subject-position' (1995: 8). In clear contrast, the ethnomusicologist Louise Wrazen describes place as 'fundamentally concrete', while space is 'socially constructed in practice' (2007: 186). Other approaches question the foundations of these distinctions, suggesting instead that the two ideas 'require each other for definition' (Tuan, 1977: 6). Indeed, there is something compelling about a sceptical conclusion advanced by the cultural scholar Paul Kendall: their meanings have 'converged in a way that it has become doubtful whether these terms can still be coherently used alongside each other' (2019: 33). Kendall justifies this stance with reference to Henri Lefebvre's ([1974] 1991) division of 'social space' into three elements whose scope may be broad enough to cover, among other things, the idea of 'place': *perceived* space, people's sensory perception of their environment, *conceived* space, the representation of the environment such as on maps, and *lived* space, the symbolic associations that make spaces meaningful in everyday life (Kendall, 2019: 31–32).

But the notion of place is not entirely redundant, not least because it is useful in distinguishing between social spaces on different *scales*. On one scale, countries, regions, cities, and tokens of similar geographical categories have enduring identities, existing as ideas on maps, as pertinent to people's sense of self and others, and as reference points in experience. The boundaries and characteristic qualities of a city as a space, however, might not primarily be *perceived* through the senses but may instead emerge predominantly in the *conceived* realm of mapped boundaries. In, for instance, the Pearl River Delta conurbation in southern China, precisely where the city of Guangzhou gives way to its adjoining neighbour city Foshan is less up for negotiation in everyday perception than it is fixed in formal representations and in the imaginations of people for whom the distinction matters. And indeed these cities may very well matter as discrete musical spaces; Foshan, for example, has claims to a key role in the origins of the region's famous dramatic genre, Cantonese opera, and this colours contemporary practice (Ng, 2015: 17–18), while the legacy of a distinct set of claims on the genre's subsequent development is at play next door in Guangzhou (Yung, 1989: 9).

On a more intimate scale, on the other hand, when people produce a social space through temporarily establishing sets of understandings and behavioural expectations applicable to a strip of activity, the boundaries within which these expectations prevail are often set with more emphasis on shared *perceptions* of the situation's materiality. The auditorium within a larger concert hall building, for example, takes on heightened qualities as a musical space when audience members enter its confines from outer rooms ahead of a concert, and musicians getting ready to play cues its doors being sealed. It seems important, then, to acknowledge disparities in *how* spaces are constructed at different scales; the production of the city and the concert hall each emphasizes different modes of spatial experience, and 'place' remains particularly useful for talking about the larger scale — the city of Guangzhou, for instance. Andrew Eisenberg suggests that 'place might be described as another modality of space' (2015: 198), and this model helps appreciate the conceptual nesting of 'place' within 'musical spaces' in chapters of the present volume.

One way of approaching a place as a kind of musical space is to consider how it is represented in musical texts, and what this reveals and fosters about how people think of the wider environments of their lives (Guy, 2009), or how people 'constitute their landscapes and take themselves to be connected to them' (Basso, cited in Wrazen, 2007: 185–186). The representation of landscape in musical texts and surrounding rhetoric has quite often been a theme of work on Western classical music from both geographers (Knight, 2006;

Revill, 2012) and historical musicologists (Grimley, 2006a, 2006b; Hicks, Uy, and Venter, 2006). But another avenue, one concerned more directly with musical activity as well as with texts, is to examine soundscapes of localness, how places are lived and made meaningful through the characteristic sounds (whether considered 'musical' or not) that are experienced walking their streets or country tracks, and the emphasis here has tended to fall more on popular musics (Brandellero and Pfeffer, 2015; Cohen, 1991, 2007; Hudson, 2006). I return to the connections between these ideas and the intimate scale of musical spaces when I talk below about 'performance'. But first it is important to widen the lens by considering the power relationships that are inextricable from musical processes of place-making — the histories of encounter between groups of people, the contestations, and the asymmetries that entwine music and place.

Power

In the substantial body of work on music emerging from the discipline of geography in the last two or three decades, a key reference point has been non-representational theory (Thrift, 2008). The overarching idea is to promote a focus upon process over product and upon how non-human material things are implicated in the forms and effects of musical practice. Arun Saldanha sums up the theory:

> ... a catch-all term for a heterodox range of approaches emphasizing material process and practice over semiotics and cognition, it now seems to stand for any (British) geographical work influenced by actor-network theory and theorists writing broadly in the wake of Deleuze.
>
> (2005: 708)

Non-representational theory's strength, for another geographer Nichola Wood, lies in how 'explor[ing] practices at the moment of their doing' can lead to 'an awareness of the social, cultural, political and emotional geographies through and in which social practices are negotiated and performed' (2012: 201). But there are also criticisms that prioritizing material process risks underemphasizing the effects of the power structures behind practice and that it offers fewer ways of linking bodily experience to matters of identity born of culture, history, and social values (Revill, 2004).

Indeed, in his work on the rave scene of Goa in India, Saldanha demonstrates that even the acts of dancing, archetypal celebrations of the immediacy of embodied practice, are shaped of specific power relations that play out on intellectual levels (2005: 716). He asks about the early morning phase of a rave that has seen through the night: 'What are the material conditions for race to become a differentiating factor when the sun starts shining on dancing bodies?' (*ibid.*: 710), noting that domestic and foreign tourists tend to end up spatially differentiated in their dancing and other activities. The two categories of people respond differently to the bright sunlight in the morning hours late on in an event, with Indian tourists usually seeking dark corners or returning to bed and white visitors continuing to dance in conspicuous positions. The tanning of skin is a marker of prestige and a display of status in the rave scene, earned through longevity of participation. But being a status primarily available to those with naturally lighter skin, it is an indication that the different physical and cultural factors behind how bodies relate to environmental conditions are implicated in musical practices when they are understood as sites of power negotiation. As such, musical spaces are fertile ground for interrogation of key contemporary issues related, for instance, to gender (Impey, 2013; Magowan and Wrazen, 2013; Spracklen, 2020), race (Bodenheimer, 2015; Stokes, 1994; Tsitsos, 2018), mobilities and circulation (Milburn, 2017; Revill, 2010), and political resistance (Hancock-Barnett, 2012; Hughes, 2016; Kong, 1995; Leyshon, Matless, and Revill, 1998).

Performance

Often conceptualized as coming prior to these formulations about power is a level of experience stemming from sensory engagements in musical encounters. Scholars have highlighted, for instance, various kinds of 'affective ties with the material environment' that arise in performance (Tuan, cited in Pace, 2017: 115) and that contribute to the symbolic meanings of Lefebvre's *lived* space (Inwood and Alderman, 2018; Leyshon *et al.*, 2016; Stokes, 2017; Watson and Ward, 2013). It is essential here for the understanding of 'performance' to be a broad one, encompassing not only what Thomas Turino (2008) has characterized as typical 'presentational' performance, but also forms with clearer 'participatory' qualities, orientated more towards collective involvement than passive contemplation. This opens up consideration of musical experiences such as those in protest chanting, group improvisation, leisure ensemble playing, and many other contexts. How do the perceived qualities of these musical spaces, their acoustical characteristics, and the arrangement of bodies afforded by features of the built environment enable sociality that intersects with wider issues in participants' lives?

Indeed, intimate musical spaces are often self-consciously shaped and manipulated, with the awareness that how they are *perceived* can be harnessed in pursuit of experiential, social, and political effects. Considering musical spaces on the level of 'performance', then, is also about exploring how creative practitioners make decisions in the rationalizing and designing of spaces for their activity, including those emphasizing new technologies and virtual environments (Milburn, 2017). Impacts are felt especially in practices that blur boundaries between 'life and art' (Kruse, 2019), those aimed at fostering creativity in service of educational and economic benefits (Burnard, 2007; Morreale *et al.*, 2014), and those linked to community-building projects (Baker, Bennett, and Homan, 2009).

This Volume

Chapters in this volume integrate insights into different scales of musical spaces, combining a concern for the structural with 'field research built on "real people" and the truths of their mutual lives' (Magowan and Wrazen, 2013: 5). Authors' varied disciplinary backgrounds inspire a range of methods. Research from musicology, popular music, and cultural studies featuring, for instance, textual analysis is complemented by ethnographic methods from ethnomusicology and geography, while authors channelling performance studies approaches and practice-based research make and analyze interventions in their roles as creative practitioners. The volume is formed of two parts, with the starting points of Part I's chapters found in larger scale issues of place and power, and those in Part II building from a look at the intimate scale, at musical spaces in performance. Chapters in each section are grouped into themes as outlined here.

Part I: (Trans)local Musical Spaces

Within this first theme of Part I, authors discuss music's various roles in negotiating and channelling local and regional identities. Jonathon Grasse explores musical spaces through the idea of 'deep regionalism', which draws together articulations of localness found in physical, social, and emotional dimensions of music in Minas Gerais in Brazil. Employing three case studies — about the *Congado* processional, the *viola caipira* Brazilian guitar, and lyrics from songs by the Corner Club popular music collective — Grasse argues that regional identities show 'formative spatialities' at the intersection of these different levels of experience. Alican Koc's chapter sets out to interrogate a rhetoric promoting the idea of North American DIY punk as formed of culturally and aesthetically distinct regional scenes. From his position within that of Oklahoma City (USA), Koc locates a sense of the scene's coherence in an affective power found in expressions of utopianism and violent escape from small-town conservatism.

The following two chapters in this section pay particular attention to translocal identities based on diasporic awareness and affinities for border-crossing forms of cultural expression. Stephen Wilford looks at music in the lives of Algerian people in the UK capital, with the notion of 'Algerian-London' made meaningful through various musical engagements. The Algerian population in this city is geographically dispersed rather than concentrated in any particular area, and thus any 'public space for the community' is apparently absent. But the pathways taken by musicians across the city and by genres in virtual transnational space are central to a place-based identity that is in constant negotiation with reference to Algerian populations in North Africa and France. Shabnam Goli's account of Iranian hip-hop takes Hamid Sefat's Persian cover of Drake's song 'Hotline Bling' as the starting point to consider the production of local place through an orientation to the transnational hip-hop world. Arguing against global/local dichotomies, Goli refocuses on the mutual play of outward-looking and place-based lyrical references, uses of language and instruments, and experiences among musicians.

Regionality in Learning and Heritage

This section looks at regional identities as negotiated in education and heritage contexts. Daithí Kearney begins with a discussion of learning and research in a higher education institution and their roles in regional identities of Oriel on the east coast of Ireland. Drawing on experiences as lecturer and ensemble director, Kearney tackles an apparent overemphasis on musical style in conceptualizations and canonizations of localness in Ireland's traditional music, countering with evidence found in various forms of wider practice. He understands regionality here as *process* found in the 'networks, infrastructure, and ecosystems through which musical culture is shaped, supported, and nurtured'. In the context of an education institution for Tibetan pupils in India, James Williams contributes discussion of the opportunities for *growth* that refugee children are afforded through music lessons and instrument pedagogy in the school. He questions how far opportunities for creativity and broader cultural encounter are compatible with an institutionalized focus on heritage preservation. Kiara Wickremasinghe then adds a postcolonial theoretical angle to considering musical spaces as sites for the promotion of 'indigenous knowledge' and anti-colonial narratives, in the experiences of Inuit and Sámi and in identities of the Arctic Region. The *joik* vocal style, in particular, is linked to manifestations of cultural sovereignty not only in musical texts, but also of processes of listening and action.

Music and Spatial Imaginaries

The final theme of Part I comprises four chapters in which felt connections and non-physical co-presences are at the heart of the *lived* symbolic meanings of musical spaces. The first two concern national imaginaries at play in a broadly conceived European art music context. Anne Macgregor's subject is the early twentieth-century Swedish composer Ture Rangström; she examines his music as connected to the granite of the lived Swedish landscape and to the intersections of regional and national identities. Musical manifestation of landscape here is not simply a matter of transposing the visual to the audible, but instead it is to add to the *experience* of landscape in a Nordic imagination. Artemis Ignatidou considers how ideas connected to Greek history are significant in the biography and career of Russian composer Dmitri Shostakovich. In the Soviet era where composers faced pressure to produce 'ideologically correct' works, a sensationalized notion of a communist 'Greek revolutionary spirit' provided Shostakovich with a sense of place that he could employ at the service of more domestic concerns, managing his relationship with Soviet cultural authorities to alleviate personal financial problems.

For Diego Astorga de Ita, it is encounter between humans and enchanted spirit worlds that is enabled in, and that produces, fandango occasions in the Sotavento region of Mexico. The genre of *son Jarocho* is associated with the production of a world beyond the everyday, with the transformation of 'abstract, Cartesian space into a phenomenological, lived place'. Musical spaces here are formed not only of human interaction with material realities, but also with an otherworldly dimension of nature. Elsewhere, the production of other worlds comes through multimedia engagement with fictional locations. Alex Jeffery's case study is the album and interconnected media *Plastic Beach* by Gorillaz, in which a fictional island is the site for fans to cross ontological boundaries between the 'real' and the 'virtual', as they insert themselves into this world through various creative practices. Among Jeffery's arguments is that fan engagement in virtual spaces is a necessary part of appreciating the reception of contemporary musical products.

Part II: Music-Making Environments

Chapters in Part II of the volume take as their starting point a microscopic focus on details of performances, of musical occasions and their soundings. James Edward Armstrong's key argument is that certain performance spaces carry significances to solo instrumentalists for more simply than their acoustical characteristics, and he proposes a deeper engagement with the 'experiential' qualities of performance rooms. With an eye on approaches from environmental psychology, Armstrong extends laboratory-style experimental research to real-life environments, giving musicians platform to elaborate on the cultural conditions and individual histories that form their subjective involvements in performing. Thomas Graves then digs down into one such element of individual and group experience. In the context of a non-professional folk 'orchestra' in the UK, he examines connections between session environments and joy and happiness. Part of the 'social conditions' underpinning the emergence of these subjective phenomena, he argues, are seating arrangements that enable particular kinds of interpersonal interaction, plus a symbolic level of lived experience that includes the history of the wider venue. Similarly, Sunmin Yoon's study of recent developments in the Mongolian long-song folk genre of *urtyn duu* shows connections between singing techniques and singers' relationships with the immediate spatial characteristics and wider environments of their lives and practices. The 'harmonizing' between members of the group Shurankhai relies upon particular ways singers conceive of and utilize the inner and outer spaces of the body and the environment, respectively, with the voice understood as projecting vertically to link the earth and the sky. This engagement with space feeds into various scales at which ideas of 'home' are constructed, ranging from the specific spot on a mountain ridge to the level of the province or the nation.

Designing Creative Spaces

This section gathers insights from practice-based research on how creative spaces are self-consciously designed and managed to foster certain effects. Opera director Benjamin Davis begins the section with a report on his experiences of staging Handel's eighteenth-century opera *Ariodante*. He sees the process as involving various kinds of mapping, with a score interpreted, a design world realized, and 'musical, physical, and conceptual spaces' navigated as part of the production. A further element is how the physical stage is embedded in a host of virtual territories of marketing, reception, and criticism. Davis thinks of the process as involving 'dialogical cartographies of culture', where these various elements collaborate to make the opera psychologically congruent for the audience. David Leahy's chapter on free improvisation performances in the UK adds to this a detailed focus on the 'shifting topography of the performance space'. Setting up public events whose spatial qualities are intended to develop an 'all-encompassing and non-hierarchical musicking practice', he observes how spatial conditions entwine with participants' collective engagements, in tandem with the sonic qualities of the environment. Leahy takes up the theme of connecting intimate relations in space with wider visions of reality, this time linking the qualities of the performances to 'egalitarian, interconnected, and collaborative values'. For Joanne Mills, too, a goal is the dissolution of barriers to active participation for audiences in contemporary art practices. She concludes the section with a chapter on the fostering of 'expanded narratives' by creations that exploit multi-sensory elements. Building on work from the 1960s by La Monte Young, Terry Riley, and their collaborators, she presents two of her own works, which harness 'active' performance space and interdisciplinary collaboration in generating immersive modes of audience engagement.

Musical Spaces and Power

Each of the three chapters in the volume's final section presents an example of a bounded musical space, highlighting how power relationships are manifest and challenged in politicized contexts. As Jelena Gligorijević puts it in the first chapter, these 'micronational spaces' each inspire particular modes of citizenship. In Gligorijević's case, the musical spaces under consideration are contemporary music festivals, whose connections to branding she interrogates to assess their potential as musical spaces of political belonging and action. The conclusion is that this potential is 'truly democratic', based on the multiple spatial and temporal trajectories of engagement that festivals can inspire. Oscar Galeev's chapter takes the intimate space of Cairo's Tahrir Square and shows how the distinct 'carnivalesque' soundscape that arises there during the January 2011 Revolution contrasts with the normal soundings of politically constrained life before it. He argues that political chants in the square make people aware of spatial relations among them that are unique to the specifics of this historical moment. Joseph Nelson's chapter applies a similar focus to Bethlem Hospital in seventeenth-century London, considering how the sounds within the building contribute to the emergence of a space of 'biopower' and resistance in the face of the sovereign state's attempts to exert control over the physical condition of those inside with mental illness. But this micro-soundscape is also in dialogue with the 'madsongs' that circulate London's streets and taverns as a means for commentators to explore the ruptures of current politics.

References

Baker, S., A. Bennett, and S. Homan. 2009. 'Cultural Precincts, Creative Spaces: Giving the Local a Musical Spin', *Space and Culture* 12(2): 148–165.

Bodenheimer, R. 2015. *Geographies of Cubanidad: Place, Race, and Musical Performance in Contemporary Cuba.* Jackson: University Press of Mississippi.

Boland, P. 2010. 'Sonic Geography, Place and Race in the Formation of Local Identity: Liverpool and Scousers', *Geografiska Annaler: Series B, Human Geography* 92(1): 1–22.

Brandellero, A., and K. Pfeffer. 2015. 'Making a Scene: Exploring the Dimensions of Place Through Dutch Popular Music, 1960–2010', *Environment and Planning A: Economy and Space* 47(7): 1574–1591.

Burnard, P. 2007. 'Reframing Creativity and Technology: Promoting Pedagogic Change in Music Education', *Journal of Music, Technology & Education* 1(1): 37–55.

Cohen, S. 1991. *Rock Culture in Liverpool: Popular Music in the Making.* Oxford: Clarendon Press.

_____. 2007. *Decline, Renewal and the City in Popular Music Culture: Beyond the Beatles.* Aldershot: Ashgate.

Eisenberg, A. 2015. 'Space'. In *Keywords in Sound*, edited by D. Novak, and M. Sakakeeny. London: Duke University Press.

Grimley, D. 2006a. *Grieg: Music, Landscape and Norwegian Identity.* Woodbridge: Boydell Press.

_____. 2006b. 'Music, Landscape, and the Sound of Place', *The Journal of Musicology* 33(1): 11–44.

Guy, N. 2009. 'Flowing Down Taiwan's Tamsui River: Towards an Ecomusicology of the Environmental Imagination', *Ethnomusicology* 53(2): 218–248.

Hancock-Barnett, C. 2012. 'Colonial Resettlement and Cultural Resistance: The Mbira Music of Zimbabwe', *Social & Cultural Geography* 13(1): 11–27.

Hicks, J., M. Uy, and C. Venter. 2006. 'Introduction: Music and Landscape', *The Journal of Musicology* 33(1): 1–10.

Hirsch, E. 1995. 'Introduction: Between Place and Space'. In *The Anthropology of Landscape*, edited by E. Hirsch, and M. O'Hanlon. Oxford: Clarendon Press.

Hudson, R. 2006. 'Regions and Place: Music, Identity and Place', *Progress in Human Geography* 30(5): 626–634.

Hughes, S. 2016. 'Beyond Intentionality: Exploring Creativity and Resistance within a UK Immigration Removal Centre', *Citizenship Studies* 20(3–4): 427–443.

Impey, A. 2013. 'Songs of Mobility and Belonging: Gender, Spatiality and the Local in Southern Africa's Transfrontier Conservation Development', *Interventions: International Journal of Postcolonial Studies* 15(2): 255–271.

Inwood, J., and D. Alderman. 2018. 'When the Archive Sings to You: SNCC and the Atmospheric Politics of Race', *Cultural Geographies* 25(2): 361–368.

Kendall, P. 2019. *The Sounds of Social Space: Branding, Built Environment, and Leisure in Urban China*. Honolulu: University of Hawai'i Press.

Knight, D. 2006. *Landscapes in Music: Space, Place and Time in the World's Great Music*. Lanham: Rowman & Littlefield.

Kong, L. 1995. 'Music and Cultural Politics: Ideology and Resistance in Singapore', *Transactions of the Institute of British Geographers* 20(4): 447–459.

Kruse, R. 2019. 'John Cage and Nonrepresentational Spaces of Music', *Social & Cultural Geography* (published online).

Lefebvre, H. [1974] 1991. *The Production of Space*, translated by D. Nicholson-Smith. Oxford: Blackwell.

Leyshon, A., D. Matless, and G. Revill. 1998. 'Introduction'. In *The Place of Music*, edited by A. Leyshon, D. Matless, and G. Revill. New York: Guilford Press.

Leyshon, A., L. Crewe, S. French, N. Thrift, and P. Webb. 2016. 'Leveraging Affect: Mobilising Enthusiasm and the Co-Production of the Musical Economy'. In *The Production and Consumption of Music in the Digital Age*, edited by B. Hracs, M. Seman, and T. Virani. Abingdon: Routledge.

Magowan, F., and L. Wrazen. 2013. 'Introduction: Musical Intersections, Embodiments, and Emplacements'. In *Performing Gender, Place, and Emotion in Music: Global Perspectives*, edited by F. Magowan, and L. Wrazen. Woodbridge: Boydell & Brewer.

Milburn, K. 2017. 'Rethinking Music Geography Through the Mainstream: A Geographical Analysis of Frank Sinatra, Music and Travel', *Social & Cultural Geography* 20(5): 730–754.

Morreale, F., A. De Angeli, R. Masu, P. Rota, and N. Conci. 2014. 'Collaborative Creativity: The Music Room', *Personal and Ubiquitous Computing* 18(5): 1187–1199.

Morton, F. 2005. 'Performing Ethnography: Irish Traditional Music Sessions and New Methodological Spaces', *Social & Cultural Geography* 6(5): 661–676.

Ng, W. C. 2015. *The Rise of Cantonese Opera*. Urbana: University of Illinois Press.

Pace, A. 2017. 'Refiguring Maltese Heritage Through Musical Performance: Audience Complicity and the Role of Venues in Etnika's Stage Shows'. In *Musicians and their Audiences: Performance, Speech and Mediation*, edited by I. Tsioulakis, and E. Hytönen-Ng, 105–119. London: Routledge.

Revill. G. 2004. 'Performing French Folk Music: Dance, Authenticity and Nonrepresentational Theory', *Cultural Geographies* 11(2): 199–209.

_____. 2010. 'Vernacular Culture and the Place of Folk Music', *Social & Cultural Geography* 6(5): 693–706.

_____. 2012. 'Landscape, Music and the Cartography of Sound'. In *The Routledge Companion to Landscape Studies*, edited by P. Howard, I. Thompson, and E. Waterson. Abingdon: Routledge.

Saldanha, A. 2005. 'Trance and Visibility at Dawn: Racial Dynamics in Goa's Rave Scene', *Social & Cultural Geography* 6(5): 707–721.

Solomon, T. 2000. 'Dueling Landscapes: Singing Places and Identities in Highland Bolivia', *Ethnomusicology* 44(2): 257–280.

Spracklen, K. 2020. *Metal Music and the Re-Imagining of Masculinity, Place, Race and Nation*. Bingley, UK: Emerald Publishing.

Stokes, M. 1994. 'Introduction: Ethnicity, Identity and Music'. In *Ethnicity, Identity and Music: The Musical Construction of Place*, edited by M. Stokes. Oxford: Berg.

_____. 2017. 'Musical Ethnicity: Affective, Material and Vocal Turns', *World of Music — New Series* 6(2): 19–34.

Tatro, K. 2014. 'The Hard Work of Screaming: Physical Exertion and Affective Labor Among Mexico City's Punk Vocalists', *Ethnomusicology* 58(3): 431–453.

Thrift, N. 2008. *Non-Representational Theory: Space, Politics, Affect*. Abingdon: Routledge.

Tsitsos, W. 2018. 'Race, Class, and Place in the Origins of Techno and Rap Music', *Popular Music and Society* 41(3): 270–282.

Tuan, Y-F. 1977. *Space and Place: The Perspective of Experience*. Minneapolis: University of Minnesota Press.

Turino, T. 2008. *Music as Social Life: The Politics of Participation*. Chicago: University of Chicago Press.

Watson, A., and J. Ward. 2013. 'Creating the Right 'Vibe': Emotional Labour and Musical Performance in the Recording Studio', *Environment and Planning A: Economy and Space* 45(12): 2904–2918.

Whiteley, S., A. Bennett, and S. Hawkins, eds. 2005. *Music, Space and Place: Popular Music and Cultural Identity*. Aldershot: Ashgate.

Wood, N. 2012. 'Playing with 'Scottishness': Musical Performance, Non-Representational Thinking and the 'Doings' of National Identity', *Cultural Geographies* 19(2): 195–215.

Wrazen, L. 2007. 'Relocating the Tatras: Place and Music in Górale Identity and Imagination', *Ethnomusicology* 51(2): 185–204.

Yung, B. 1989. *Cantonese Opera: Performance as Creative Process*. Cambridge: Cambridge University Press.

Part I

(Trans)local Musical Spaces

Chapter 1

Musical Spaces and Deep Regionalism in Minas Gerais, Brazil

Jonathon Grasse
Music Department, California State University, Dominguez Hills,
1000 E. Victoria Street, Carson, CA 90747, USA
jgrasse@csudh.edu

1.1 Introduction

In researching music and regional identity in Minas Gerais, Brazil, I engage three interrelated notions of musical space: the geography of cultural territory's physical places; the spaces of social development within and between historical communities; and the spaces of consciousness hosting identities of place as emotional, cognitive experiences enabling what some have theorized as neuro-phenomenological 'enactment' (Schiavio, 2014). This chapter focuses on applying theories of musical spaces while briefly engaging other important aspects of three case studies illustrating the defining role these spaces play in considering what I term 'deep regionalism'. These musical spaces, equated with the places of self

Musical Spaces: Place, Performance, and Power
Edited by James Williams and Samuel Horlor

ISBN 978-981-4877-85-5 (Hardcover), 978-1-003-18041-8 (eBook)
www.jennystanford.com

and social identities, frame the music depicted here with notions of empowerment: individual, social, historical, and political. The three select case studies in Minas Gerais are: the *Congado* popular Catholic processional of contemporary African-descendant communities in the city of Ouro Preto; the ten-string Brazilian guitar (*viola* or *viola caipira*), an instrument/object of regional iconicity; and lyrics extolling pathos of regional place and history in songs from the region's Corner Club popular music collective (Clube da Esquina). Each case study engages the array of musical spaces in unique ways.[1]

Encountered at various places and times throughout the year, the often-termed Black Catholicism of the *Reinado* festival, or *Congado*, is also manifest in Ouro Preto during Epiphany week, when the ensemble Ouro Preto Moçambique Group of Our Lady of the Rosary and Saint Iphigenia parades with invited groups visiting from other towns and cities. They follow a route from the Saint Iphigenia Church in the Alto da Cruz neighbourhood in the direction of the Chico Rei mine. Groups typically include extended families with children and elder participants, reflecting their basis in community and neighbourhood. In 2016, the Epiphany week festival titled 'The Faith that Sings and Dances' attracted *Congado* groups invited from many regions of Minas Gerais (also simply Minas). A person or thing from Minas is still called 'miner' (*mineiro/a*), and the heritage of Baroque-era mining wealth was enabled through the enslavement of black Africans and their descendants. Though the procession's destination is the site of Chico's old gold mine, the gates to the Church of Our Lady of the Conception serve as both a practical and picturesque compromise on the part of organizers; the masses of musicians, dancers, and spectators would overwhelm the narrow street running past the mine's entrance, a stone's throw away.

A regional foundation myth posits Chico Rei (Francisco the King) as having been the very first King of Congo (*Rei do Congo*), a title still bestowed to others as part of an election rite during or in conjunction with many *Reinado* festivals. For this very festival, the

[1]Fieldwork on the Ouro Preto *Congado* festivals occurred in January of 2015, 2016, and 2018 and includes interviews with organizers and participants, and videographic documentation. Fieldwork examining the *viola* includes interviews with luthiers, *viola* players (*violeiros*), and scene events in Belo Horizonte, Nova Lima, and Sabará in 2012, 2015, and 2016. Research into the Corner Club includes interviews with collective members beginning in 2006.

Moçambique community's king is joined by royal representatives of visiting groups as they bestow blessings on procession participants at the gate of the Church of Our Lady of the Conception. In a legend shared by countless *Congado* practitioners, Chico Rei purchased his freedom, and that of his son and other members of his African ethnic group, by saving what gold he could from his mining work. He later purchased the mine, earning the title in recognition of his leadership and having overcome slavery in such grand fashion. Before the dawn of the eighteenth century, the first of hundreds of thousands of slaves were brought to this highland interior of Brazil to work in the gold, diamond, and gem mines for their owners and the Crown of Portugal, and later as forced labour in the coffee fields of southern Minas. 'The Faith that Sings and Dances' is a popular Catholic festival of a compound nature, dedicated to the Virgin Mary and the group's patron Saint Iphigenia of Ethiopia, two figures joined by Chico Rei in the nearly equally passionate devotion they inspire. Several festival events are held at the Chico Rei House of Culture, a community centre dedicated in part to the organization of social activism. Saint Iphigenia Church, the heart of the hosting group's community, and dramatically perched atop one of Ouro Preto's many hills, was constructed by slaves between 1733 and 1785, partially funded by Chico Rei.

The colourful procession, characterized by pounding percussion, graceful call-and-response praise songs, bright costumes and banners, and enthusiastic spectators, winds through steep streets and past landmarks. The pageantry links iconic physical locations symbolizing Minas' primary historical modes of regional identity: gold mining and the spectacular colonial-era architecture now protected by UNESCO. The Alto da Cruz neighbourhood in Ouro Preto's eastern edge, from which both the community and the parade emerge, dates to the village known as Antônio Dias founded at the dawn of the eighteenth century, and bordering Ouro Preto, when it was known as Vila Rica, Portugal's wealthy colonial gem in the New World. It was near Antônio Dias that gold was discovered in 1697 and that slaves and the working poor were forced to live. Today, the parade route almost traverses what were these two worlds: one, of the poverty-stricken, marginalized black workers;

the other, of a picture postcard of a cultivated colonial city reflecting the European elegance of political and social power. Brazilians typically embrace the symbolic picturesque beauty of Ouro Preto's renowned landmarks as a national treasure, leaving behind as a seemingly forgotten legacy the marginalized districts still called home by descendants of forced labour. Here, the Chico Rei legend and the legacy of urban marginalization engage musical spaces of both physical territory and the development of historical communities; celebration of the historical figure's meaningfulness, mingling with the religiosity of public devotion, re-inscribes cultural places still marked in 2018 by difference and confrontation.

The *Moçambique* ensemble's call for unity among visiting groups, and the festival's role in cultural resistance, positions the religious event, in part, as a confrontation to the general marginalization of Afro-Mineiro communities, while reawakening attention to historical legacies and to their spaces of racially segregated social development. The festival, a spectacular display of public devotion to Catholic icons, also contests Brazil's prevailing racial injustices in a theatrical 'desegregation' of the old city's streets and landmarks and thus a marking of cultural territory. This aspect of black pride inherent in *Congado*, a facet not lost on the general Brazilian public, was emphatically shared by that year's *Rei do Congo* in an interview during the festival (Benifácio, 2015 interview in Ouro Preto). At play here too are the notional social spaces of freedom — Chico Rei's freedom and the promise of freedom — forming the figurative grounds for contemporary commemoration. These 'spaces of freedom', augmenting Ouro Preto's physical cultural territory, emanate from social arenas of confrontation etched into history and into notions of collective difference as a set of social places inhabited by communities. Music's formative spatialities created by the *Reinado* tradition in Minas confront, on their own terms, specific histories and legacies of hope and suffering while expressing and celebrating both Catholic religiosity and Chico Rei's heroic role in contemporary *mineiro* black consciousness. Geographic, historical, and cognitive layers of formative spatiality further link these Afro-Mineiro communities to their place in a particular frontier of the African diaspora.

1.2 Musical Space in Consciousness

Ouro Preto's culturally symbolic physical locations, and the historical arenas of the *Reinado's* figurative spatialities of confrontation, are near-empty ideas without the participatory consciousness in which individuals enact identity of place as a cognitive musical experience in real time, mapping in some way to these meanings. 'Musical sense-making is not a passive representation of elements of the musical environment. Rather, it is a process of bringing forth, or enacting, a subject's own domain of meaning' (Schiavio, 2014: 142). Intentionality and purposefulness on the part of *Congado* participants (performers, audience, and community members), as a musical space, form a key element to deep regionalism, an identity of place made real by neuro-phenomenological enactment. Similar neuro-phenomenological processes hold true, as key to meaningfulness, in the following descriptions of the *viola* as an icon of Minas' rural past and of regionalist themes in the music of the Clube da Esquina popular music collective. If we are to say that these examples offer potential for deep regionalism, then we are saying that participants cognitively enact such sentiments, positioning themselves towards identities of place through music. Aesthetic and formal realms attract and encourage such participation through style, sound, and structures.

Music is neither expressive nor symbolic without participants consciously enacting experiential meaningfulness of those expressions and symbols. Enactment leads to processual, musical spaces of consciousness hosting identities of place. Here, I theoretically concretize cognitive experience as being elemental to deep regionalism; through performer and audience enactment of meaningfulness, music enables mindful engagement of both the objective physicality of cultural territory and the discursive arenas of historical social implications. Sense-making brought about in musical encounters is, furthermore, socially conditioned and linked to expectations inherent in broader experiences with the world. Theories of 'embodiment' suggest how affective encounters made beautiful, powerful, and sensual by music augment and facilitate the formation of identities of place enhanced by physical places and historical space. Notions of embodiment locate regionalism within a participant's sense-making of experiential meaningfulness with music. In the twenty-first-century ethnomusicological literature, embodiment acquires interrelated meanings (Ciucci, 2012; Gibson and Dunbar-Hall, 2000; Guy, 2009; Magowan and Wrazen, 2013; Schultz, 2002). I draw from across these sources to form a threefold neuro-phenomenological concept of embodiment, emplacement, and emplotment: in lifting ourselves up into music (enacting), we place ourselves into relationships with geographies of cultural territories and into narratives of music's 'formative spatialities' (after Leyshon, Matless, and Revill, 1995: 424–425) of social development. *Congadeiros*, I suggest, *perform* territoriality and historical space as part of experiencing identity of place.

1.3 *Viola* as Icon of the Rural Past

The metallic timbres of the guitar-like *viola's* five double-course strings have become increasingly iconic of Minas Gerais' rural past, as they also have in other Brazilian regions. Colonists, the clergy, Indians, and slaves in proximity to larger colonial-era settlements were eventually exposed to the instrument, known variously as the 'ten-string' *viola* (*viola de dez cordas*) and the 'country hick' *viola* (*viola caipira*). Portuguese immigrants pouring into Minas Gerais during the eighteenth-century gold rush strengthened its presence there. Telling is the etymology of *caipira*: it is a word in the Tupi language for 'those who cut and clear the trails' (*caa* = forest, *pir* = cut). The original *caipiras* were the hardened labourers controlled by 'flag-bearing' colonizing settlers known as flag bearers (*bandeirantes*), the first Europeans to settle in the south east interior, doing so with the help of enslaved coastal Tupi Indians. Folklorist Luis da Camará Cascudo defined the *caipira* as 'a man or woman of little education not living in a city. A rural worker, of the river banks, of the seashore, or of the sertão' (Travassos, 2006: 128). Reconstructions of pre-modern, rural life remain elemental themes of the 'country music' (*música caipira*) associated with them and their social spaces. This musical space greatly informs *mineiro* regional identity, as it resonates with the regional, physical places of cultural territory and with participatory enactment.

Caipira is also the land itself, strongly related in the Brazilian psyche to the *sertão*, a truncation of *desertão*, a vast uninhabited place. Another aura of the *sertão* stems from the earliest, most

Eurocentric days of the Portuguese empire. It is portrayed as Godless and lawless in its lack of Church and Crown, a region home to cannibalistic pagans without faith or promise. These were seen as inhospitable zones far from Western-style settlements, distanced from civilized ideas, and a place in which the devil did his work. In cities, towns, and *fazendas* (farms), the *pardo* (today *mulatto*) offspring of blacks and whites began forming a majority before the end of the eighteenth century, and parcel farming and sharecropping (*parceira* and *meiação*) allowed the rural poor to live on privately owned farms. These positions are still handed down today through generations of labourers via landowners. This rural life, racially mixed and sprinkled with African-descendant communities, also became associated with the *caboclo* culture of the *caipira*, the mixed-race majorities of the backlands. A strong-arm oligarchy grew as economic opportunities for commoners vanished in a post-mining-boom economic decay. However, the *mineiro sertão* is celebrated too as a vast poetic space by nativist writers such as Guimarães Rosa, whose novel *Grande Sertão: Veredas* remains a landmark of the twentieth-century Brazilian literature. In this *mineiro* heartland, the so-called 'illiterate maestros' (*maestros analfabetos*) (Faria and Calil, 2003) created the repertoire of *caipira* music.

Today a roots instrument cultivated by luthiers and collectors as well as performing musicians, the *viola* comes wrapped in notions of this cultural territory's physical places. The *caipira* experience, within an unforgiving social structure, and as depicted in the tragic themes of sentimental song traditions, is fictively buoyed in musical space by respect for nature, family, love stories, a satisfying taste for independence, and unique strains of folk Catholic religiosity. The *viola* is found in the widely practised folk Catholicism of the *folia de reis* procession and in many *Congado* groups. The secular *viola* famously channels romantic notions of the natural world and the self (Freire, 2008), its cultural emblems drawing from poetics of wilderness, isolation, and, throughout the twentieth century, shifts between poetics of big-city modernity and rural traditions. It is in the struggles and victories of survival that *mineiro* individuality and independence proudly arise as dimensions of *sertão* existence. These themes of *mineiro* identity remain relatively unconnected to the rich, powerful, colonial-era fabric of the European-derived 'Minas Baroque' (*barroco mineiro*), with its legacies of gold wealth, diamonds, and lavish urban architecture as seen in sumptuous eighteenth-century churches in towns such as Ouro Preto.

As industrialized, urban Brazil developed at the dawn of the twentieth century, images of both the *caipira* culture and the *viola* emerged even more strongly as belittling stereotypes of the backwoods/backwards poor, the uneducated rural masses (*povo*) of the interior. Many grew to see the *viola* as an Iberian artefact repackaged as 'backwoods' through the prism of urbanized modernity (Travassos, 2006). The *viola* engages these notions representing this cultural territory, navigating the spaces of historical social confrontation, and symbolizing contestations that define aspects of Brazilian society. Wedded to a fictive countryside curated by twentieth-century social change, the *viola* emerged in new contrasts to burgeoning urban popular music styles, such as the guitar-dominated *choro* and *samba*. The '*viola* song' (*moda de viola*) arose as a generic term for repertoire performed by a vocal duet (*dupla*) singing in thirds while accompanied by *viola*, with an optional guitar. The cities of Rio de Janeiro, São Paulo, and, at the start of the twentieth century, Belo Horizonte, swelled with rural immigrants subscribing to these associations. In Minas, as elsewhere in Brazil, the unofficial 'songbook of the *caipira*' (*cancioneiro caipira*) presents memories and myths of lyrical battles for countryside soul and rural authenticity. This ethos was clearly showcased in an event I attended, Belo Horizonte's first Minas Festival of Viola in 2013, which was organized by the Brazilian Viola Caipira Institute (IBVC) and featured original songs by finalists drawn from all corners of the state. The IBVC overlaps with smaller organizations throughout Minas Gerais, such as FENACRUPE (National Viola Festival of Cruzeiro dos Peixotos) of Uberlândia, a city in the state's western section, known as the Triangle, in which agriculture and cattle industries dominate. FENACRUPE's festival routinely draws audiences of over 10,000.

Yet to 'consider the place of music is not to reduce music to its location, to ground it down into some geographical baseline, but to allow a purchase on the rich aesthetic, cultural, economic and political geographies of musical languages' (Leyshon, Matless, and Revill, 1995: 424–425). A human landscape of independence, suffering, hope, survival, and storytelling opens windows into the heterogeneous spaces of contested social development within and between hinterland communities; the *viola* becomes the voice, a 'musical language' of these spaces, articulating those pasts and resonating in its reception by today's Brazilian audiences. Antique, nineteenth-century *violas* handmade by the fabled luthiers of Queluz, MG (a city since renamed Conselheiro Lafaiete) have been rescued, refurbished, and heralded as cultural emblems by sophisticated collectors, just as master *mineiro* luthiers such as Virgílio Lima and Max Rosa design and construct improved *violas* for the demands of instrumental masters and the burgeoning high-end *viola* market (interviews in 2015 with Lima in Sabará and with Rosa in Nova Lima). Today, the *viola* taps into the unidentifiable beauty of *música caipira's* composite disposition of empowerment enjoyed by the mythologized personae of the lonely and isolated *caipira* that is both safe (nostalgia for a rural home and small community) and dangerous (the vulnerability of the isolated individual in the wide-open natural world, lacking political power or social standing). The instrument, laden with symbols of the places of subjugation and forgottenness, engages the peculiar pathos of those sometimes-sweet stories etched into song depicting the individual within a rustic simplicity set against a massive backdrop of natural wonder and devastating social change, where feelings of loss are not just for former lovers, but also for traditional ways of life transformed by market economies, industrial agribusiness, and a political world that has betrayed the *caipira*.

1.4 Corner Club: A Popular Music Collective and Its Odes to Regional Identity

The *viola* narrative fluidly relates to music spaces, its timbral identity, and stylistic associations facilitating transformational processes of memory and belonging, and also of transcending boundaries, of crossing over to other places, and of forgetting, leaving, and departing. This empowerment by spaces physical, historical, and emotional also characterizes popular urban music based upon expressions of *mineiro* regional identity. Emerging from Belo Horizonte (also BH) during Brazil's eclectic popular music scenes of the mid-1960s, singer, songwriter, and guitarist Milton Nascimento spearheaded a collective of musicians and poets known as the Corner Club (Clube da

Esquina). The 1972 EMI-Odeon double LP titled *Clube da Esquina* is regarded as one of Brazil's most important post-*bossa-nova* albums. The release came at the culmination of the decade that saw Milton's rise from obscurity to winning nationally televised music festivals, a period during which the collective became further linked to its *mineiro* roots, and which mapped on to turbulent years of a brutal military dictatorship (1964–1985). The LP launched and furthered careers of many musicians and poets, among them Milton's primary album collaborator Lô Borges. Their two names appearing on its back cover, the LP's wordless front cover features only a full-bleed photo of two hauntingly anonymous boys sitting beside an iconically rustic dirt road, a fictively rural mirroring of the urban Milton-Lô collaboration: the elder is black, the younger is not. Both LP title and the collective's moniker are carried over from a song of the same name, the earliest Nascimento-L. Borges collaboration. The 'corner' is a celebration of the physical location, in BH's Santa Teresa district, of the intersection of Rua Divinópolis and Rua Paraísopolis, street names taken from small interior towns in diverse regions of Minas Gerais. Today, the corner is also the epicentre of a series of inscribed silver plaques, placed by the Corner Club Museum in partnership with the City of Belo Horizonte and mounted on buildings and sites that played important roles in the collective's development.

Deep regionalism's multiplicity of musical space (physical, historical, emotional/cognitive) is here illustrated textually by poetic reference to *mineiro* cultural territory, the physical layout of the city of Belo Horizonte, and the regional lineages found among the collective's members and their families. Musicians Wagner Tiso, Beto Guedes, Toninho Horta, Tavinho Moura, and Nivaldo Ornellas, among others, each with fascinating family legacies traversing Minas geographically, culturally, and historically, brought their diverse aesthetic input to group-orientated recording sessions, musically augmenting the poetic textual identities found in many of the song lyrics: urban, rural, historical, modern, regional, national, Latin American, and global. Many lyrics of the Clube da Esquina's most successful songs were penned by one of three young poets Márcio Borges (Lô's older brother), Fernando Brant, and Ronaldo Bastos. This section focuses on textual analysis, forgoing discussion of the great extent to which the collective's music supports and extends those meanings.

By the mid-1970s, the collective released a string of well-received recordings interlaced with jazz, folk-rock music, classical music, art rock arrangements, and social protest sensibilities. Some songs showcase Minas Gerais as a unique cultural territory. In addition to themes of modernity and contemporaneousness, some songs invoke regionalist themes covered in the *viola* section of this chapter: rustic simplicity, reverence of nature and tradition, and both a questioning and a valorization of isolation, travel, and escape. Lyrics speak yearningly to national and universal issues from within this guise of mythical *mineiro* landscapes, of introspective, regional heritage, and from within both expansive and intimately composed sound worlds. During the dictatorship, with citizenship itself threatened, Clube da Esquina songs elevated contexts of *mineiro* identity to the equivalency of a Brazil writ large. Notable production values of sonic musical space often portray the contradictory *mineiro* tapestry of Baroque golden age brilliance, the self-conscious doubt of the post-mining-boom decay (*decadência*), and the cleaved identity of BH's gleaming modernity cast against a vast, isolated hinterland and the often dark yet celebratory religiosity of popular Catholicism. My three notions of musical space emerge once more: geographical place, historical space, and cognitive processes of identity of place in individual consciousness. In these songs, space is empowerment: personal, creative, social, and political.

Lô and Márcio Borges are joined by lyricist Fernando Brant on the telling rock song 'Para Lennon and McCartney' (For Lennon and McCartney), featured on the 1970 EMI-Odeon LP *Milton*, and an anthem masterfully navigating late '60s rock aesthetics in a conjoining of *mineiro* identity and the collective's desire to be heard beyond the region's borders, beyond Brazil. This shout-out to the world openly addresses the Beatles, fans, critics, and all listeners, with the message that Minas Gerais is 'on to' what is happening in the rest of the world:

Eu sou da América do Sul,	I am from South America,
Eu sei, vocês não vai saber	I know you will not know that
Mas agora sou cowboy,	But I am now a 'cowboy',
Sou do ouro, eu sou vocês,	I am made of gold, I am you,
Sou do mundo, sou Minas Gerais.	I am of the world, I am Minas Gerais.

A clear message emerges, to place Minas Gerais within South America's cultural array, and with a hip cowboy delivering a regionalist manifesto on the international stage. A vigorous rock soundtrack powerfully conveys the sentiments. Embedded in the song's sheen, its strong delivery, is BH's own modernity, and this developing world's tropical modernism, in which the collective's appreciation for their regional heritage ('I am made of gold' is also a nod to the Baroque past) is a spatial-cultural condition most outsiders 'will not know'. And only when placing itself on the world stage, in contexts of other continents and hemispheres, does the Luso-Brazilian voice, in Portuguese, associate so strongly with Hispanic, Spanish speaking neighbours ('I am from South America').

The song 'Os Povos' (The People) is ostensibly about the *mineiro* past, lost in a moody remembrance of struggle, and of light and dark. Mountainous Minas is portrayed in Márcio Borges' lyrics as existing at the edge of the world, isolated, with a hidden historical pain. The lines eerily map onto the oppressive decay of citizenship during the military regime. Musical devices are contoured to these meanings, the accompaniment's plodding determination reflecting at once both a vision of a troubled Minas landscape in its post-mining-boom decay and for the stifling oppression of the regime; the 1972 album came at the time of the dictatorship's darkest years, nicknamed the 'Leaden Years' (*Anos de Chumbo*).

Na beira do mundo	At the edge of the world
Portão de ferro	An iron gate
Aldeia morta, multidão	A dead village, multitudes
Meu povo, meu povo	My people
Não quis saber do que é novo	No longer want to know what is new
Eh! Minha cidade	It is my city
Aldeia morta, anel de ouro, meu amor	Dead village, gold ring, my love
Na beira da vida	At the edge of life
A gente torna a se encontrar só	We return to meet in solitude (we turn to find ourselves alone)
...	...
Ah, um dia, qualquer dia de calor	Ah, one day, whatever hot day
É sempre mais um dia de lembrar	Is always one more day to remember
A cordilheira de sonhos que a noite apagou	The mountain range of dreams that the night erases

The programmatic accompaniment could be that of a burro mindlessly treading the paths of Old Minas, or of an ox cart wheel turning undeterred by change, unfazed by Milton's lonely voice coaxing the listener back through time in melodically simple, matter-of-fact declamations bathed in reverberation. Breathy human voices whisper softly percussive 'chuh-chuh' and 'wah' vocal sounds, accenting an arpeggio's rhythmic scheme. But listeners also recognized within this a hypnotic churning of boredom, the deadly acquiescence to dictatorship. Finally, a cry — 'Ah, one day...' — sharply contrasting the matter-of-fact, purposefully drab vocal delivery of the opening; Milton is calling out to gain attention on solitude's behalf. But it is an echo chamber. No one listens. It is of no use contesting this near-tragic loneliness of the *mineiro* interior from which a voice of the past calls out; or is it emanating from within a 1972 police dungeon's interrogation cell? The simplicity of the musical gestures is no longer simple; spaces of identity are being drawn from Minas' history of decay and tragic current events. This is Minas, 'my people', and a Brazil lost to both time and brutality.

Lô and Márcio Borges contributed the curiously sardonic 'Ruas da Cidade' (City Streets) to Nascimento's 1978 sequel LP *Clube da Esquina 2*. With a deliberately unsettling chord progression reminiscent of material from Lô's first solo album, the so-called Tennis Shoe album of cult fame (*Lô Borges*, 1972, EMI-Odeon), the song's ostensible attempt at celebration quickly, intentionally, fades. Names of indigenous ethnicities, appropriated in the original 1890s city plan for streets criss-crossing downtown Belo Horizonte, fail in their homage and instead confront 300 years of settlers paving over these ethnicities, now anonymously collapsed beneath modern BH. Rather, the streets become well-trodden, metaphoric cemeteries. The pseudo-solemn, ceremonial naming has utterly failed to honour the rich cultural histories of the indigenous peoples who once called the region home. That is what the song is about — they are all in the ground:

Guacurus, Caetés, Goitacazes	Guiacurus, Caetés, Goitacazes,
Tubinambás, Aimorés	Tupinambás, Aimorés [names of indigenous ethnicities]
Todos no chão	All in the ground
A parede das ruas	The wall of streets
Não devolveu	Do not give back
Os abismos que se rolou	The abyss from which they turned away
Horizonte perdido no meio da selva	Lost horizon in the middle of the jungle
Cresceu o arraial	The village grew

Such was the case for the Guiacurus, forced westward from Minas during colonial times, into what are now the states of Goias and Mato Grosso do Sul, adopting horses along the way to augment their warrior-like resistance to Portuguese expansion. The violent decline of Native American populations, a forgotten history in Brazil even as it brutally continues today, came in the form of slavery, disease, displacement, and massacres. But such details are not the teleological goal of Lô and Márcio Borges' literary gesture and would have weighed down the song. 'Ruas' is neither a documentary nor a commemorative ballad. Rather, a morbid, ironic deadpan turns the non-celebration of these peoples, thoroughly unconvincing in the flaccid boosterism of nineteenth-century city planners (and resonating with the *Indianismo* literary movement of nineteenth-century nationalist Brazilian intellectuals), into a questioning of those very intentions: the song jabs at the notion of street names doing justice to genocide. Meanwhile, the queasy yet pedestrian-sounding song takes us on a journey through those streets of Belo Horizonte (beautiful horizon), now *Horizonte perdido* (lost horizon):

Passa bonde, passa boiada	Street car passes, cattle pass
Passa trator, avião	Tractor passes, an aeroplane
Ruas e reis	Streets and kings
Guajajaras, Tamoios	Guajajaras, Tamoios
Todos no chão	All in the ground
A cidade plantou no coração	The city planted in the heart
Tantos nomes de quem morreu	Many names of those who died
Horizonte perdido no meio da selva	Lost horizon in the middle of the jungle
Cresceu o arraial	The village grew

The two most convincing follow-ups to the 1972 LP are the Nascimento albums *Minas* (1975) and *Geraes* (1976). Thanks to a friend's observation, Milton was inspired by the first title, if not for obvious reasons, then partly due to its matching the first letters from his name: MI-lton NAS-cimento, with its paired concept album companion *Geraes*, assuming an archaic spelling while completing the ode to regional identity in blatant fashion. *Minas* sold hundreds of thousands of copies in Brazil, still ranking as Nascimento's best-selling LP. The song 'Fazenda' (Nelson Angelo) appears on *Geraes*, a nostalgic meditation on the gentle, family-based weekend country house. Visited with the extended family, and far enough from the city to constitute 'country', yet close enough to visit frequently, the *fazenda* typically boasts a small orchard, pond, or some livestock. 'Fazenda' celebrates via first person narrative the natural beauty and local delicacies laced with familial, generational experiences. There are no apologies given for the warmth of family space as a *mineiro* treasure. Life is good.

água de beber	drinking water,
bica no quintal	backyard spring
sede de viver tudo	everything with a thirst to live
e o esquecer era tão normal	to forget troubles was so normal
que o tempo parava	that time stopped
tinha sabiá, tinha laranjeira	we had the *sabiá* [a bird], the orange grove
tinha manga-rosa	we had the *manga-rosa* [mango]
e na despedida	and during our farewell
tios na varanda	uncles on the veranda
jipe na Estrada	a Jeep driving away
e o coração lá	and our hearts left behind
...	...
e a meninada	and the children
respirava o vento	breathed the wind
até vir a noite	until night fell
e os velhos falavam	the old folks spoke
coisas dessa vida	things about this life
eu era criança	I was a child then
hoje é você	today it is you
e no amanhã	and tomorrow
nós	us

The romantic arrangement, with surging emotions delivered by Milton's unmistakable, trademark voice, speaks directly to a nostalgia balanced with confidence in a future that holds these dreams intact for subsequent generations. Beto Guedes' high vocal line of the backing chorus is heard as an echo of sorts, a memory in falsetto calling as witness to rustic traditions. The varied, subtle intricacies of Angelo's arrangement artfully mask a rock ballad, complete with driving rhythms and soaring strings. The gently unsettling harmonic modulations leave a question mark as to the relationships we might have with these sweet memories. Above all, this is an ode to a *mineiro* sense of place where home is beautiful in every way, as it is spatially synonymous with privacy, family, and belonging.

1.5 Conclusion

The three case studies discussed here are regional traditions, each offering narratives of deep regionalism engaging the geography of cultural places, the heterogeneous spaces of social development within and between communities, and the processual spaces of consciousness hosting identities of place as cognitive experiences enabling what some have termed enactment. As with the *Congado's* territoriality and Afro-Mineiro historical narratives, and the *viola's* iconicity of the rural past, downtown BH's streets represent historical confrontation and conflict. In speaking to power, 'Ruas da Cidade' accomplishes this in its questioning of hollow local boosterism. Likewise, the equating of the normalization of a military dictatorship to a curmudgeonly historical population in a dead village at the end of the world, as in 'Os Povos', relies upon music's formative spatialities. Here, 'the mountain range of dreams that the night erases' casts Minas as hopeful space lost in time. 'Para Lennon and McCartney' thrusts this almost secretive, golden place onto the international stage. As claimed here, regional identity for many *mineiros* is linked to *Congado*, *viola*, and the Clube da Esquina. For many other *mineiros*, foreign imports are a passionate stand-in for how music articulates their regional identity: rap, hip-hop, death metal, and virtually all global trends have found their way here as localized currency. They are simply less traditional, yet carry their respective symbolic, formative spatialities.

References

Ciucci, A. 2012. '"The Text Must Remain the Same": History, Collective Memory, and Sung Poetry in Morocco', *Ethnomusicology* 56(3): 476–504.

Gibson, C., and P. Dunbar-Hall. 2000. 'Nitmiluk: Place and Empowerment in Australian Aboriginal Popular Music', *Ethnomusicology* 44(1): 39–64.

Guy, N. 2009. 'Flowing Down Taiwan's Tamsui River: Towards an Ecomusicology of the Environmental Imagination', *Ethnomusicology* 53(2): 218–248.

Leyshon, A., D. Matless, and G. Revill. 1995. 'The Place of Music: Introduction', *Transactions of the Institute of British Geographers* 20(4): 423–433.

Magowan, F., and L. Wrazen. 2013. 'Introduction: Musical Intersections, Embodiments, Emplacements'. In *Performing Gender, Place, and Emotion in Music: Global Perspectives*, edited by F. Magowan, and L. Wrazen, 1–14. Rochester: University of Rochester Press.

Schiavio, A. 2014. *Music in (en)action: Sense-Making and Neurophenomenology of Musical Experience*. PhD thesis, The University of Sheffield.

Schultz, A. 2002. 'Hindu Nationalism, Music, and Embodiment in Marathi Rashtriya Kirtan', *Ethnomusicology* 46(2): 307–322.

Travassos, E. 2006. 'O Destino dos Artefatos Musicais de Origem Ibérica e a Modernização no Rio de Janeiro (ou Como a Viola se Tornou Caipira)'. In *Artifícios & Artefactos: Entre o Literário e o Antropológico*, edited by G. Santos, and G. Velho, 115–134. Rio de Janeiro: 7Letras.

Other Media

Faria, L., and M. Calil. 2003. *Clube da Viola: Raizes 2* (CD liner notes, BMG).

Freire, P. 2008. *Violeiros do Brasil: Músicas e Conversas com Artistas da Viola Brasileira* (DVD liner notes, Projeto Memôria Brasileira).

Chapter 2

'Trapped in Oklahoma': Bible Belt Affect and DIY Punk

Alican Koc
Department of Art History and Communication Studies, McGill University, 845 Sherbrooke St W., Montreal, QC H3A 0G4, Canada
ali.koc@mail.mcgill.ca

2.1 American Hate

This story begins in the back room of a Chicago skate shop in the summer of 2014. Hundreds of us are assembled in the damp heat of the room for the main gig of a local DIY (do-it-yourself) punk festival, chatting over cans of cheap beer and sweating profusely as we wait for bands to play. A new band whose name I have not yet learned has begun setting up on stage, and they look a little different from the others.

Despite the massive span of North American DIY punk and its abundance of small regional scenes, there is a general sense of aesthetic uniformity within the greater community. Over the years, this has been achieved through the proliferation of online blogs,

Musical Spaces: Place, Performance, and Power
Edited by James Williams and Samuel Horlor

ISBN 978-981-4877-85-5 (Hardcover), 978-1-003-18041-8 (eBook)
www.jennystanford.com

popular zines such as *Maximum Rocknroll,* independent music festivals, and the movement of distinct sounds and styles through the touring circuit. The increased access that participants in the DIY punk network have to the wider scene contributes to a quickly shifting dialectic of subcultural styles in which dominant aesthetic trends routinely shift. More importantly, this access to regional scenes throughout the continent allows participants to geographically map out dominant trends by region, distinguishing, for example, between the wacky cartoonishness of Midwestern 'egg punk' bands, such as Warm Bodies and Big Zit, and the pummelling post-apocalyptic sounds of Japanese and Swedish punk influences in Pittsburgh raw punk bands, such as Eel and Ratface. In the geographic imagination of a style-savvy North American punk rocker, each city or region on the continent seems to have its own localized sound and style.

Despite my involvement in this community for years, this band's aesthetic feels unplaceable within a particular geographic locale, subgenre, or style. The band's guitarist, bassist, and drummer all have wild, long hair — an anomaly in this festival, particularly in the heat of the summer. Even more terrifying is the band's singer, a huge guy with a big frizz of bleach-blonde hair covering part of his beet-red face. Furiously pacing back and forth in front of his band as he waits for them to finish setting up, I see on his face a boiling point of years of pent-up rage ready to be released at the unsuspecting onlookers. As the band begins to play, the singer charges at the audience bellowing at the top of his lungs, grabbing the first person in front of him by the face and throwing them to the ground — a daringly violent move in the show's otherwise convivial atmosphere. The band thrashes around wildly, swinging their long locks as they force out a pummelling and deranged strain of rock 'n' roll driven hardcore punk with a sense of primal rage that I have rarely seen or heard before. As the set comes to a close, the guitarist begins violently slamming his guitar against his amplifier, filling the room with a screeching feedback that amplifies the sense of tension building in the room since before the band even began playing. The set finally ends, and I am astounded by the performance. Attempting to conjure up words again, I ask my friend if they know the name of the band that just played. 'American Hate', replies someone standing behind me in the crowd. 'Where are they from?' I ask. 'Oklahoma City'. Somehow everything begins to make sense.

Having known next to nothing about the state of Oklahoma, let alone its punk scene, something about American Hate's furious performance and sonic aesthetic captivated me. Upon discovering the existence of a thriving DIY music scene in a city that I barely knew about, I became consumed by an anthropological curiosity surrounding the lifeworld that spawned this strange music. After hours spent chatting with the band about their local music scene, and their experiences living with the conservative politics of their home state under Republican governor Mary Fallin, I realized that I had begun to form a romanticized idealization of a bizarre place in the centre of America in which a suffocating climate of religious conservatism in the prairie heat gave way to an aesthetic of pure American hatred — an artistic condensation of the country's legacy of violence planted right in its heartlands.

2.2 'Something in the Water'

The expression 'something in the water' has become a cliché in the DIY punk circuit, used to describe the rapid and unforeseen emergence of cultural and aesthetic activity within a specific region. My fascination with this overused expression is twofold. Firstly, the phrase implies some degree of formal aesthetic unity across music emerging from the region in question. Music journalism in particular is rife with romanticized accounts of musical space in which there is an implied correlation between the affective experience of living in particular places and the forms of musical expression that emerge from them. Secondly, the joking tone of the expression seems to point to some other phenomena responsible for the spontaneous eruption of creative energy from a particular locale. The question thus becomes: What is happening in these places to contribute to the rapid emergence of a specific aesthetic if it is not something in the water of these cities? Put another way: What is the relationship between the circulation of affect generated by lived experience in a particular place and the cultural aesthetics emerging there? This question will be the central concern of this chapter, which will draw upon ethnographic research conducted in Oklahoma City's DIY punk scene to examine the relationship between the scene's sonic aesthetics and the affective experience of living in the American Bible Belt.

2.3 'Can You Feel It?': Exploring the Intersections of Aesthetic and Affect Theory

What exactly is affect? Few words seem to have been used so much to connote so little over the past few years. The answer depends on who you ask. Drawing on Deleuze and Guattari's adoption of the term from Spinoza, Brian Massumi calls affects 'virtual synaesthetic perspectives anchored in (functionally limited by) the actually existing, particular things that embody them' (1995: 96). In Massumi's work, 'affect' refers to a set of presubjective bodily sensations that are distinguished from emotion due to their ability to resonate prior to cognition and social qualification (*ibid.*: 88). Following the staunch anti-representationalism of Deleuze and Guattari, Massumi treats affects as singularities largely abstracted from their point of emergence in the world (*ibid.*: 94). For Deleuze and Guattari, art is one of these points of emergence, a 'bloc of affects' that has never been representational (1994: 164, 173). Rather than attempting to trace particular relationships between art and affects anchored in the actual, the Deleuzo-Guattarian approach to aesthetic theory tends to view art's affects as imminent to knowledge and meaning and focus instead on potentialities of being through artistic expression to set off new powers of thought (McMahon, 2002: 4; O'Sullivan, 2001: 126).

While demonstrating an overdue sensitivity towards the autonomy of relations between art's affectivity, content, and form, Deleuze and Guattari's theory of expression has received criticism for ignoring the role that affect plays in mediating culture (Anderson, 2014; Hutta, 2015; Mazzarella, 2009). As William Mazzarella writes in 'Affect: What is it Good for?',

> Rather than expending vast amounts of energy recuperating the constitutive instability and indeterminacy that attends all signification... would it not be more illuminating to explore how this indeterminacy actually operates in practice as a dynamic condition of our engagement with the categories of collective life?
>
> (2009: 302)

Indeed, a number of scholars have more recently begun examining the dynamism of processes of representation, signification, and symbolism within the field of affect studies. J. S. Hutta writes: 'as the quarrel around affect's non-representational alterity to language is becoming increasingly tired, attention is shifting to how language and affect are mutually implicated' (2015: 296). I suggest here that Raymond Williams' elusive 'structure of feeling' concept might serve as a tool for making sense of more collective expressions of affect. Williams defines his concept as 'a kind of thinking and feeling which is indeed social and material, but each in an embryonic phase before it can become fully articulate and defined exchange', which has 'a special relevance to art and literature' (1977: 131, 133). Existing in an elusive relationship with the cultural forms through which they are felt, Williams' 'structure of feeling' concept refers to the amberization of collective feeling within the still sticky and fluid resin of aesthetic form. Subtly interweaving material and historical processes, affects, and creative expression, Williams' concept functions as a powerful theoretical tool for trying to identify the proverbial something in the water without resorting to linear causal relations.

2.4 Arrivals

My arrival in Oklahoma City to do fieldwork in the summer of 2016 was admittedly not my first time in the 'Sooner State'. Following the festival in Chicago, I quickly began to develop a close friendship with the members of American Hate, who booked a show for my band in the summer of 2015 and visited my home town of Toronto where I booked a show for them just two months later. After being astonished by American Hate's wild performance in Chicago and hearing hours' worth of anecdotes about furious hardcore punk shows taking place in the garages and basements of homes in the capital of the conservative Christian state, I was excited to be playing in Oklahoma City. I imagined the city's scene as an almost uniform entity, composed of American Hate and a number of similar sounding bands, transforming the conservative political and social climate of their home state with violent bursts of noise and hardcore punk coupled with dreary images of life in Oklahoma. I was particularly excited by the fact that the city's punk scene appeared to be on the cusp of a renaissance. American Hate's singer Ross had recently begun booking a small DIY music and art festival in Oklahoma City in the spring of 2015. Despite its humble size, the first Everything Is Not OK fest drew sizable crowds of spectators, bands, and visual artists from around the country and generated a substantial amount of hype amongst punks on the internet, ultimately shining a spotlight on the city's scene and creating a fertile environment for younger members of the scene to start bands and begin circulating their art.

As we approached Oklahoma City for the first time during our tour in August of 2015, I looked over the city's small skyline, below the massive and gleaming Devon Energy Center, and began wondering just what sort of a place I was in. All across the city's massive quasi-suburban sprawl were signs of Oklahoma's religious conservatism: billboards with images of cute babies that read 'Cherish Life', church signs with slogans like 'Abortion is Murder', and a looming white cross erected beside the highway north of the city. That night, we played at The Shop, a former car repair shop in a remote industrial part of the city, owned by the father of local punk siblings Becky and Nora, which had been converted into a DIY concert space. In the warm glow of Christmas lights hung from the walls and rafters, the beaten and dust-covered tractors lining one side of the room struck me as a tongue-in-cheek nod to Oklahoma's rural prairies, while the eerie vintage dolls that hung from ropes tied to the rafters felt like an oppositional response to the pro-life billboards and signs I had seen throughout the city. Despite the relatively small turnout at the show, our performance was well received, and upon being asked by Ross to play his upcoming fest, we happily agreed.

During the following months, my romanticized idealization of Oklahoma City's punk scene continued to grow. Just two months after Oklahoma City police officer Daniel Holtzclaw was convicted of sexually assaulting thirteen African-American women across the city, his crying face from a picture taken during his court date appeared on the cover of local powerviolence band Crutch's second demo tape, this time with a noose drawn around his neck. With their almost inhuman speed and crushing metallic hardcore riffs, Crutch struck me as a sonic embodiment of the rage experienced by punks in Oklahoma City. With lyrics like 'sick of your bigoted mentality/ you lost your way in evolution/now I'm feeling violent/backward hillbilly fuck/your superiority is shit/bigot racist sexist pile of shit' or 'crossbearers, saints and saviors/sin-makers, death bringers/no resolution on their stained glass stage/two thousand years of empty promises/their mandates set to control and confine/turn the tables, burn their books, and see the end of faith/resist control', Crutch were directly addressing the bigotry and narrow-mindedness of Christian conservatives in their home state (Crutch, 2016).

As I geared up for our return to Oklahoma City for the festival, I began ransacking the internet for more punk bands from Oklahoma that sounded and felt like American Hate and Crutch. A particularly excellent resource was *Terminal Escape*, a cassette blog whose author had grown up in Oklahoma before moving to California. There, I found a seemingly boundless plethora of tidbit-like descriptions of Oklahoma's punk scene over the decades, describing it as 'arguably the most socially conservative state in the country, with little to fall back on but a history of white oppression and unemployment' (The Wizard, 2012b). Amid descriptions of Oklahoma as a 'geographical handicap', and as 'culturally void', a number of reviews in *Terminal Escape* wrote of bands which pointed a 'stiff middle finger to the Oklahoma bible belt culture that weaned and corrupted them' and attributed the use of manic, frustrated hardcore punk vocals to living in Oklahoma (The Wizard, 2010, 2012a, 2014). As I listened to the new bands coming out of Oklahoma City and read these reviews, I began to associate the anguished sounds of Oklahoma City punk with a particular feeling of suffocation and rage towards the Christian Right in the Bible Belt. It seemed almost perfect. Perhaps a bit too perfect.

2.5 'You Are Now Entering America's Corner'

I arrived in Oklahoma City to do my fieldwork in July of 2016. During this time, I stayed at 'American House', the residence of Ross, Taylor, Tim, and Nora — the respective singer, guitarist, and bassist of American Hate and one of the front people of Cherry Death. Staying at 'American House' turned out to be the best opportunity to learn about punk in Oklahoma. As the oldest member of Oklahoma City's DIY punk scene, Ross functioned as something like the scene's big man, running a local festival, routinely booking shows for out-of-town bands, starting a small local cassette label, and playing in a number of the city's bands. Moreover, through his years in the local scene, Ross had collected an encyclopaedic archive of local fanzines, demo tapes, and show flyers dating back decades. Within days of my arrival, Ross gave me a stack of local zines and tapes to look at, and he began introducing me to locals within the scene.

The small grassroots network of North American DIY punk can sometimes feel like a secret society. The Germs' iconic slogan 'what we do is secret' resonates in the world of contemporary DIY punk (The Germs, 1979), which is generally kept within tight-knit communities and can sometimes come off as hostile, cold, or otherwise unwelcoming to outsiders. The multitude of reasons for this secrecy mostly surrounds the issue of authenticity. While cultural theorists such as Dick Hebdige have famously theorized punk's ability to signify chaos and opposition at every level of its style, the subculture's relatively short history is rife with stories of the co-option and commodification of the subculture's striking aesthetic sensibility and youthful angst (Hebdige, 1979: 113).

Not only is such commodification considered a disgrace to punk's oppositional aesthetic, but it also has tangible effects on the lives of participants within the subculture. In his influential essay '"Gentrification" and Desire', Canadian geographer Jon Caulfield notes how the emergence of marginal groups seeking to improve their lives in lower income inner-city neighbourhoods constitutes the first wave of gentrification in urban space (1989: 623). According to Caulfield, these 'marginal gentrifiers' become the cultural vanguard for more mainstream and middle-class groups to arrive into these neighbourhoods, driving up housing prices while diluting the marginal groups' avant-garde values and desires for a unique place to live (*ibid.*). Considering the centrality of punk culture to the practice of marginal gentrification described by Caulfield, and punk's aestheticized critique of mainstream bourgeois culture, it seems evident why punks have had an averse attitude towards the 'discovery' of their scene by outsiders and 'normies'.[1] In order to prevent their scenes from becoming overridden by mainstreamers, punks often avoid contact with people outside of their subculture and have devised elaborate methods of ensuring that their events are only known by punks within their scene. Perhaps the most famous of these is 'Ask a Punk', a generic address posted on flyers and Facebook

[1] A term used to describe 'normal' people with mainstream values.

events for shows happening in houses, parks, abandoned buildings, or any other DIY spaces in which the presence of police or other outsider groups is unwanted.

Despite being an outsider to the scene, and of all things, a nerdy graduate student, my fortune in being able to meet and instantly establish rapport with a vast number of local scene members for my research was boosted by a number of factors. One of these factors is that I was already known as a punk musician who had visited Oklahoma City twice to perform with my band. Not only was I a member of a somewhat known band from out of town, but my year-long friendships with several members of American Hate ensured that I was in with the right crowd as well. Under the angelic wing of the scene's 'big man', Ross, I was immediately introduced to everybody I was not already familiar with operating within the city's small but tight-knit hardcore punk scene. In addition to this, Taylor, Tim, and Nora, whose interests exceeded the confines of hardcore punk, introduced me to a number of talented musicians who used the scene's DIY punk network as a venue for more avant-garde sounds, ranging from noise rock, post-punk, psychedelic, and indie rock bands to solo projects focused on ambient music and noise.

After Ross drove me home on my first night, the two of us began a long conversation on punk in Oklahoma. As we chain-smoked cigarettes in the dim light of the house's front porch, Ross recounted stories to me of growing up as a young punk in the city. In stark contrast to my conception of Oklahoma City as a Mecca for Christian Republicans, Ross spoke of it as a cultural oasis within Oklahoma. He had recently designated the city as 'Freak City, U.S.A.' and described it as a place in which punks and other marginal groups were less likely to be harassed by rednecks, jocks, and police for looking different; a city in which DIY spaces and shows were never in jeopardy; and a refuge for dozens of young, disenfranchised freaks from across Kansas, Arkansas, Oklahoma, and North Texas to escape the banality of their conservative small towns. When I subtly began trying to ask Ross about American Hate's relationship to the abrasive hardcore punk sound that I had attributed to the place, Ross matter-of-factly mentioned that there had never really been a specific Oklahoma punk sound. Within a matter of hours of my visit beginning, my sensationalistic vision of the pulverizing sound I associated with Oklahoma City and the affects of enraged freaks living in the heart of the Bible Belt continued to crumble. It began to seem like I had gotten it all wrong. Not only was Oklahoma City a far cry from the neoconservative dystopia I had imagined, but the sound that I had previously associated with the uncontrollable anger of Bible Belt punks was also beginning to look like a fiction, or at least the tip of the iceberg of a scene whose eclectic range of sounds may have been as wide as Oklahoma City's sprawling city limits. I began to think that my desire to find a perfect representation of a Bible Belt punk structure of feelings, and its embodiment in a regional music scene, was more akin to the sensationalistic fantasies of record collectors or music writers than the findings of an anthropologist. Still, I could not shake my strange attraction to the tremendous affective resonance of the city's scene.

2.6 Moving Towards the Church of Freak City

As I began paying closer attention to the diverse array of sounds in Oklahoma City and chatting with local musicians about their experiences in their home state, Ross' designation of the city and its DIY punk scene as 'Freak City' began to come to life. Virtually everybody I spoke to described the scene as a kind of refuge they had stumbled upon after leaving some sort of *elsewhere*. For many people, this elsewhere was a geographical one, generally one of the small towns bordering Oklahoma City. During my first conversation with Ross, he had mentioned leaving his home town of Edmond — a small town to the north — due to the police harassment he was experiencing for dressing differently.

In addition to being an oasis for frustrated youth fleeing the conservatism of their small towns, Freak City as a scene also seemed to be a sanctuary from a *cultural* elsewhere for many members. For almost everyone I spoke to, this cultural elsewhere was their home. While none of the local punks I encountered described an active hatred towards their parents and families, almost all mentioned conflicts arising from the rift between the conservative values of their parents and the progressive ideals associated with their punk subculture. Perhaps the strongest example of this was a conversation I had with Jerry, a close friend I had made during my first visit to Oklahoma, who was the most active transgender member of the city's scene. During our conversation, Jerry noted that despite loving her parents, she did not feel the need to agree with them or to continue living her life as a conservative Southern boy. For Jerry and everybody else in the scene who I spoke to, Freak City punk provided a chosen family characterized by an inclusivity rarely found in the rigid structure of family life in their home state.

While the Freak City scene seemed to be a place in which its alienated members created a family-like sense of community, I soon came to realize that the scene was not necessarily the first place in which its participants had sought refuge. During many of my conversations, the DIY scene was portrayed as an oasis that had been stumbled across after many botched attempts to feed a subcultural thirst for community in a number of mirages. For many members of the scene, these mirages were other alternative music communities in Oklahoma City that offered more aesthetic coherence in their formal styles and yet seemed incapable of properly ridding themselves of the problematic ideologies and behaviours of their society. This came up extensively during my conversations with The Garters' vocalist Labangry, who described to me how she and other women had fled to the local DIY scene after experiencing the intense misogyny of the city's more mainstream hardcore scene. Similarly, Jerry mentioned her feelings of alienation from the drug use and hookup culture within the city's queer scene.

For Jerry, Labangry, and almost everybody else I spoke to, the title of Freak City seemed to have a dual meaning. As a geographical space, Freak City referred to Oklahoma City, a somewhat more progressive city in Oklahoma in which disillusioned youth from nearby areas could escape the banality and conservatism of small-town life in the Bible Belt. Yet perhaps more importantly, Freak City also seemed to function as a cultural space referring to the city's DIY punk scene, which functioned as a creative network as well as a community for people who felt alienated by their families and other social networks to find solace in one another. Upon asking Ross what Freak City was, he mentioned this dual meaning to me:

> This. It's here. It's Oklahoma City. It's a place where you can be a fuckin' freak and it's chill and you know, Freak City is our community. I think the way that we're perceived here is not the way that we are but it's like when you go to shows at The Shop, that's Freak City, that's a bunch of people who have found three hours of this given week or month or day where they can exhale, they can feel safe being who they are and talk about life or not talk about it, however they want to be. And nobody is going to be like, 'Look, you're different'.

Alienation was perhaps the most prevalent theme I noticed throughout my conversations with scene members and in the work of local bands. Looking through the stack of local demo tapes Ross had lent me, I came across an American Hate cassette whose insert simply read, 'Thanks to anyone who has ever felt alone'. The message running throughout the scene was clear: *you don't have to feel alone,* and all of the members of the community seemed to feel that the do-it-yourself ethos and progressive politics of punk created a space in which alienated freaks could come together and express themselves.

Considering the pervasive hatred towards Christian conservatism that I found almost ubiquitously amongst punks in the scene, I was interested to find that a number of members likened the social organization of their community to that of a church. As my friend Colin told me when I asked what they sought to achieve with their participation in the scene, 'It would be... a place where cultural hierarchy is negated, and it's just a pure horizontal space. I want it to be like a church, I guess'. Not only was Colin's idealization of a church interesting to me, but I was also intrigued by the utopian sensibility of Colin's description of their vision for the scene. As the church bells of Freak City rang throughout the sprawling land mass of Oklahoma City, permeating through the lazy midday heat, I began to wonder if their violent, feedback-ridden tune was merely a distorted take on an old utopian hymn.

2.7 Cruising Dystopia/Slamming Utopia

Considering its oppositional aesthetic centred around chaotic imagery, jarring stylistic choices, and violent and abrasive soundscapes, punk's beer-soaked and sweat-stained mosh pits seem like an unlikely place to turn in the pursuit of cruising utopia.

Yet, throughout my conversations with members of the Freak City punk scene, I could not help but notice an undeniably utopian affect underlying their feelings towards the scene. Several members described the DIY punk community as a blessing and a utopian breath of fresh air and acceptance in the sweltering heat of a red state that seemed predisposed to breathing down the necks of its inhabitants. Punk's counter-intuitive existence between utopianism and dystopianism has been noted by a number of scholars, including José Muñoz. In an essay on The Germs, a notoriously violent Los Angeles punk band, Muñoz writes, 'I propose that we can see the negation that is negativity as something that can be strangely utopian while simultaneously dystopian. It can conterminously represent innovation and annihilation' (2013: 98). For Muñoz, punk serves as a space in which negativity towards one's surroundings may correspond to utopian affect — a space in which the destruction of the present world clears space for a potential utopia to be built in its place.

Like Muñoz, I suggest here that the utopian feelings of expression and belonging which bled out of the damaged sounds in Oklahoma City's punk scene can also be read as simultaneously representing innovation and annihilation. Living in the heart of the American Bible Belt, the participants of the Freak City scene I spoke to seemed to have no option but to adopt a sense of militant optimism. It was this bitter optimism that gave the scene its creative fuel, charging its participants to simultaneously destroy the things around them that they hated, while building a better world in its place. At the end of shows at The Shop, like at the unhinged after-parties that took place during Everything Is Not OK, after the firecrackers had been lit and thrown around the room, inanimate objects smashed in the mosh pit; after the state-endorsed brand of cowboy Christianity had been lit aflame by the drinking and the screaming, the sweating, and the slamming; after the room filled with ghostly echoes of erupted frustrations towards jocks, rednecks, parents, teachers, cops, and politicians, the venue would hollow out, and the residents of Freak City would silently get to work sweeping up cigarette butts, beer cans, and debris to prepare the space for another day.

If the correlation between the feelings circulating amongst alienated artists living in Oklahoma City and the aesthetics of the Freak City punk scene did not immediately make itself evident to me, it was perhaps largely because my scope had been too narrow. As I eventually learned, it was not that the Freak City scene had not produced pummelling hardcore punk that reflected the dystopian feelings of alienation and frustration experienced by punks in Oklahoma, but that these feelings and sounds only accounted for a minute fraction of the affective/aesthetic output of the scene. The reality of it seemed to be that Freak City was a space far more complex in its affective and aesthetic boundaries than I had imagined, closely resembling Oklahoma City in its vast, underpopulated, and decentred sprawl.

If the designation 'Freak City' is to be understood as an imagined space of expression built out of the simultaneously utopian and dystopian affects circulating amongst punks in Oklahoma City, then what did this sound or look like? Thus far I have discussed the fury of American Hate's live performance in Chicago, the frenzy of destruction in late-night shows at The Shop, and a space of love, acceptance, community, and expression that had blossomed out of this. Yet I suggest that the latter utopian feelings were not merely cosy caresses and comfortable pillow talk following the frantic release of pent-up emotions within the scene. Instead, these utopian affects animated the aesthetic make-up of Freak City with the same intensity as the scene's destructive, dystopian feelings, creating a fluid relationship between the two. Put another way, the utopian feeling of Freak City was not only achieved through the sonic annihilation of Christian conservatism, and the liberating feeling of expressing frustration, but it was also reflected in the music itself in all of its incredible diversity.

One of the best examples of the utopian affect of Freak City was Cherry Death, the psychedelic quasi-supergroup led by Tim and featuring a rotating cast of musicians from in and around the city's scene, including Nora and Taylor. Knowing Tim only as the bassist of American Hate, I had been intrigued by how vastly different Cherry Death sounded from the belligerent hardcore produced by his other band. On 'Bite My Nails', the first track from the band's eponymous EP (Cherry Death, 2015a), the seemingly quotidian nervous tic referenced in the song's title erupts into a hopelessly beautiful pop ballad, carrying the listener off into a rosy and enchanting lifeworld of its own. Within seconds of the song's opening chords ringing out, the skies seem to open up, lifting the listener into the warm embrace of a golden horizon of flowers and heartbreak. On 'Puppet Dance', an instrumental track on the band's *Stone Shake Golden Mile EP* (Cherry Death, 2015b), a simple country-style walking bassline keeps a steady pace through a heart-stopping piece of folky Americana, its dazzling guitar licks conjuring visions of the rolling green fields of America's heartland.

It was not just Cherry Death's music that had astounded me in its utopian feel, but the way in which Tim produced the band's music as a pure labour of love. I could see it on his face as we sat on the porch, drinking beer and engaging in casual conversation, with Tim smiling shyly to himself while he wrote new songs on his guitar. Early on in my stay in Oklahoma City, I participated in a recording session with Cherry Death as they were putting the finishing touches on what would soon become their newest LP, *Saccharine*. It was a sunny afternoon and I was walking into a liquor store in the city's northwest end to pick up some beers to quell the heat when Taylor pulled up in his car and told me that Cherry Death was recording that afternoon. When I arrived at the house, I was immediately invited by Tim to participate in the recording session and could not refuse. Tim handed me some shakers and told me to play along. Upon asking him what to play, Tim smiled at me and told me to just do my thing. As the sky dimmed on a quiet residential street, ten of us, most of whom had never before heard the song we were now recording, clustered around in a wide bungalow, jamming out the slow and dreamy instrumental track I now know as 'Brilliant Love' (Cherry Death, 2016). To me, Tim's faith in his friends to stumble in from the street and record on his upcoming album, his ability to record new material live off the floor with an ensemble of musicians who did not know the song, and to finish it by the second take was a perfect testament to the utopian feeling of Freak City.

The utopian and dystopian affects circulating through Freak City's virtual lifeworld did not merely exist as black and white, however. Throughout my stay in Oklahoma, I witnessed the complex and fluid interplay between the two, echoing Muñoz' mention of the coterminous relationship between annihilation and innovation (2013: 98). I had first felt this upon initially meeting the members of American Hate after their Chicago performance in the summer of 2014. Having been somewhat terrified by the band's violent and destructive performance, I could not help but notice the glowing smile on Ross' face as he smoked in the back alley of the show space drenched in sweat, members of the audience surrounding him with congratulations on the band's incredible set. Back at The Shop, I would see fresh new faces beaming as they slammed back and forth to the tough hardcore punk of bands like Life as One and Leashed. Everywhere there was frustration, anxiety, and an appetite for the destruction of the menacing world of the Bible Belt in Freak City; it seemed to be coupled with a desire to create better worlds of community, belonging, and diverse forms of self-expression. Inspired by the infectious energy surrounding me, I drove to a lake north of the city with Tony and Crutch's front man Garett on one of my last days in town. Sitting on the bank of the gleaming river in the hot midday sun, I remember rubbing Oklahoma's famous red dirt on my jeans, hoping that it would never wash off.

2.8 It Starts with a Feeling

Upon returning home from Oklahoma, I was glowing with the energy of the local scene I had spent my time with, yet I was also somewhat confused over how to theorize my observations on punk in Oklahoma. My initial premise of a hateful structure of feeling in the American Bible Belt, and its manifestation within a specifically violent local hardcore punk sound, had proven to be largely fictive. Not only had there never been a specific local sound and style to punk in Oklahoma, but the violent feelings I had associated with Bible Belt punk only seemed to account for half of the picture, leaving out the vibrant sense of utopianism that animated the scene. Since these observations seemed a far cry from the juicy and sensationalistic story that I had been chasing — of an emerging regionalized sound fuelled by the anguished cries of alienated punk rockers in the heart of the American Bible Belt — the 'so what' question loomed in my mind.

My answer to the question is as follows: it starts with a feeling. Even before condensing into aesthetically coherent forms, genres, and sounds, music begins as the expression of affects and emotions. Rather than communicating meaning through language, music functions through what Suzanne Langer terms 'implicit meaning', an 'articulate but non-discursive form having import without conventional reference' (1953: 29). Occasionally, a tangle of feelings specific to a particular place in time starts to solidify into something vaguely aesthetically coherent, but whether this is a structure of feeling, a singularity, or a record collector's fantasy is a question for another day. The emphases on becoming and singularity that characterize Brian Massumi's affect theory certainly resonate with the uncodified intensities of the Freak City scene, yet I argue that this does not make research on the codification and signification of aesthetic form any less relevant or interesting. The relationships between the feelings articulated through aesthetics, the cultural and material circumstances through which these feelings emerge, and formal modes of articulation are incredibly complex, but the indeterminacy of these relations should not deter research on dynamic processes which ultimately shape the recognition of categories in social life. Only time will tell whether the recent growth of the Oklahoma City DIY scene will become recognized as footprints in a dusty red dirt road towards a regionalized sound, but the difference it made in the lives of its participants is undeniable.

References

Anderson, B. 2014. *Encountering Affect: Capacities, Apparatuses, Conditions*. Burlington: Ashgate.

Caulfield, J. 1989. '"Gentrification" and Desire', *Canadian Review of Sociology and Anthropology* 26(4): 617–632.

Deleuze, G., and F. Guattari. 1994. *What is Philosophy?* Translated by H. Tomlinson, and G. Burchell. New York: Columbia University Press.

Hebdige, D. 1979. *Subculture: The Meaning of Style*. London: Methuen.

Hutta, J. S. 2015. 'The Affective Life of Semiotics', *Geographica Helvetica* 70(4): 295–309.

Langer, S. 1953. *Feeling and Form: A Theory of Art Developed from Philosophy in a New Key*. New York: Scribner.

Massumi, B. 1995. 'The Autonomy of Affect', *Cultural Critique* 31: 83–109.

Mazzarella, W. 2009. 'Affect: What is it Good for?' In *Enchantments of Modernity: Empire, Nation, Globalization*, edited by S. Dube, 309–327. London: Routledge.

McMahon, M. 2002. 'Beauty: Machinic Repetition in the Age of Art'. In *A Shock to Thought: Expression After Deleuze and Guattari*, edited by B. Massumi, 43–48. London: Routledge.

Muñoz, J. 2013. 'Gimme Gimme This... Gimme Gimme That: Annihilation and Innovation in the Punk Rock Commons', *Social Text* 31(3): 95–110.

O'Sullivan, S. 2001. 'The Aesthetics of Affect: Thinking Art Beyond Representation', *Angelaki: Journal of the Theoretical Humanities* 6(3): 125–135.

Williams, R. 1977. *Marxism and Literature*. Oxford: Oxford University Press.

Other Media

Cherry Death. 2015a. 'Bite My Nails' (from the album *Bite My Nails*, self-released).

_____. 2015b. 'Puppet Dance' (from the *Stone Shake Golden Mile EP*, self-released).

_____. 2016. 'Brilliant Love' (from the album *Saccharine*, City Baby).

Crutch. 2016. 'Sin-Makers' (from the album *Crutch S/T 2016*, self-released).

The Germs. 1979. 'What We Do is Secret' (from the album *(GI)*, Slash Records).

The Wizard. 17 January 2010. 'Multiple Choice' (blog, *Terminal Escape*). Available online at https://terminalescape.blogspot.co.uk/2010/01/

_____. 27 March 2012a. 'Autonomy' (blog, *Terminal Escape*). Available online at https://terminalescape.blogspot.co.uk/2012/03/

_____. 28 March 2012b. 'Grasseaters' (blog, *Terminal Escape*). Available online at https://terminalescape.blogspot.co.uk/2012/03/

_____. 7 July 2014. 'Subsanity' (blog, *Terminal Escape*). Available online at https://terminalescape.blogspot.co.uk/2014/07/

Chapter 3

Musical Pathways Through Algerian-London

Stephen Wilford
Faculty of Music, University of Cambridge, 11 West Road, Cambridge, CB3 9DP, United Kingdom
stw31@cam.ac.uk

3.1 Introduction

London's Algerian diaspora community has grown significantly over the past two decades as individuals and families have moved from North Africa and France to the UK to study and work. As the city's Algerian population has expanded, so has a local music scene that embraces everything from ensembles of *andalusi* and *chaabi* performers to rappers and *raï* singers.

In this chapter, I suggest that the musical practices of Algerians in London are closely entwined with the city in which they live and that through musicking Algerian performers and audiences construct a

Musical Spaces: Place, Performance, and Power
Edited by James Williams and Samuel Horlor

ISBN 978-981-4877-85-5 (Hardcover), 978-1-003-18041-8 (eBook)
www.jennystanford.com

shared notion of 'Algerian-London' (Small, 1998).[1] Music, I argue, plays a vital role in bringing together people from across the city and creating a sense of social cohesion, while simultaneously demarcating a shared cultural identity for Algerians within the public sphere. While this identity is unique to London, it is shaped by an ongoing cultural dialogue with other Algerian populations in North Africa and France. It is through this interaction and negotiation of the local and transnational, I claim, that music produces a distinct, if fluid, sense of belonging and identity for Algerians in contemporary London. While this shared identity is complex and multi-layered, ensuring that the very idea of an Algerian community within the city remains in a state of constant negotiation, music nevertheless produces social interaction and a feeling of kinship unique to the local diaspora.

I draw upon the works of two scholars to provide a framework for this chapter. Firstly, I am influenced by Ruth Finnegan's notion of urban 'pathways', which she explicates in her classic ethnographic study of music-making in Milton Keynes in the UK. While the pathways that Algerian musicians and audiences in London travel might not be as tangible or established as those featured in Finnegan's work, they certainly correspond with her suggestion that such pathways are 'not all-encompassing or always clearly known to outsiders, but settings in which relationships (can) be forged, interests shared, and a continuity of meaning achieved in the context of urban living' (1989: 306). Finnegan claims that such pathways embody a 'symbolic depth' and while they are not necessarily bounded and fixed,

> They form broad routes set out, as it were, across the city. They tend to be invisible to others, but for those who follow them they constitute a clearly laid thoroughfare both for their activities and relationships and for the meaningful structuring of their actions in time and space.
>
> (1989: 306, 323)

It is by traversing such pathways across the city, I suggest, that Algerian music and musicians move around London and construct a

[1]I employ Small's term 'musicking' throughout the chapter to articulate the collective processes of musical production, listening, and circulation that take place among Algerians in the city and beyond. As Small notes, this is 'not so much about music as it is about people, about people as they play and sing, as they listen and compose, and even as they dance' (1998: 8).

sense of shared cultural identity. I also extend Finnegan's pathways to include the transnational flows of music that circulate between the UK, France, and North Africa and that play an important role in demarcating Algerian-London from diaspora communities elsewhere.

Secondly, my thinking is shaped by Ayona Datta's notion of a 'geographical turn' within diaspora and transnationalism studies, from which she claims that 'the movement of goods, ideas, people, and capital across (and beyond) nation-states allows us to explore more situated politics of power that shape migrants' relations to other spaces and scales' (2013: 89). I am drawn to Datta's plea to move our thinking about diaspora and urban space beyond notions of forced migration, asylum, and a desire to 'return', instead thinking of cities as spaces in which new and vibrant diasporic cultural identities are constructed. I suggest that this is the case for the Algerian community in contemporary London, for whom music plays such a vital role in producing a unique shared identity and negotiating an understanding of the city, and this chapter aims to address Datta's call for 'framing cities as laboratories in the making of identities, difference, otherness, and the production of home and belonging during the movement/immobility of people' (*ibid.*).

The chapter is based upon four years of extensive fieldwork with Algerians in London between 2011 and 2015. The quotations included throughout the chapter are drawn from personal interviews with interlocutors and have been anonymized where appropriate to protect the identity of individuals.

3.2 London's Algerian Community

The Algerian community in London is far smaller than those of many large French cities, but still forms one of the largest Algerian populations outside of North Africa. Official statistics regarding the number of Algerians living in the city remain unclear, but numbers have undoubtedly grown significantly since a 2004 report estimated a population of 20,000–25,000 (Communities and Local Government,

2009).[2] Many individuals moved to London in search of work, to study at one of the city's universities, or to escape the violence of the political situation in Algeria in the 1990s (Collyer, 2003).[3] By the early 2000s, a significant Algerian community had emerged on and around Blackstock Road in Finsbury Park, in the north-east of the city, but negative media reports at the time suggesting links to terrorism meant that many people subsequently elected to move away from the area.[4] The effects of this negative stereotyping have never entirely disappeared, with one Algerian musician stating that 'some people think [of Algerians that] "they are trouble-makers, they are terrorists"'. Others are concerned with the lack of interest in Algerian culture among the general public in London, and a local radio station owner argues that

> Until now it has been as if Algeria is just an alien country to everyone. They don't know anything about it. Either the media has decided not to talk about it, or it has focused on politics rather than culture. We have got to a stage where we have to be proud of who we are, and we have to reconnect with who we are.

He adds that

> A lot of people [in London] don't actually know about Algerian culture. Maybe they know that when it comes to music, it's *raï*, and when it comes to food, it's couscous. So there is a tremendous area that needs to be explored, and to be shown and exposed. But unfortunately it's really understated, and people don't know anything about it.

---

[2]Making accurate statistical comparisons between the Algerian diaspora communities in the UK and France is difficult, but the French 2011 census recorded over 465,000 Algerian-born individuals residing in France. This figure does not include those born in France who consider themselves Algerian.

[3]Between late 1991 and early 2002, a violent conflict occurred between the government and an armed Islamist opposition. Estimates of deaths caused by the conflict vary significantly, from 50,000 to 200,000. With much of the country engulfed by violence in the early 1990s, many elected to leave Algeria. Major cities across northern Algeria were affected by the conflict, but the majority of those arriving in London had migrated from the capital city of Algiers. For a detailed discussion of this conflict, see Evans and Phillips (2007).

[4]An example of a headline in the British media at this time, from an article in the *Daily Telegraph*, is '100 Known Algerian Terrorists Came to this Country as Asylum Seekers' (Laville, 2003).

This apparent lack of knowledge about Algerian culture in London is a concern for many in the community, and musical performances offer one way of increasing public awareness.[5] There is a strong sense of nationalistic identity felt by many Algerians which can be explained, in part, by the efforts of the Algerian government to promote nationalism and Arabization after independence from French colonial rule in 1962.[6] A large majority of those now living in London grew up in postcolonial Algeria and were exposed to this state-promoted nationalism at an early age, and while notions of a shared Algerian identity do not go unquestioned, this helps to explicate the heightened sense of collective belonging among Algerians in the city and the desire to publicly promote their culture.[7]

While strong connections to North Africa endure, there are also many other transnational links, particularly with Algerian communities in France. Despite a fractious shared history resulting from the legacy of colonialism, France remains home to a significant Algerian diaspora, and those that have moved from France to the UK differentiate themselves from other members of the local community. A female music promoter that I interviewed, who was born in north-eastern France, believes that differences are apparent in the musical tastes of those born in Algeria and France, and she claims that 'many French-Algerians might listen to *raï*, but they're not very likely to listen to *ma'luf* or *gnawa*. Some would... but we are more Westernized in the way that we listen to music. We listen

---

[5]Other opportunities for social interaction among the community include book clubs, political groups, public lectures on topics relating to Algeria, and small cultural organizations. But these are rare in comparison to the number of musical activities that exist within the local community.

[6]The promotion of Arab culture and language has been problematic for many in Algeria. In 1980, the 'Berber Spring' protests were a result of political unrest and a questioning of the predominance of a monocultural Arab identity for Algeria. By disputing Arabization, Martin Evans writes that the Berber Spring 'challenged one of the cornerstones of Algerian nationalism' (2012: 357).

[7]A 2007 report claimed that almost 50 percent of Algerians living in London were aged 30–39 and noted that 'most Algerians have therefore arrived in the last 12 years. These figures must be increased further because Algerians really started applying for asylum in significant numbers from 1995' (IOM, 2007: 13).

to pop music, we listen to reggae, we listen to R&B'.[8] Here she points to the perceived differences that coalesce broadly around notions of 'Western' modernity and cultural 'authenticity', in which those who were born in Algeria feel a stronger connection to socially respected musical traditions, while those born in Europe generally engage more eagerly with 'hybridized' musical forms which are often produced and recorded in France.

There is an ongoing tension, therefore, between a desire for cultural unity and a shared sense of community and internal differences and diversity within the local Algerian population. Music plays an important role in negotiating and overcoming some of these anxieties, bringing the local community together, forming transnational connections, and producing a shared Algerian-London identity.

3.3 Algerian Music in London

Algerian music in London is performed, disseminated, and heard within a variety of public and private contexts. These range from spaces restricted almost exclusively to Algerian audiences, such as the cafes scattered across the city that regularly host performances of *chaabi* and *raï*, to more public contexts, such as concerts, festivals, and weekly broadcasts by the city's only Algerian radio station. While there is a desire to expose Algerian culture to the city's wider public, audiences in each of these contexts remain predominantly Algerian.

The social interaction facilitated by collective acts of musical performance is fundamental to overcoming the physical dispersal of Algerians across London and the lack of public space for the community. One musician, who arrived in London in the late 1990s, complains that 'it's not like with the Turkish, if you go to Dalston. You find Pakistanis in Whitechapel. But we haven't got that special area...

[8]*Raï* is a form of popular music that emerged in the western Algerian city of Oran in the early twentieth century and found international success in the 1990s, particularly among the diaspora in France. *Ma'luf* is a form of *andalusi* high art music from the east of the country, while *gnawa* is a form of spiritual music performed in the west of Algeria. It is important to note that *raï* is considered socially disreputable, while *ma'luf* and *gnawa* are generally more respected.

our own area'. Without a sense of grounded belonging in a particular area of the city, music therefore provides something around which the community can coalesce, and a local radio station owner claims that

> It doesn't matter what you earn. If you are Algerian, that's fine. Most of the people I know, we don't have that differentiation between job titles, or you have papers or you don't have papers. For us, if you are Algerian, you are Algerian. We get on fine, there are no worries. We have that solidarity.

While some challenge these assertions, highlighting the diversity of the local community along lines of education and class, there are many contexts within which music undoubtedly fulfils a unifying function. Perhaps the most salient are the handful of Algerian cafes found across the city, often on high streets in predominantly residential areas. These cafes are spaces almost exclusively reserved for men and host performances of styles such as *raï* and *chaabi*.

Chaabi emerged in the working-class cafes of urban Algiers in the early twentieth century and continues to provide the soundscape for neighbourhoods in the Algerian capital, such as the iconic Casbah. Mustapha Harzoune (2013) claims that 'Algiers would not be Algiers without chaabi', while James McDougall, describing the history of the Algerian cafe, writes that

> In both urban and rural areas, cafes were vital and multifunctional spaces of Algerian male sociability, both preserving social ties and exclusions, and... incubating a new 'civil society' where work, music, news, football and politics were all organised and expressed.
>
> (2017: 108)

Chaabi plays an important role in the lives of a small network of musicians in London, many of whom will travel significant distances in the evening or at weekends to meet and perform together. Reflecting its roots in the urban working-class neighbourhoods of Algiers, *chaabi* in London remains the preserve of groups of male performers who meet in cafes scattered across the city. Rostomia, a cafe run by two Algerian brothers on Goldhawk Road in west London, is typical of such venues and acts as a meeting point for a number of the city's Algerian musicians. Ali, a darbuka player who lives in south London, regularly travels on his own musical pathway across the city, on his motorbike, to perform there and explains that 'I'm from Algiers. So most of my interest was in it, because every time there's a wedding party you'll see bands playing music and it will be *chaabi* music. You go out and get in the car with my dad and it's *chaabi* music. So that's what I was really interested in'. These musicians are mostly men in their late 30s or older, who perform the more traditional *chaabi-malhûn* repertoire made popular by the likes of Mohamed El Anka.[9] Performances can last for a number of hours, and while they help to sonically demarcate the cafe as Algerian, they often garner relatively little interest from the cafe's regular clientele.[10] It is not uncommon to see a group of musicians sitting at the back of Rostomia and playing *chaabi* together but having little interaction with the other people present, and thus while the music serves to bring the performers together, its influence upon the wider community is somewhat limited. Such musical performances are certainly not without criticism, and one Algerian musician highly active on the local scene claims that 'the *chaabi* people' do not understand the lives of young Algerians in London, arguing that 'if you do a *raï* festival and a *chaabi* festival, you will find more people at the *raï* than the *chaabi*. This is the truth. Because those *chaabi* people are still stuck in the tradition, in the old clothes'.

Chaabi's influence upon the wider community is limited by the age and gender demographic of many of its performers, and the cafes in which it is primarily performed are not spaces in which women are generally welcome. A female musician highlights the restrictive nature of this situation, noting that 'I think in these areas people still meet in cafes and the cafes that are Algerian are only male. It's not a rule, but that's what tends to happen. And so for women, it's not as easy I think to meet with the community'. Therefore, while *chaabi* plays an important role in the lives of a number of musicians in the

[9]*Chaabi-malhûn* is closely identified with the working-class neighbourhoods of Algiers and developed from the local poetic *malhûn* traditions (Morgan and Nickson, 2006: 7). Elsner writes that a traditional *chaabi* lyric 'used the vernacular (*malhûn*) and was formally related to the *qasīda* in its strophic structure and equal rhymes over several lines and half-lines of verse' (2002: 475).

[10]Tony Langlois writes that traditional *chaabi* songs 'consist of lengthy narrative songs sung by a single performer interspersed with vociferous choral sections involving the ensemble' (Langlois, 2014).

city, it acts as a constrained musical subculture within the broader local diaspora community.

Given the limited number of performers and restricted audiences for traditional musics like *chaabi*, larger public events and festivals provide an important public space in which Algerian music-making can take place. Whereas smaller venues, such as cafes, are often dispersed across the city and in areas with relatively low rental costs, these larger events commonly take place in the centre of London, allowing easy access to them via public transport.[11] Musicians and audience members trace pathways across the city to coalesce and share musical experiences. Events and festivals provide opportunities for local Algerian musicians and offer a public space for community interaction.

The Algerian Cultural Festival was held in October 2012 to celebrate the fiftieth anniversary of Algerian independence from French colonial rule and took place at Rich Mix, a community arts venue in Shoreditch, east London. Organized by a group of young professional Algerians living in the city, the Festival programme explained that

> The Algerian Cultural Festival (ACF) was born out of a desire to both commemorate a momentous anniversary in Algerian history, an important date little talked about in Britain, and to promote the country's rich cultural heritage, under-stated, under-covered and often misunderstood abroad... Fragmented and lacking cultural references, the Algerian community in the UK is largely absent in Britain's showcase of international folklore... Algerian cultural manifestations are seldom seen across the UK, and initiatives to connect and enable expression and exchange are few and far between.
>
> (Anonymous, 2012)

The festival brought together a broad range of musicians and ensembles from across London and was considered a momentous event among many within the local community. Houria, a musician and artist who moved to London from Algeria in the late 1970s, reflected that

[11]Areas that are home to well-established Algerian cafes include Finsbury Park (north-east London), Hounslow (south-west London), and Shepherd's Bush (west London).

> I think that this cultural event is really vital and important in terms of bringing Algerians together. This kind of cultural festival in London has never happened, for years. I have been here since 1977 and we were a tiny Algerian community in London... These kinds of event, like the Algerian Cultural Festival, it is really something very valuable and it should be done every month!

Karim, one of the event's organizers, believes it was important for the festival to provide a platform for local Algerian musicians and he claims that

> It was for the local community and that is exactly what I wanted. I said 'look, if you want to celebrate something, let's give an opportunity to local bands'. Because every time there is a gig, we have flown in someone [from France or Algeria] to be the star for the community. Let us say 'within the community we can have fun with what we have!'

This focus upon local performers was extremely important and meant that the festival was not simply a representation of Algerian culture, but specifically of the music of the Algerian-London community. It offered a space within which a shared community identity was constructed and negotiated and featured performances of a range of musical styles, from the more traditional *andalusi*, *chaabi*, and *ma'luf* to hip-hop, *raï*, and *gnawa*-influenced blues-rock. The demographic of the musicians and audience at the festival was similarly heterogeneous, and the event therefore simultaneously highlighted the diversity of the local Algerian population and produced a sense of community cohesion and unity through shared acts of musicking.

While the focus of the festival was primarily local and featured acts based in London, it drew an audience from across the UK who were eager for an opportunity to reengage with Algerian culture. Karim spoke to individuals at the festival who had travelled to London on buses from Manchester and Glasgow to attend and recalls one particular conversation with an audience member:

> He said 'the festival is for Algeria, and independence', and they wanted to be part of it... People were there for the fiftieth anniversary of independence. I don't think they came only for the music. They came for music, and a night out, and to be Algerian, and to be proud for one evening.

The Algerian Cultural Festival therefore provided an important space within which a sense of community and cultural reengagement was produced through shared musicking. Although a festival of this scale is yet to be repeated, it marked an important moment in which community cohesion appeared possible to many within the local diaspora. While such events focus upon local music-making practices in London and the development of a defined Algerian-London culture, the collective identity of the local community is also shaped by factors beyond the city.

3.4 Algerian Music and the Internet

The physical distances that separate members of the local Algerian community within the city, and the lack of established venues for musical performances, mean that the internet has taken on a particular significance for Algerian musicking in London.[12] Online interactions, especially via social media, enable musicians and audiences to organize musical performances locally, while engagements with Algerian populations elsewhere in the world serve to shape and delineate Algerian-London culture as unique.

At the most fundamental level, internet communication enables individuals to discuss music and coordinate performances. Samir, a percussionist who is very active on the local scene, articulates this through his own frustrations at the lack of performance opportunities for Algerian musicians in the city. He is a highly active user of social media (particularly Facebook) and declares: 'Thank God that there are social things, on the internet, so you can communicate, leave a message, even share an event'.

While Samir uses social media to interact with friends, the Al-Andalus Caravan employ the internet to facilitate their ensemble's musical activities. The group is the city's leading proponents of *s'ana*, a school of traditional North African *andalusi* music that is associated with the city of Algiers, and they meet each Sunday afternoon to

---

[12]For example, members of the Al-Andalus Caravan (discussed below) travel from north-east and south London, as well as other towns and cities in the south-east of England (Reading and Woking), to attend weekly rehearsals in central London. Tewfik, the ensemble's Director, spends around two and a half hours driving to and from these rehearsals each week.

rehearse in a church hall in Pimlico, central London.[13] Their director Tewfik, a trained and highly respected musician, explains that

> It is possible in fact to use the technology for what we do. So Facebook is a very important point of contact. To look at pictures, and listen to records, spread messages, inform people of what's going on and what's happening... At the end of the day, technology was intended to help people.

The Caravan describe themselves as an association, a term that connotes a collective desire to protect the traditions of *andalusi* music through performance and teaching. Jonathan Glasser writes that Algerian 'amateur associations are the preeminent spaces of initiation for novices, of musical advancement for initiates, and of intensive sociability for all' (2016: 175). The Caravan fulfil this function for Algerians from across the city, and the association member Yasmine suggests that

> I think it's a very nice thing to have for the [Algerian] community [in London]. It's not that this is important for the association, but that this is important for the community. People who have started from scratch and have someone like Tewfik to teach them from scratch, to teach them about Andalusian music.

Tewfik runs two teaching and rehearsal sessions each Sunday for novice and more experienced members, respectively, but his time with the musicians is limited. The ability to interact and share information online is therefore vitally important, and he uses the association's Facebook page to share audio and video files that serve to teach repertoire and techniques to the musicians. His primary concern, however, is exposing the Caravan's members to the sound world of *andalusi* music and he notes that 'what I want them to do is to listen to this music because by listening to music, you will record it and you will have that flavour. I focus so much on the flavour. There's no point in playing music without giving the flavour of it'.

---

[13]*S'ana* is one of the three schools of *andalusi* music performed in Algeria along with *gharnati* in the west of the country and *ma'luf* in the east. Given its connections with the capital city of Algiers, it is perhaps the most prestigious of the three and is most popular among the educated wealthy classes of Algerian society. *Andalusi* music is based upon the concepts of *tab'* (comparable to Western ideas of mode) and *nubat* (a song cycle).

This engagement with music, he claims, has clear benefits for the well-being of the musicians, and he describes their musical activities as 'a good way for relaxation... it's made a lot of changes, not only for myself, but for the other guys as well... This type of music just reminds you how life can be beautiful... And you will say "yes, work is important, but we work to live, not live to work"'.

At the same time, social media allows the Caravan to interact with associations in other parts of the world. The term association does not simply denote a discrete and locally sited ensemble, but describes groups within a transnational network of music-making that share a focus upon performance, education, and nostalgia (Shannon, 2015). Through online interaction, the Caravan connect with other ensembles in this network, and Yasmine notes that

> There are a lot of [videos on the Caravan's Facebook page] that are of associations in Algeria, which is good because it kind of creates a community across borders, on one page... What's good about it is that with the people who are in the associations in Algeria or France, we could exchange ideas or videos of what they think is good, and we can go and talk about it.

The internet therefore plays a dual role for the Caravan, enabling rehearsals, performances, and teaching locally, while constructing connections with associations globally.

The Caravan's membership remains relatively small, but the internet also facilitates processes of musical engagement among the wider local community. One way in which this occurs is through the broadcasts of Rihet Bladi, the UK's only Algerian radio station, which is run by restaurant owner and music enthusiast Djamel from a small room in Brixton, south London. He created Rihet Bladi to imitate similar radio stations in the UK that represent diasporic communities from other nations and states that 'I wanted to liaise with the Algerians here in the UK. Plus to promote the Algerian culture, through the music and the shows we do, and to actually engage some Algerians'. The station's weekly Sunday evening broadcasts include features and interviews (with Algerians in the UK, France, and North Africa) but dedicate much of their time to Algerian music, whether playing pieces from Djamel's extensive record collection or notifying the community of forthcoming performances. Djamel's rules for the station are to never discuss politics and to only play music that he deems appropriate for a family listenership. This excludes much repertoire from certain genres, such as *raï*, but Djamel claims that

> I have a lot of families listening to me on a Sunday. They have their dinner when the radio starts, they put it on in the background and listen. They are here in London, here in England. And they say to me that it just reminds them of home.

The station's broadcasts embrace a similar diversity of music to that performed at the Algerian Cultural Festival, and Djamel is keen to point out that 'you can't just play *chaabi*, you can't just play *hawzi*. I think every audience has to have its share'. He also believes that his radio shows are bound up in personal memories and nostalgia and serve to ameliorate the daily stresses of life in London for his listeners, adding that

> I need to give him [the Algerian listener] something to take his mind off everything and bring him back to memories of his family when he was little. We put on something that makes him go 'oh this song reminds me of this, this reminds me of that'. The listener starts to go back. And it becomes, the radio, part of his childhood and his memories, and he will cherish it. So that is exactly my idea, to give them something to bring them back to where they used to be. Things that with being here in London, they forgot all about.

While some members of the community are critical of Djamel and his station, suggesting that it is self-serving, many describe the importance of individual and collective memories, and music seems particularly central to such processes of remembering. While the Caravan draw upon, and engage with, the shared nostalgia of *andalusi* music through online interactions, Djamel's radio broadcasts evoke personal memories for his individual listeners. In both cases, we might understand music as taking on the role of what Pierre Nora calls *lieux de mémoire* (sites of memory), physical and metaphysical spaces 'where memory crystallises and secretes itself' (1989: 7). Nora's *lieux* are sites that can evoke and reproduce collective memories and serve to challenge the 'breaking apart' of the contemporary and historical. Similarly, music in this context draws together the past and present for London's Algerian community and creates specific meanings around musical performances and recordings.

Musicians in London also employ the internet to distribute their music to Algerian audiences in other parts of the world. The Papers are a group based in London that were formed by brothers Yazid and Massyl. Their sound combines a number of musical influences but is not sonically identifiable as 'Algerian', yet they have been embraced by the local community and appeared at the Algerian Cultural Festival. Yazid explains that

> I wanted the Algerians to be proud of us... Even if it's not necessarily your music. As long as you think it's good, it's cool and you like it, and you see that we are Algerian. If I saw something and I was on the fence a bit with it, but then I saw that there was an Algerian flag, I'd feel like out of solidarity I should like it.

Although the group has been welcomed into the local music scene, he believes that their music remains more popular with Algerian listeners outside of London, suggesting that

> The local Algerian community [in London] that we know, I've found is either students or have their own tastes. A lot of them are into politics, and musically it's not really connecting just yet. Maybe certain tracks that are more political might work with them. But because it's not specifically an Algerian style, it's difficult.

The group share their music online, through Facebook, YouTube, and other platforms, but he adds that they have struggled 'finding Algerian groups, and trying to push it on that side, they weren't that receptive'. In contrast, they have gained more positive responses from Algerian audiences outside of the UK, and Yazid believes that this is because of their Algerian-London identity. He explains that

> We actually tried to push it more in Algeria, back home, because we figured that they have their Algerian rap out there, *chaabi* or whatever, and ours is a bit of a different flip on things. It's European, it's Western music, but we're Algerian, there's Algerian in there.

For The Papers, having a unique identity provides a degree of cultural capital, and by differentiating themselves from traditional forms of music, and Algerian musicians elsewhere, they have been able to garner interest in their music via online distribution.

Through online interaction with music and musicians across the city, and throughout the transnational diaspora, a unique Algerian-London cultural identity is constructed that draws upon a shared past and individual experiences in present-day London. Social media and file-sharing platforms facilitate performances, rehearsals, and active listening, while musicians and sounds flow along sonic pathways that traverse the city and connect the UK with France and North Africa.

3.5 Conclusion

Ayona Datta identifies a tension within urban diaspora and transnationalism scholarship that she calls the 'structure-agency divide' (2013: 91). This term, she explains, describes the tendency for research to fall into two broad categories: 'transnationalism from above' and 'transnationalism from below', or those studies focusing upon macro structures of control and those engaging with the individual experiences of members of diaspora communities. This chapter attempts to understand the experiences of Algerians in London, and their involvement in musicking practices, from a place between these two binary positions. On the one hand, Algerian culture in London is structured by the city and is shaped by the sense of dispersal and marginalization that many members of the local community encounter. Algerians often find themselves living and working across London, with few opportunities to engage with one another, and little knowledge or interest in Algerian culture within the wider public sphere. It is here that music plays an important role in simultaneously facilitating social interaction and providing a sonic and visual identity for Algerian culture in the city.

On the other hand, Algerian musicians and audiences are actively involved in the development and mediation of a vibrant Algerian-London culture. Music is performed and experienced in a variety of contexts, from small cafes to festivals and from rehearsal rooms to online radio shows. It is shaped by a shared history and frequent nostalgia for North Africa, and by connections with Algeria and France, but it is also unique to the local community in London. Given the relatively small size of the local diaspora, musicians and audience members often know each other personally and perform alongside one another. Musics that might remain discrete in Algeria or France, such as *andalusi* and *raï*, appear together on festival programmes and radio broadcasts, producing a vibrant and diverse local music scene.

The local Algerian music scene in London is not static however, and both musical sounds and musicians travel along pathways that are real and imagined, local and transnational. These pathways are sometimes obligatory, in the sense that musicians and audience members are required to traverse the city in order to experience and perform music together. But they are also voluntary, in the ways that individuals elect to join ensembles, share audio-visual materials, and listen to local radio broadcasts. Expanding upon Finnegan's original concept, I suggest that sounds, and their associated meanings, also travel along 'pathways' that flow between the UK, France, and Algeria, connecting people, places, and memories. It is through these local and transnational musical flows, and the encounters that they produce, that musicking enables Algerians in the city to construct, negotiate, and reify a unique and vibrant shared Algerian-London culture.

References

Datta, A. 2013. 'Diaspora and Transnationalism in Urban Studies'. In *A Companion to Diaspora and Transnationalism*, edited by A. Quayson, and G. Daswani, 88–105. Chichester: Wiley Blackwell.

Evans, M. 2012. *Algeria: France's Undeclared War*. Oxford: Oxford University Press.

Evans, M., and J. Phillips. 2007. *Algeria: Anger of the Dispossessed*. New Haven and London: Yale University Press.

Finnegan, R. 1989. *The Hidden Musicians: Music-Making in an English Town*. Cambridge: Cambridge University Press.

Glasser, J. 2016. *The Lost Paradise: Andalusi Music in Urban North Africa*. Chicago: University of Chicago Press.

Harzoune, M. 2013. 'Salah Guemriche, Alger la Blanche, Biographies d'une Ville', *Hommes & Migrations* 3(1297): 170

Langlois, T. 2014. 'Algeria'. *Grove Music Online*. Available online at https://doi.org/10.1093/gmo/9781561592630.article.42951

McDougall, J. 2017. *A History of Algeria*. Cambridge: Cambridge University Press.

Morgan, A., and C. Nickson. 2006. 'Algeria, Raï: Rocking the Casbah'. In *The Rough Guide to World Music: Africa and the Middle East*, edited by S. Broughton, M. Ellingham, and J. Lusk, 6–21. London: Penguin/Rough Guides.

Nora, P. 1989. 'Between Memory and History: *Les Lieux de Mémoire*', *Representations* 26: 7–24.

Shannon, J. H. 2015. *Performing Al-Andalus: Music and Nostalgia across the Mediterranean*. Bloomington: Indiana University Press.

Small, C. 1998. *Musicking: The Meanings of Performing and Listening*. Middletown, CT: Wesleyan University Press.

Other Media

Anonymous. 2012. *Algerian Cultural Festival*. Rich Mix, London, 20th October 2012 (festival programme, Algerian Cultural Collective).

Collyer, M. 2003. *Explaining Change in Established Migration Systems: The Movement of Algerians to France and the UK* (working paper, Sussex Centre for Migration Research). Available online at www.sussex.ac.uk/migration/documents/mwp16.pdf

Communities and Local Government. 2009. *The Algerian Muslim Community in England: Understanding Muslim Ethnic Communities* (report, commissioned by the Department for Communities and Local Government). Available online at https://webarchive.nationalarchives.gov.uk/20120919192752/http://www.communities.gov.uk/publications/communities/algerianmuslimcommunity

IOM [International Organization for Migration]. 2007. *Algeria: Mapping Exercise, London, September 2007* (report, commissioned by the International Organization for Migration). Available online at www.iomuk.org/doc/mapping/IOM_ALGERIA.pdf (last accessed 26 September 2012)

Laville, S. 17 January 2003. '100 Known Algerian Terrorists Came to this Country as Asylum Seekers' (newspaper article, The *Daily Telegraph*). Available online at http://www.telegraph.co.uk/news/uknews/1419075/100-known-Algerian-terrorists-came-to-this-country-as-asylum-seekers.html

Chapter 4

Dancing to the Hotline Bling in the Old Bazaars of Tehran

Shabnam Goli
School of Music, University of Florida, Gainesville, FL, USA
shgoli@ufl.edu

4.1 Introduction

Since its emergence in the 'hoods of the Bronx in the late 1970s, hip-hop has been closely linked to the concept of place. Local abbreviations for 'hoods and telephone area codes have been major localizing devices used by hip-hop artists to unite marginalized members of society around the notion of locality and indigenous issues (Sarkar and Allen, 2007: 125). In response to socio-political maladies including poverty and inequality, a defiant youth subculture emerged which gave voice to African Americans and other minorities in US urban centres and provided them with new spaces for expressions of identity (Forman, 2000: 65).

Musical Spaces: Place, Performance, and Power
Edited by James Williams and Samuel Horlor

ISBN 978-981-4877-85-5 (Hardcover), 978-1-003-18041-8 (eBook)
www.jennystanford.com

Hip-hop's potential to empower the marginalized and its mobility via mediated technology (Nooshin, 2011: 2) transformed the initially local genre of hip-hop into a globally spread and hybrid domain of cross-cultural creativity and communication. Today, hip-hop music can be traced in locations as geographically, culturally, and socio-politically removed as Palestine (Maira, 2008), Turkey (Solomon, 2005), Egypt (Williams, 2010), Germany (Bennett, 1999), Japan (Condry, 2001), and Iran (Breyley, 2014; Johnston, 2008; Nooshin, 2011, 2017), all sharing musical and cultural features of the global hip-hop scene while simultaneously reflecting their own localities.

Outside the US, as Tony Mitchell stresses, hip-hop scenes rapidly moved away from adopting the US musical idioms and forms towards adapting them to meet local needs (2001: 3). Borrowing Roland Robertson's (1995) concept of glocalization to describe the complexity of the intersection between the local and the global, Mitchell discusses processes of localization and indigenization of the global form of hip-hop in diverse locations. The 're-emplacement' of the global form (Solomon, 2005: 51) is achieved through embedding music in specific places and commenting on indigenous realities in locally meaningful ways.

Rather than dealing with locality as a politically, socio-economically, and culturally confined phenomenon, the postmodern world has created a fluid concept of translocality, of being here and there. In such a context, investigating the complexities of hip-hop culture, with its postmodern rejection of metanarratives spread all around the world via new media and technologies, not only provides us with a more nuanced view of the concept of translocality, but also defies dichotomous understanding of global and local and modern and traditional. With the constantly shifting meaning of the concept of locality in today's increasingly interconnected world, where bounded notions of culture and place are contested by rapid advances in telecommunication technologies, hip-hop connects youth across the globe through common taste in music and collective expressions of identity.

In this chapter, I investigate contemporary Iran's rapidly flourishing hip-hop scene in light of Robertson's (1995) concept of glocalization. Informed by the enlightening research of scholars such as Laudan Nooshin (2005, 2011, 2017), Bronwen Robertson

(2012), Nahid Siamdoust (2017), and G. J. Breyley (2014), I examine the complexity of the interaction between local and global forces in the Iranian hip-hop scene. Focusing on the Persian cover of Drake's 'Hotline Bling' by a well-known and highly controversial hip-hop artist Hamid Sefat, I argue that the postmodern nature of hip-hop (its rejection of metanarratives, its tendency to cross boundaries, and celebration of pluralism (Pasler, 2001)), as well as its global appeal, mobility via mediated technology, and accessibility of production equipment (Nooshin, 2011), has shaped it as a translocal site for sociocultural contact between young Iranians and the transnational hip-hop nation. Iranian hip-hop artists contest the confined notion of locality in terms of cultural and geopolitical boundaries. Moreover, they challenge the state's control over the cultural domain. Since the revolution, the state has rejected Western cultures and values in an attempt to fight cultural imperialism and protect its own ideology (Siamdoust, 2017: 5–6). Iranian hip-hoppers have not only claimed membership of transnational hip-hop culture but have created new forms and meanings influenced by global sounds and informed by local realities.

Functioning as what Mary Louise Pratt (1987) termed a 'contact zone', the translocal hip-hop scene in Iran has shifted from imitating the Western form in its early years in diaspora (Nooshin, 2011: 9) to an indigenous medium of expression with strong musical and extra-musical ties to its local setting. Instead of simply imitating and absorbing Western forms and norms, as Iranian authorities claim and warn against using Jalal Al-Ahmad's concept of Westoxification or West-struckness (Siamdoust, 2017: 10), Iranian hip-hop artists use their music as a tool to give voice to local concerns and to empower young Iranians who share the experience of suppression and underrepresentation with their African American counterparts, though in different forms.

From highly political commentaries to rich kids' bragging about their hedonistic lifestyles, Iranian hip-hoppers comment on their lives in modern-day Iran. They utilize the global template to send out localized messages and to transcend geographical, political, and cultural barriers. Examination of Persian hip-hop as a contact zone between local and global musical forces, and in light of Robertson's (1995) notion of glocalization, emancipates us from bounded conceptions of locality (and identity) and illuminates the intricacy of cross-cultural interactions in today's world.

Exploring Iran's hip-hop scene beyond the binary of the local and the global, theories of Westoxification,[1] cultural imperialism, and cultural resistance cast light on the ways through which Iranian youths challenge the state's control of space and culture, forge new translocal, transnational identities, and, most globally impactful, expose modern Iran to the outside world. Defying the media-imposed stereotypes of not only Iranians but also Muslims, hip-hoppers raise awareness about life in other locales and thus play a key role in increasing cross-cultural understanding, which is a major component of a peaceful future.

Moreover, as Nooshin (2017) observes, the representation of Iranian popular music in romanticized discourses of resistance and freedom in global academic and journalistic writings is reductionist and serves wider regimes of orientalist representation. It is thus vital to study Iranian hip-hop beyond the dichotomy of the local's active resistance against or passive absorption of the global. Briefly discussing the situation of hip-hop music in today's Iran and the theoretical paradigms informing this study, I explore how Persian hip-hop functions as a bridge, a contact zone, and a translocal sphere encouraging pluralism and rejecting metanarratives of Westoxification and cultural indigenization against imperialism.

The increasingly growing global presence of hip-hop culture and music has inspired multidisciplinary scholarly inquiries since the 1980s. Groundbreaking studies by Tricia Rose (1994, 2008) and ethnomusicological research on hip-hop have explored issues of race (Perry, 2016), gender (Rabaka, 2011), musical borrowing and copyright (Olufunmilayo, 2006), youth identity (Alim, 2011; Pardue, 2008; Perullo, 2005), and locality (Flores, 2000; Mitchell, 2001). Hip-hop in the Middle East has also been investigated as a window into locales geographically and culturally distant from hip-hop's birthplace. Much attention has been given to introducing music scenes and their particularities or interpreting music as a medium for oppositional youth self-expression and social commentary. Growing scholarship also highlights the struggle between local and global forces in the hip-hop scene (Johnston, 2008; Nooshin, 2011).

---

[1]Westoxification is the destructive plague of being struck by the West, as proposed by Jalal Al-e Ahmad (1982) in his study of Iran in the face of modernization.

While such research offers critical insights into local scenes, the emphasis on the confrontation of the local against the global (Nooshin, 2017), whether submissive or defiant, has led to some crucial blind spots, depriving us of a profound understanding of the on-the-ground realities. The examination of hip-hop scenes reveals that the interaction between local and global cultures is more complex than an either/or question. Rather than re-enforcing such bipolar and reductive readings of the late modern world's socio-economic and cultural movements, Robertson (1995) suggests we examine the ways through which both tendencies have become features of life in the postmodern world.

4.2 Globalization, Glocalization, and Translocality

Globalization theories, the *Zeitgeist* of the 1990s (Rosenberg, 2005), predicted that interconnectedness of the late modern world would eventually transform human society, leading to the complete erosion of the local in the face of the global and replacing the sovereign state with a multilateral, multinational, global system of governance. A decade later, however, globalization faced criticism as an all-encompassing and ambitious theory that distinguishes local and global as opposing poles. Highlighting that locality cannot be understood outside of the global notions of identity, particularism, and universalism, Handler (1994) maintains that even the most extreme assertions of nationality and ethnicity are in fact informed by global terms of identity and particularity. Locality, thus, is shaped, defined, and promoted from the outside.

Considering the local not as a counterpoint to the global but an inseparable aspect of globalization, Robertson argues that accepting globalization as a process that overrides locality is problematic due to its neglect of the complex notion of translocality (1995: 26–30). As Marjorie Ferguson (1992) also stresses, the overuse of the term 'globalization' in both academia and the public sphere has led to the myth of globalization as a powerful homogenizing force, obliterating not only locality but even history around the world. The future world, in Ferguson's view, is increasingly pluralistic (*ibid.*: 81).

Maintaining that globalization — the compression of the world in the general sense — involves not only the incorporation but also the invention of locality, Robertson proposes to replace the term globalization with 'glocalization' (1995: 40). Glocalization clarifies that the two seemingly opposing trends of homogenization and heterogenization are in fact interpenetrative and complementary. The challenge, as Robertson observes, is to pinpoint how these forces interact with and influence each other. Hybridized cultural artefacts, including hip-hop music, embody such complex processes of glocalization. The transnational and postmodern character of hip-hop music (Manuel, 1995) creates a translocal contact zone in which global and local forces communicate in a dialogic manner.

Globally spread, localized hip-hop music rejects the common myth of the local fading into the global and illuminates the dialogic relationship between the two tendencies. Examination of hip-hop highlights the multifaceted processes of blending the universal and the particular, and it suggests that understanding globalization as a tension between a 'McWorld' and a 'Jihad World' (Barber, 1996) is reductive. Discourses of cultural imperialism and Westoxification also fail to address the complexity of the encounter. 'Cultural messages' sent out by imperial powers — containing Western, mainly American culture — are not received and interpreted monolithically around the world (Tomlinson, 1991). Even the most universalistic artefacts and messages, whether it be Shakespeare (Billington, 1992), a McDonald's sandwich, or a hip-hop song, are received and interpreted differently in distinct localities. Moreover, incorporating local messages in the global template casts light on local participants' agency in the construction of new meanings.

Persian hip-hop is a prime example of the multifaceted encounter of the local and the global. It rejects the simplistic readings of the local resisting global forces or struggling to stay alive in the face of complete eradication and obliteration. By going beyond decoding localized messages and paying closer attention to the liberating impact of the new locality created via Persian hip-hop, I inquire about the ways Iranian post-revolutionary youth constructs a new locality that transcends barriers, is outside the control of the state, and is hidden to outsiders. Hip-hop in Iran functions as a site for cultural participation with some far-reaching effects, including resisting stereotypes and portraying the new face of Iran.

4.3 Hip-Hop in Iran

Since its emergence in the Bronx in the late 1970s, hip-hop has swept across the globe, securing a central position in global youth culture. As a 'vehicle for global youth affiliations and a tool for reworking local identity all over the world' (Mitchell, 2001: 1–2), hip-hop has developed a local meaningfulness enclosed in the global form. In Palestine, hip-hop has functioned as a site for youth political expression, education, and solidarity. An aesthetic form used by Palestinian youth, hip-hop has played a key role in raising awareness about the Palestinian condition (Maira, 2008). In Japan, it has been re-interpreted to fit the Japanese context and way of life (Condry, 2001). Having initially emerged in the Iranian diaspora in California in the 1990s (Nooshin, 2011: 9), Iranian hip-hop soon reached a prominent position among the youth and in Iran's alternative music scene. By the mid-2000s, under the influence of President Khatami's liberalizing policies and relaxation of regulations on popular music production, and by taking advantage of technological advances and the internet, a flourishing local hip-hop scene was established inside Iran (*ibid.*).

Iran's vibrant hip-hop scene has attracted much scholarly and journalistic attention since its emergence in the mid-2000s (Nooshin, 2011; Robertson, 2014; Siamdoust, 2017). While Johnston (2008) introduces the scene and conditions of music production, consumption, and distribution, Nooshin focuses on the ways hip-hop has become localized and how it is used as a means of youth empowerment and expression of local messages (2011: 30). Siamdoust's (2017) study of music in Iran provides a comprehensive account of the roots and history of Persian rap as well. Introducing a typology of Persian rap (street rap, gangsta rap, party/love rap), Siamdoust discusses the style and repertoire of some of the major Iranian hip-hop artists, such as Yas, Hichkas, Shahin Najafi, and Justina. Siamdoust investigates hip-hop through the lens of youth empowerment and expression of resistance to sociocultural metanarratives (*ibid.*: 240–250).

To better understand the situation of hip-hop in today's Iran, it is important to have an understanding of Iran's larger popular music scene, expanded throughout the world via the Iranian diaspora and constantly impacted by the political atmosphere inside the country. Soon after the Islamic revolution, Iranian authorities focused on strengthening Iranian-Islamic national identity and creating solidarity on the basis of shared citizenship and the state's ideology (Siamdoust 2017: 9). Having labelled popular music as 'youth decaying' Western influence (*ibid.*: 6) and highlighting its ties to the Pahlavi regime's sociocultural attempts to modernize and Westernize the country (Johnston, 2008: 103), authorities banned popular music for nearly two decades, only accepting certain forms of revolutionary and religious songs (Siamdoust, 2017: 2). Consequently, pre-revolutionary musicians were compelled to abandon their careers or migrate to the West.

The first instances of Persian hip-hop emerged in the US Iranian diaspora in the late 1980s/early 1990s in songs such as 'Shagered-e Avval' (Top Student) by pop star Shahram Shabpareh with a rap section by female vocalist Nahid. The genre later developed in the works of the band Sandy, with its lead singer Shahram Azar (Nooshin, 2011; Siamdoust, 2017: 235). With the arrival of the internet and satellite TV broadcasting, soon exile-based Persian hip-hop, along with Tupac, Dr. Dre, and Eminem, entered Iran. Thus, the accessibility of satellite TVs and the internet in urban areas not only introduced Western hip-hop to Iranian society, but it also reconnected the exiled to their homeland.[2]

It was almost two decades later that hip-hop appeared in the 'undergrounds' of Iran. While the Khatami administration's relaxation of regulations on music permits facilitated musical activities, it did not lead to the emergence of a democratic music scene.[3] The fact that music production and performance still required an officially issued permit pushed many young artists to the peripheries. Women still faced prohibition from solo singing, and lyrical contents and musical outputs were closely examined. Under such circumstances, 'underground music' or *musiqi-ye zirzamini* appeared as a youth sensation sweeping across major urban centres in the late 1990s.

Authorities' rejection of hip-hop, even linking it to Satanism (Siamdoust, 2017: 235) due to its associations with American

[2]For more information about Iranian exile television channels, see Hamid Naficy (1993).

[3]All kinds of artistic and journalistic productions in Iran require official permits issued by the Ministry of Culture and Islamic Guidance. Several scholars including Hemmasi (2010), Johnston (2008), Nooshin (2005, 2011), and Youssefzadeh (2000) have covered the process of applying for and acquiring permits ('*mojavvez*' in Farsi).

culture, its direct language and candid content, and the 'un-Islamic' appearance of musicians and fans, made the genre a major component of Iran's underground music scene. By the end of the 2000s, hip-hop had become a significant element of Iranian youth culture, progressively emerging in the undergrounds of Tehran and other major urban centres, such as Mashhad and Isfahan. Soon, the consumption of Western hip-hop was replaced by the production and active participation of the Iranian youth in global hip-hop culture via internet forums, chat rooms, and numerous weblogs (Johnston, 2008: 104).

In the first place, hip-hop linked the Iranian diaspora, and its Americanized popular music, to the homeland, creating a larger, more interconnected music scene transcending geographical and state-imposed boundaries. Moreover, hip-hop became a window through which American culture and values — fashion and lifestyle, gender roles and codes, rebellious youth expressions of identity, partying and drinking — entered the urban centres of Iran. Although a language barrier hindered full understanding, the flow of values and meanings transformed the young sector of society. Taking the beats and rhythms of Western hip-hop and blending them with local ingredients including language, dialect, and Iranian musical instruments and melodies, Iranian hip-hop covers a wide range of topics and issues, from rejecting the norms of Iranian society/culture to criticizing the state and its control over the cultural sphere and the music scene.

Unlike in other Western styles of Persian popular music such as rock, the youth-dominated hip-hop scene in Iran is led by the middle and lower classes of society (Nooshin, 2011: 10). The accessibility of its means of production and consumption has made hip-hop a music of the streets. The scene, however, is not as monolithic as Nooshin depicts. While prominent rappers such as Hichkas (Sourush Lashkari) and Yas have devoted their careers to addressing issues of the lower socio-financial sectors, bands such as Zedbazi and Tik Tak have focused their music on rich kids' hedonistic lifestyles and partying. Thus, it is vital to understand that Iran's hip-hop scene is vastly diverse with musicians and audiences from different walks of life and with a variety of styles, contents, and audiences. The most prominent commonalities in the scene include the age spectrum of producers and consumers (teens to 30s), the use of Farsi language, and the inclusion of Iranian instruments and melodies.

Many studies of Iran's underground music (Breyley, 2014; Nooshin, 2005; Rastovac, 2009; Robertson, 2012; Zahir, 2008) have focused on Iranian underground music's defiant character, framing it as a Western-influenced reaction to the state's censorship and a major youth medium for socio-political commentary. What has been left relatively unattended is the complexity of musicians' agency in not merely rejecting the state's undemocratic policies and regulations on musical activities and other aspects of life, but also in bridging the gap between Iran and the outside world, particularly the US.

While underground musicians have used a variety of styles and genres from rock and heavy metal to jazz, blues, and hip-hop to give voice to their generation's concerns, they have also taken influential steps in connecting the Iranian youth to global youth culture. They not only receive but also send messages. With the use of social media and the internet, Iranian hip-hop musicians and fans have taken an active role in global hip-hop culture. Through sharing songs and discussing Persian hip-hop on different social media platforms and forums, Iranian hip-hop fans in Iran and in the diaspora talk about their identity, life, and concerns. Persian hip-hop thus provides a lens through which to see and learn about modern-day Iran. As Nooshin (2017) highlights, it is time to step away from monolithic and reductionist readings of Iran's popular music and attempt to cast light on on-the-ground realities.

4.4 Persian 'Hotline Bling'

A few days after Canadian hip-hop artist Drake released his award-winning hit 'Hotline Bling', the online world was bombarded by numerous parodies on social media.[4] One popular parody widely disseminated among Iranians replaced the song with a famous Persian 6/8 dance tune by female singer Fattaneh. Perfectly matching Drake's dance moves, it created a comical juxtaposition of a Western artist dancing to a low-brow Persian dance tune. The widespread circulation of the clip led to the song's popularity inside Iran. By the time Hamid Sefat released the Persian cover, 'Hotline Bling' had become a major hit, allowing Sefat to reach a larger audience on the back of the song's established popularity.

[4]The song brought Drake the American Music Award 2016 and two Grammy Awards for Best Rap Song and Best Rap/Song Performance 2017.

Recently emerged, controversial rapper Hamid Sefat has gained tremendous fame in Iran's hip-hop scene in the past few years. Born in 1993, Sefat has established himself as a rebellious rapper with a unique persona. His distinctive appearance with a dense, black beard, frequently appearing in military-style apparel as well as Western hip-hop fashion, has played a key role in establishing his fame (Figure 4.1). His reverence for the Iranian martyr, military commander, and the first defence minister of post-revolutionary Iran, Dr. Mostafa Chamran, has shaped his persona as a partisan rapper. He devoted the song 'Che' to Chamran and other Iranian martyrs.[5] Sefat has repeatedly introduced himself as a '*Yaghi*' (rebel), which to him also describes Chamran, a guerrilla character with religious and nationalistic values. Another significant aspect of his persona is his short motto '*Gholam-e Nanam*' (['I'm] My Mum's Slave'). Appearing as a zealous young man, Sefat encouraged others to respect and protect their female family members to the extent of controlling them.

Figure 4.1 Hamid Sefat. Caption translation: 'At a time when music in Iran is dominated by repetitive and commercial pop music, we must be united, we must create quality music, we must transcend the boundaries, Iran's music is drowning in a swamp, new thoughts must be valued. Otherwise what you see as pop music in Iran is not presentable anywhere in the world' (credit: Sefat's Instagram page).

[5]Ebrahim Hatamikia's 2014 movie, *Che*, is also devoted to the life of Mostafa Chamran. Sefat's song brings the movie to mind and builds its message in reference to the movie. Intertextuality is a major aspect of the postmodern nature of rap as it reaches out to various texts to create meaning.

Sefat's rapper persona is similar to two other significant Iranian rappers Hichkas and Amir Tataloo. On the one hand, his street, rebel personality is reminiscent of the '*luti*' character of Soroush Lashkari, whose rap name Hichkas (Nobody) depicts his appreciation of the humility of the *luti* and of staying down to earth. Historically, the *luti* was a chivalrous man of a certain neighbourhood of Tehran who fought for justice and valued selflessness, modesty, and honour (Siamdoust, 2017: 247). The power of a *luti* is manifest in his gang and their obedience towards him. Hichkas's *luti* character is revealed in his claiming of the streets, pursuit of justice, projection of nationalism and manliness, and honouring of *namous* (female family members) and *vatan* ('homeland') (*ibid.*: 248). Sefat depicts such *luti* personality traits in using street slang, showing pride in the country, exerting protective control over his mother (*namous*), and mimicking the old *luti* appearance with his gang of men in black suits and hats.[6]

Contrary to Hichkas, who is greatly respected and loved by Iranian hip-hop fans, Tataloo is widely ridiculed and criticized on social media and in the public sphere for his political sidings, supporting the right-wing, conservative cleric Ebrahim Raisi in the presidential election in 2017 as well as for his hypocritical promotion of Islamic and state values, including making the hijab compulsory, expanding nuclear power, and controlling women under the labels of *gheyrat* ('zeal and honour-based jealousy') and *namous*. Sefat shares his support for the state's ideology and Islamic values with Tataloo, establishing both as 'good boy' characters of Persian hip-hop. Despite working within the framework of the Iranian state by staying away from addressing political issues on the one hand and venerating the state's ideological values including its respect for the martyrs on the other hand, Sefat has not been given an official permit. As an underground, *zirzamini* artist, Sefat has stated that acquiring the permit is accepting censorship and control and is against the

[6]Protecting one's *namous* ('honour') has historically been a significant aspect of manhood in Iranian culture. Two major components of *namous* include female members of the family (mother, sister, daughter, and wife) and the homeland (*vatan*). Street rappers commemorate protection of *vatan* and *namous* and encourage their fans to do so as well. In the song 'A Bunch of Soldiers', Soroush Lashkari (Hichkas) sings: 'We sacrifice ourselves for four things, *namous, vatan, khunevadeh* ('family'), and *rafigh* ('friend').

defiant essence of rap. Having been active in the rap scene since his teenage years, Sefat's songs address the emotional challenges of youth, loneliness, love, and family relations, the importance of God and the protection of the homeland, and *namous*.

In August 2017, Sefat was arrested on suspicion of murdering his stepfather and spent some time in jail awaiting trial (Shahrabi, 2017).[7] The incident exaggerated the *luti* aspect of his character and mobilized a great number of hip-hop fans and musicians who were keen to see him released (Figure 4.2). This ignited debates on his persona as a hip-hop musician with a 'good boy' or *luti* character. His Instagram page posted prayers and contributed to a change in his public image from that of a rebel rapper to one of a devoted Muslim respecting Sharia law and Islamic occasions. He acquired a quite unconventional position among Iranian hip-hop enthusiasts. While many hip-hop fans support his religious and traditional stance, others ridicule him for his paradoxical amalgamation of Western values and Islamic ideology.

Figure 4.2 Sefat behind the scenes of the 'Hotline Bling' video (credit: Sefat's Instagram page). Comments 2–5 all express hope for his release and comeback to the music scene. Note the motto on his shirt, which in Farsi says: '*Gholam-e Nanam*' ('My Mum's Slave').

With over 5 million views on Radio Javan (The Radio [for] the Youth), the main platform for Persian music broadcasting, the Persian cover of Drake's 'Hotline Bling' ignited heated debates on Westoxification and the absurd replication of Western artefacts,

[7]He was released in 2018 paying bail and with further investigation showing that his assault was not the sole reason for his stepfather's death.

as well as on issues of originality and 'authenticity'.[8] Some music fans vehemently rejected what they saw as a meaningless cover of the song and its lack of creativity and authenticity, whereas many hip-hop enthusiasts appreciated the Farsi version and enjoyed the possibility of singing Farsi words to the familiar beat.

Despite the fact that cross-cultural musical borrowings, non-Iranian influences, and the covering of foreign songs have been intrinsic to Persian pop since its inception in Tehran in the 1960s (Hemmasi, 2010; Nettl, 1972), debates on cultural imperialism and Americanization of the world inspired a more attentive examination of the song in light of glocalization and realities of the postmodern world. The trifold delivery of the song's meaning in lyrics, images, and music illuminates the complexity of cultural 're-emplacement' (Solomon, 2005).

The lyrics provide a translation of the original song's text, expressing feelings about the leaving of a beloved. The assertive male point of view is highlighted in the word choice, articulating dissatisfaction rather than agony and melancholy. Sefat addresses the beloved:

> *To midoonesti man bi to tanhãm, midoonesti tanhãm man, khandidi o gofti, boro dast as saram bar dãr, ya'ni delet bã mã ni, hamash boodim sar-e kãr, ya'ni delet bã mã ni, to mãro mikhãy pas chikar?* (x2). *Delam ba inke jã moondesh pish-e to, dige nemitoonam bemoonam bãt, khaste shodam man az harfãt, khandehãt az chesham hattã oftãd. Delam bã inke jã moonde bood pish-e to, khoob bood amã yek dafe bad shod, bad shod o bãzio balad shod, hãlã begoo ki beinemoon sad shod*?[9]

[8]Radio Javan (RJ) is the most well-known and widely used platform for the production, dissemination, and consumption of Persian music inside and outside Iran. Based in the United States, RJ has housed and supported both young and established musicians. Initially a music broadcast website (https://www.radiojavan.com/), RJ soon created a music blog, hired numerous DJs, hosted cultural festivities and parties, recorded its own radio and television programmes, and eventually became the most significant broadcaster of Persian music. While RJ has its own politics, its relative openness has freed marginalized musicians from the need for securing a governmental permit to release music inside Iran. Hosting all permitted, underground (*zirzamini*), and exiled musicians, RJ has created a relatively democratic virtual Persian music scene. It is, however, vital to note that Iranian authorities require permitted musicians to remove their music from RJ.

[9]The letter 'ã' is used to represent the long vowel 'a' as in the word 'car' to distinguish it from 'a' as in 'apple'.

> You knew I would be lonely without you, you smiled and said 'go and leave me alone', this means your heart is not with 'us', 'we' have been played with, this means your heart is not with us, why would you need 'us' anyways? Although my heart is still with you, I can no longer be with you, I am sick of your words, your smile is no longer precious to me. Although my heart is still with you, it [your heart] was good, but it suddenly became bad, became bad and learned the game. Now tell me, who came between us?

In the second verse, he adds:

> *Delam bã inke jã moonde bood pish-e to, amã kheli vaghte ke mordi barãm, nemikhãm bargardam be ghablanãm... heif delam az to roo dast khord... dar o nadaram ro bãz havas bord...*

> Although my heart is still with you, it's been long since you've been dead to me. I don't want to go back to my previous me. What a pity! you played with my heart. I lost my all to longing [for you]

In the bridge, he asks the beloved to either come back or stay away forever. He sings:

> *Bargard ammã khoob Bedoon ãshegham tã jonoon, yã bemoon ghãne sham, yã ke boro khãheshan*

> Come back but be aware that I love you to death, either stay and convince me or go and leave me alone please

The lyrics are localized in terms of the point of view and choice of words. As Samy Alim highlights, 'hip-hop artists vary their speech consciously to construct a street-conscious identity, allowing them to stay connected to the streets' (2002: 288). Sefat uses 'we' instead of 'I' to refer to himself (the first-person plural). This is an indication of the empowered male stance as is common in Iranian culture. By referring to himself as 'we', Sefat draws on the dialect of the lower sector of Tehran, especially in the vernacular language of *lutis* and *lãts* (tough guys). In his analysis of music videos in exile, Naficy explains that the tough guy characters in black suits and fedora hats represent the robust men of old Tehran who devoted their lives to protecting the vulnerable, the poor, and, most importantly, women (1998: 61–62; see also Breyley and Fatemi, 2016: 16).

The *luti's* image is associated with a strong, jealous, and protective man who exerts his power and control over females, an ideology taking women as precious 'objects' to be owned and protected. In the language of *lutis* and *lãts,* 'I' is replaced with 'we' as a form of expressing power, as if one person has the power of a collective. Furthermore, the short word '*ni*' ('there is not', 'it is not') used in the dialect of *Tehrani* ('that of Tehran') instead of the word '*nist*' (the negative form of the verb 'to be') highlights the linguistic tie to the locality. By utilizing the local dialect and cultural capital (Bourdieu, 1986), Sefat embeds Tehran in the song. The complex localization of the song in the linguistic domain, taking place in the use of the *Tehrani's* and *Luti's* dialect, can be particularly difficult to pinpoint by outsiders.

The video clip further localizes the song through different techniques of videography, editing, and effects (Sefat, 2016). The song begins by showing Sefat sitting on an old chair and getting a haircut on the street. A very rare image in today's metropole of Tehran with modern and luxury hair salons, the scene is a marker of temporal intertextuality. Black and white images make the haircut situation look more vintage and remote in time. As soon as the beat enters, the video switches to colour and Sefat appears among his gang strolling through old alleys of Tehran Grand Bazaar. Surrounded by bazaar workers, the gang lip-syncs the song as they move their bodies in a way similar to Western rappers dancing in American hip-hop videos.

The video reflects the juxtaposition of the East and West, tradition and modernity, on multiple levels. The scene showing young bazaar workers crowding around Sefat and his friends juxtaposes the poor and the wealthy: the workers and the Western-looking young men on hoverboards. The 'gangsta' rap feel of the music video intensifies the paradoxical co-presence of two worlds. Moreover, the location of the video, the Grand Bazaar of Tehran, strengthens the feeling of the concurrence of the local and the global. A historic landmark of Iran's capital with numerous shops, banks, restaurants, and mosques, the Grand Bazaar of Tehran reminds of Tehran's history, lower social sectors, downtown life, and the *luti* culture. In the alleys of Tehran's bazaars, urban workers and rural immigrants meet the Western-looking, rich kids singing Farsi and dancing to a Western beat. The confrontation of the two is reminiscent of the juxtaposition of the East and West.

Hoverboards are particularly important as they manifest the consumerist aspect of global youth culture. Uniform patterns of youth consumption around the world — fashion trends, music tastes, and media habits — manifest a transnational, market-based ideology at the centre of which exists a dialectic between structures of difference and their adaptation in local contexts (Askegaard and Kjeldgaard, 2006: 232). Global youth culture is best understood in terms of glocalization as it exhibits creolization and appropriation of globally spread features. Such creolized and appropriated glocal youth culture depicted in the song video — in fashion, hairstyles, hoverboards, and body movements — casts further light on the glocalizing processes at work.

As a cover song, musical aspects do not change much. Nonetheless, the localization of the song takes place in the use of two instruments: the violin and the Iranian frame drum, *daf*. The violin, absent in Drake's version, is used in the choruses and becomes noticeable towards the end of the song where the violin player enters the music video too. The violin is an important melodic instrument in Persian popular and traditional music (Breyley and Fatemi, 2016: 35; Nettl, 1972: 222), particularly due to its timbre being close to the Persian instrument *kamancheh* and the freedom it provides Iranian musicians to play microtones. The *daf* is also used in the choruses and plays a prominent role in the rhythm section producing the beat by giving it a Persian flavour through its distinct tone colour. As Johnston stresses, the incorporation of Iranian instruments and references to Persian poetry have been effective tools in creating the distinct sound of Persian hip-hop (*Rap-e Farsi*) and reflect the musical adaptation of the global template (2008: 108).

As the close examination of the song shows, the postmodern character of hip-hop music — intertextuality, bricolage through sampling and musical borrowing, pastiche, parody, and double consciousness — creates a democratic contact zone, bringing distant locales and cultures together. In the transnational context of hip-hop, all have access to membership and a voice, a possibility provided by a key aspect of the postmodern condition, the rejection of metanarratives (Lyotard, 1996).

There are many arts to this contact zone. In Iran, hip-hop music has linked the young generation to global youth culture. Besides, through Persian hip-hop songs, modern Iran is exposed and connected to the globally dispersed Iranian diaspora and, more broadly, the whole world. According to Swedenburg (2001), Muslim hip-hop artists play a pivotal role in challenging Islamophobia and racism in France and England. Khabeer's (2007) study of American Islamic hip-hop reveals that this too serves two major objectives: (1) preserving the Islamic identity of Muslim youth and (2) educating non-Muslims about Islam and Muslims. These two studies are great examples of the benefits of the contact zone. It is through the localization of the globally meaningful template that Muslim hip-hoppers aim at destroying stereotypes and cultivating new discourses on Muslims' culture and identity. A very similar process is detectible in the Persian hip-hop scene.

4.5 Conclusion

Complex processes of musical indigenization, adaptation, and local re-emplacement inherent in the global presence of hip-hop have been widely studied in recent years (Baker, 2006; Kahf, 2007; Mitchell, 2001; Nooshin, 2011). Created at the intersection of the local and the global and serving both as a canvas and template (Baker, 2006: 236), hip-hop is an epitome of glocalization. The cultural value of hip-hop cannot be assessed without considering the local setting in which it is used as a 'mode of collective expression' (Bennett, 1999: 78). In Iran, hip-hop has developed as a globally conscious, yet locally meaningful, medium for the expression of socio-political commentary and youth active participation in civil society (Nooshin, 2011). Moreover, it has linked the young generation to global youth culture through reworking of beats, song forms, contents, and covers. As Iran's 'fastest growing popular music genre' (*ibid.*: 5), hip-hop has been utilized by local agents (both producers and consumers) to create a space for marginalized voices and forge globally influenced local identities.

In this chapter, I have claimed that the multifaceted processes of glocalization and the postmodern nature of hip-hop have given it the capacity to play more complex roles in the current context of Iran. Taking advantage of hip-hop music's inherent flexibility, Iranian hip-hoppers have (1) challenged the state's tight control of space, (2) introduced the face of modern Iran to the outside world (an underground Iran hidden to outsiders) and thus defied not only the representation of Iran as a promoter of war and terrorism and its people as oppressed, but also the anti-Iranian, anti-Muslim rhetoric in the West and global media,[10] and (3) rejected the simplistic readings of hip-hop music as a manifestation of Westoxification and cultural imperialism by depicting multifaceted processes of cultural interaction at work. As the case study of the Persian 'Hotline Bling' shows, Persian hip-hop, even in its most evident borrowing of Western culture, is much more complex than meaningless imitation.

By incorporating the vernacular Farsi of Tehran, locating the video in a historical landmark of Iran's capital, The Grand Bazaar, incorporating *daf* and prominent appearances of the violin on the one hand and, on the other hand, showing the hoverboards, phones, and hip-hop fashion trends representing global/American consumption patterns — all added to Drake's beat and dance moves — the Persian 'Hotline Bling' re-emplaces the song in Tehran and reveals how hip-hop rejects the metanarratives of cultural purity against cultural imperialism. More nuanced interpretations of hip-hop sharpen our understanding of 'on-the-ground' realities and socio-cultural interactions and can consequently impact policies and practical approaches towards the production and consumption of hip-hop.

References

Al-e Aḥmad, J. 1982. *Weststruckness: Gharbzadegi*. Lexington, KY: Mazda Publishers.

Alim, S. H. 2002. 'Street-Conscious Copula Variation in the Hip Hop Nation', *American Speech* 77(3): 288–304.

_____. 2011. 'Global Ill-Literacies Hip Hop Cultures, Youth Identities, and the Politics of Literacy', *Review of Research in Education* 35(1): 120–146.

Askegaard, S., and D. Kjeldgaard. 2006. 'The Glocalization of Youth Culture: The Global Youth Segment as Structures of Common Difference', *Journal of Consumer Research* 33(2): 231–247.

[10]While the representation of Iran in the global media is not monolithic, the hostage crisis in the 1980s, the country's support for other Shia' states in the region, and political adversity with Saudi Arabia, Israel, and the US have led to a notable anti-Iranian rhetoric in the world. Representative news stories include Browne (2016), Ahronheim (2017), and Daoud (2015).

Baker, G. 2006. '"La Habana Que no Conoces": Cuban Rap and the Social Construction of Urban Space', *Ethnomusicology Forum* 15(2): 215–246.

Barber, B. R. 1996 *Jihad vs. McWorld*. New York: Times Books.

Bennett, A. 1999. 'Hip Hop am Main: The Localization of Rap Music and Hip Hop Culture', *Media, Culture and Society* 21(1): 77–91.

Bourdieu, P. 1986. 'The Forms of Capital'. In *Cultural Theory: An Anthology*, edited by I. Szeman, and T. Kaposy, 81–92. John Wiley & Sons.

Breyley, G. J. 2014. 'Waking up the Colours: Memory and Allegory in Iranian Hip Hop and Ambient Music', *Australian Literary Studies* 29(1–2): 107–119.

Breyley, G. J., and S. Fatemi. 2016. *Iranian Music and Popular Entertainment: From Motrebi to Losanjelesi and Beyond*. London and New York: Routledge.

Condry, I. 2001. 'Japanese Hip-Hop and the Globalization of Popular Culture'. In *Urban Life: Readings in the Anthropology of the City*, edited by G. Gmelch, and W. Zenner, 357–338. Prospect Heights, IL: Waveland Press.

Ferguson, M. 1992. 'The Mythology about Globalization', *European Journal of Communication* 7(1): 69–93.

Flores, J. 2000. *From Bomba to Hip-Hop: Puerto Rican Culture and Latino Identity*. New York: Columbia University Press.

Forman, M. 2000. '"Represent": Race, Space, and Place in Rap Music', *Popular Music* 19(1): 65–90.

Handler, R. 1994. 'Is Identity a Useful Cross-Cultural Concept?' In *Commemorations: The Politics of National Identity*, edited by J. R. Gillis, 27–41. Princeton: Princeton University Press.

Hemmasi, F. 2010. *Iranian Popular Music in Los Angeles: Mobilizing Media, Nation, and Politics*. PhD thesis, Columbia University.

Kahf, U. 2007. 'Arabic Hip Hop: Claims of Authenticity and Identity of a New Genre', *Journal of Popular Music Studies* 19(4): 359–385.

Khabeer, S. A. 2007. 'Rep that Islam: The Rhyme and Reason of American Islamic Hip Hop', *The Muslim World* 97(1): 125–141.

Johnston, S. 2008. 'Persian Rap: The Voice of Modern Iran's Youth', *Journal of Persianate Studies* 1(1): 102–119.

Lyotard, F. J. 1996. 'From the Postmodern Condition: A Report on Knowledge'. In *From Modernism to Postmodernism: An Anthology*, edited by L. E. Cahoon, 481–514. Cambridge, MA: Blackwell Publishers.

Maira, S. 2008. '"We Ain't Missing": Palestinian Hip Hop—A Transnational Youth Movement', *The New Centennial Review* 8(2): 161–192.

Manuel, P. 1995. 'Music as Symbol, Music as Simulacrum: Postmodern, Pre-Modern, and Modern Aesthetics in Subcultural Popular Musics', *Popular Music* 14(2): 227–239.

Mitchell, T. 2001. *Global Noise: Rap and Hip-Hop outside the USA*. Middletown, CT: Wesleyan University Press.

Naficy, H. 1993. *The Making of Exile Cultures: Iranian Television in Los Angeles*. Minneapolis: University of Minnesota Press.

_____. 1998. 'Identity Politics and Iranian Exile Music Videos', *Iranian Studies* 31(1): 51–64.

Nettl, B. 1972. 'Persian Popular Music in 1969', *Ethnomusicology* 16(2): 218–239.

Nooshin, L. 2005. 'Underground, Overground: Rock Music and Youth Discourses in Iran', *Iranian Studies* 38(3): 463–494.

_____. 2011. 'Hip-Hop Tehran: Migrating Styles, Musical Meanings, Marginalized Voice'. In *Migrating Music*, edited by J. Toynbee, and B. Dueck, 92–111. London: Routledge.

_____. 2017. 'Whose Liberation? Iranian Popular Music and the Fetishization of Resistance', *Popular Communication* 15(3): 163–191.

Olufunmilayo, A. B. 2006. 'From J.C. Bach to Hip Hop: Musical Borrowing, Copyright and Cultural Context', *North Carolina Law Review* 84(2): 547–645.

Pardue, D. 2008. *Ideologies of Marginality in Brazilian Hip Hop*. New York: Palgrave Macmillan.

Pasler, J. 2001. 'Postmodernism'. *Grove Music Online*. Available online at https://doi.org/10.1093/gmo/9781561592630.article.40721

Perry, M. D. 2016. *Negro Soy Yo: Hip Hop and Raced Citizenship in Neoliberal Cuba*. Durham: Duke University Press.

Perullo, A. 2005. 'Hooligans and Heroes: Youth Identity and Hip-Hop in Dar es Salaam, Tanzania', *Africa Today* 51(4): 75–101.

Pratt, M. L. 1987. 'Arts of the Contact Zone'. In *Ways of Reading: An Anthology for Writers*, edited by D. Bartholomae, and T. Petrosky, 33–40. New York: St. Martin's Press.

Rabaka, R. 2011. *Hip Hop's Inheritance: From the Harlem Renaissance to the Hip Hop Feminist Movement*. Lanham, MD: Lexington Books.

Rastovac, H. 2009. 'Contending with Censorship: The Underground Music Scene in Urban Iran', *The McNair Scholars Journal of the University of Washington* 8: 273–296.

Robertson, B. 2005. 'Persian Pop Music: At "Home" in Exile and in "Exile" at Home', *Context: Journal of Music Research* 29–30: 31–41.

_____. 2012. *Reverberations of Dissent: Identity and Expression in Iran's Illegal Music Scene.* New York: Continuum.

Robertson, R. 1995. 'Glocalization: Time-Space and Homogeneity-Heterogeneity'. In *Global Modernities,* edited by M. Featherstone, S. Lash, and R. Robertson, 25–44. London: Sage Publications.

Rose, T. 1994. *Black Noise: Rap Music and Black Culture in Contemporary America.* Hanover: University Press of New England.

_____. 2008. *The Hip-Hop War: What We Talk about When We Talk about Hip-Hop and Why It Matters.* New York: Basic Books.

Rosenberg, J. 2005. 'Globalization Theory: A Post Mortem', *International Politics* 42: 2–74.

Sarkar, M., and D. Allen. 2007. 'Hybrid Identities in Quebec Hip-Hop: Language, Territory, and Ethnicity in the Mix', *Journal of Language, Identity & Education* 6(2): 117–130.

Siamdoust, N. S. 2017. *Soundtrack of the Revolution: The Politics of Music in Iran.* Palo Alto, CA: Stanford University Press.

Solomon, T. 2005. 'Living Underground is Tough': Authenticity and Locality in the Hip-Hop Community in Istanbul, Turkey', *Popular Music* 24(1): 1–20.

Swedenburg, T. 2001. 'Islamic Hip-Hop vs. Islamophobia'. In *Global Noise: Rap and Hip-Hop outside America,* edited by T. Mitchell, 57–86. Middletown, CT: Wesleyan University Press.

Tomlinson, J. 1991. *Cultural Imperialism: A Critical Introduction.* Baltimore: Johns Hopkins University Press.

Williams. A. 2010. '"We Ain't Terrorists But We Droppin' Bombs": Language Use and Localization in Egyptian Hip Hop'. In *The Language of Global Hip Hop,* edited by M. Terkourafi, 67–95. London and New York: Continuum.

Youssefzadeh, A. 2000. 'The Situation of Music in Iran since the Revolution: The Role of Official Organizations', *British Journal of Ethnomusicology* 9(2): 35–61.

Zahir, S. 2008. *The Music of the Children of Revolution: The State of Music and Emergence of the Underground Music in the Islamic Republic of Iran with an Analysis of Its Lyrical Content.* Master's thesis, University of Arizona.

Other Media

Ahronheim, A. 15 September 2017. 'Iran Pays $830 Million to Hezbollah' (news report, *The Jerusalem Post*). Available online at https://www.jpost.com/Middle-East/Iran-News/Iran-pays-830-million-to-Hezbollah-505166

Billington, M. 1992. 'The Reinvention of William Shakespeare' (magazine article, *World Press Review* 39(July): 24–25).

Browne, R. 3 June 2016. 'State Department Report Finds Iran is Top State Sponsor of Terror' (news report, CNN). Available online at https://www.cnn.com/2016/06/02/politics/state-department-report-terrorism/index.html

Daoud, D. 2015. 'Meet the Proxies: How Iran Spreads its Empire through Terrorist Militias' (magazine article, *The Tower*). Available online at http://www.thetower.org/article/meet-the-proxies-how-iran-spreads-its-empire-through-terrorist-militias/

Sefat, H. 2016. 'Hotline Bling' (music video, Avang Music). Available online at https://www.youtube.com/watch?v=GZOK3CTAiIs

Shahrabi, Sh. 24 August 2017. 'The Rapper Who Killed His Stepfather' (news report, *Iranwire*). Available online at https://iranwire.com/en/features/4782

Regionality in Learning and Heritage

Chapter 5

Performing Local Music: Engaging with Regional Musical Identities Through Higher Education and Research

Daithí Kearney
Department of Creative Arts, Media and Music,
Dundalk Institute of Technology, Dundalk, Co. Louth, Ireland
daithi.kearney@dkit.ie

5.1 Introduction

Regional differences in Irish traditional music are challenged by processes of globalization but supported by an apparent tribalism and localism amongst Irish people and potential economic valuing of regional traditions. Local musical traditions not only underpin regional identities, particularly in parts of the west of Ireland, but they also create networks that enhance a sense of community underpinned by intangible cultural heritage. Many students who undertake undergraduate music studies at Irish institutions engage in the study of regional musical styles, requiring them to critically listen to selected performers, often from regions in the west of

Musical Spaces: Place, Performance, and Power
Edited by James Williams and Samuel Horlor

ISBN 978-981-4877-85-5 (Hardcover), 978-1-003-18041-8 (eBook)
www.jennystanford.com

Ireland. This can create or reinforce a limited canon that places an emphasis on historical recordings. Understanding both regions and traditions as processes, the canon must be revised in the context of new modes of learning and engagement with tradition and communities that are shaped by new technologies and virtual spaces, and by new geographies of the tradition that relocate music-making nationally and internationally. Students become part of the musical community of the region in which they undertake their studies and can most readily immerse themselves in local cultures through observation and participation beyond the classroom.

I am a lecturer in music at Dundalk Institute of Technology (DkIT) and was the director of the traditional music ensemble from 2011, when I was appointed, until 2019. Located on the east coast of Ireland, the promotional material concerning music in the Institute has evoked a connection with the Oriel region. Often neglected in the narratives of Irish traditional music, recent research by academics and performers has not only developed a narrative of musical traditions in this part of Ireland but has also led to students and staff at the Institute staging themed music performances that have engaged the local community in the process of musical regionalization. This chapter critically examines the impact of music research and related performances on an understanding and awareness of a local or regional musical heritage in Dundalk and its surroundings in the past decade, and it explores the implications for a spatial understanding of Irish traditional music in the twenty-first century that engages with communities of musical practice (Kenny, 2014, 2016).

The field of ethnomusicology has evolved to focus on process over product, as 'interest shifted from pieces of music to processes of musical creation and performance — composition and improvisation — and the focus shifted from collection of repertory to examination of these processes' (Myers, 1993: 8). Education and transmission are also important in the development of music cultures, and academic institutions play an integral part here. But these processes also lead to products in the form of new collections, publications, audio recordings, and performances or the reinforcement of musical canon through pedagogical practice. As musician and scholar Jack Talty notes, 'Since canonicity is frequently constructed (and occasionally challenged) through pedagogy, faculty should be conscious of their influence' (2017: 104). Thus, this chapter is largely self-reflective, focusing on one cultural region and an Irish academic institution that engages with the study and performance of Irish traditional music, but I am cognisant of developing pedagogies and philosophies in higher education elsewhere in Ireland and internationally (see Hill, 2009a, 2009b; Talty, 2017). I draw upon the work of Ted Solís (2004) and Simone Krüger (2009), whose books provide themes for critiquing the activities and impact of the ensemble at DkIT from an ethnomusicological perspective. Recent doctoral theses engaging with folk or traditional music in higher education include those of Elise Gayraud (2016) on English folk music revivals and Jack Talty (2019), who focuses on Irish institutions (including DkIT) with reference to some other European counterparts. Talty's thesis, in particular, creates a lens through which I can engage in critical reflection on my own practice and institution. Drawing on these themes, in this chapter I consider in particular how students can engage with the music of a place. This chapter is further informed by a survey of graduates, and engagement with the local community, notably through an ongoing study focused on the Oriel Traditional Orchestra (OTO), an ensemble-in-residence established at DkIT, and interviews with musicians, music teachers, and other stakeholders. These sources help balance the weight of my positionality within the institution.

5.2 Location

Co. Louth is located on the east coast of Ireland, with the town of Dundalk approximately halfway between the major cities of Dublin and Belfast. The Oriel region comprises parts of Co. Louth and the surrounding counties Meath, Monaghan, Armagh, and Down, and definition of its geographical scope is based upon the area of the ancient kingdom of Orialla or Airialla. The concept of Oriel as a musical region is presented in Pádraigín Ní Uallacháin's seminal study, *A Hidden Ulster* (2003), which places particular emphasis on the language and song traditions of the region. The location of the Irish border through this region from 1922 has implications for the imagination and performance of cultural identity, but a number of cultural projects have also benefitted from cross-border and peace funding since the 1990s in particular. DkIT is located approximately 10 km from the border and has engaged in various cross-border and local cultural initiatives. Despite the prominent use of the term Oriel, this is often done without in-depth critique and, with greater engagement with the musical traditions, an understanding of other competing identities in South Louth and Co. Monaghan is emergent.

Programmes of study in the arts and humanities seek to educate students to think critically and challenge orthodoxies, and it is important to look beyond a narrow canon of sources and examples, such as the narrative that locates Irish traditional music and exemplary performers in the tradition on the west coast of Ireland. Despite a rich local musical tradition, the Oriel region does not feature significantly in the narratives of Irish traditional music in the twentieth century and does not form part of a canon for the study of Irish traditional music. Its location on the east coast, impact of English conquest, and the development of major urban centres contrast greatly with the rural, seemingly untouched west of Ireland. Research and discourse demonstrates an emphasis in Irish traditional music studies on counties along the west coast (Kearney, 2009a, 2009b; Ó hAllmhuráin, 2016; O'Shea, 2008), influenced by the development of a 'myth of the west' and issues of music and identity in Northern Ireland (Vallely, 2008) and a romantic nationalist focus on western places, but also recognizable in other studies relating to competitions and the activities of Comhaltas Ceoltóirí Eíreann (Kearney, 2013).[1] Flute player and academic Niall Keegan asserts: 'In traditional music today there are discourses and vocabularies that are privileged above others... Such terminologies are built around issues of regional style and past, privileged practice' (2011: 40). Critically responding to these narratives of tradition with references to music and musical figures beyond the canon is an important aspect of academic study.

The location of DkIT in the Oriel region provides an opportunity for engagement with its local musical traditions. However, it is important not to seek to create new canons through the rejection

[1]Comhaltas Ceoltóirí Éireann (CCÉ) is an organization committed to the promotion of Irish traditional music, song, dance, and the Irish language. It operates a network of branches in Ireland and internationally and organizes a number of music festivals, competitions, and workshops.

of established knowledge or to become insular in the approach to teaching and research. Faculty and students engage with, and bring their learning into, the communities of musical practice outside of the institution. The relationship between these two communities is part of the development of regional musical practice, and each can inform and influence the other.

5.3 Understanding Regions

Presenting at an early academic conference focused on the concept of regional styles in Irish traditional music, Co. Louth singer and collector Seán Corcoran stated:

> ... concepts of place and region have long had a powerful role in the history of Irish thought. These concepts have been largely ideological constructs with little correlation with cultural distribution patterns, and have been widely accepted in fields of folk-music and folklore studies, where they are linked with various related concepts, like 'remoteness' and 'authenticity'.
>
> (1997: 25)

Despite the perceived acceptance of 'place' and 'region' in Irish traditional music, the discourse on regional styles is quite recent and, in some instances, attempts to force the identification of regions with particular aspects of musical style, a connection that can be overly romanticized or easily problematized. Seán Ó Riada is an important figure in this context, particularly through his radio series *Our Musical Heritage*, first broadcast in 1962 and later published as a book in 1982. His influence on the study and performance of Irish traditional music extends far beyond that. The radio series and book focus on a small number of regions, not including Oriel. *Our Musical Heritage* began an examination and discussion of stylistic differences in Irish traditional music based on a regional model that was already being eroded by changes in technology and society during the 1960s. Nevertheless, the doomsayers who predicted the extinction of regional styles and the homogenization of Irish traditional music (Ó Bróithe, 1999; Ó Riada, 1982) have been challenged by a desire amongst practitioners and listeners to engage with diversity in the tradition (Dowling, 1999; Kearney, 2012), some of which remains connected at some level with a sense of place and motivated in part by a desire to commercialize regional identities in Irish traditional music within the tourism, recording, and entertainment industries (Kneafsey, 2003; Laffey, 2007; Vallely, 1997). The relevance of regional identities in Irish traditional music today relates to both the continuing importance of local music activity and the influence of marketing and commercialism on regional styles and identities (Ó hAllmhuráin, 2016; O'Shea, 2008).

The initial focus of my own postgraduate studies was the concept of musical style, but as my understanding developed, I became more interested in the importance of regional identities expressed in relation to music and in the networks, infrastructure, and ecosystems through which musical culture is shaped, supported, and nurtured. As Gregory Dorchak writes in relation to Cape Breton fiddle traditions, 'to think of cultural practices only via stylistic terms can hamper the ability of a tradition to adapt to the inevitable changes that occur within a community' (2008: 153). Thus, this chapter is not about musical style, although regional musical styles are a component of what I am discussing. More than this, I am referring to what Corcoran terms 'tribalism' (1997), cognisant of the problems he presents when discussing the processes that shape a regional understanding of Irish traditional music. Musical sounds are not abandoned as we listen to traditional music, but they are contextualized by a geographical narrative that considers the wider social and cultural contexts.

As I have stated elsewhere, 'local contexts remain important for the transmission, performance and consumption of Irish traditional music', but 'local distinctiveness is challenged by changing social and economic conditions, technology and the distances that many musicians travel to take part in musical events' (Kearney, 2012: 1). Whereas early attempts to identify regions in Irish traditional music focused on aspects of musical style, Sally Sommers Smith states that 'regional styles, and indeed dance music in its entirety, are no longer geographically bound' (2001: 115). Yet it is not unusual to meet a musician who emphasizes their connections to a place, read a review that interprets a recording in the context of a regional style, or supervise an academic project that seeks to highlight the musical heritage of a particular region. Thus, a new understanding of regions is required. As geographer and musician Deborah Thompson notes in a study of Appalachian musical traditions:

> Like the rest of space, regions are now conceived as multiple, shifting, and contingent, with porous boundaries if they are 'bounded' at all. The processual, historically contingent nature of a region and its entanglement with various networks of social relations makes it hard to characterize or describe, as it is constantly changing and evolving, with different parts changing at different rates and continually forming new webs of connection.
>
> (2006: 67)

Irish traditional music has become a globally performed art form, and that many of its participants have no hereditary links with the country provokes questions about the connections between music, place, and concepts of authenticity (see also Keegan, 2011: 40). Place and tradition must be understood as processes and, to this end, Mats Melin notes the internal and external forces that shape a dance tradition and the 'paradoxical concept of continuity and change in tradition and issues of selectivity, creativity and ongoing reconstruction within tradition' (2012: 132). Regional identities, though often based on the construction of historical raison d'être or foundation myth, are also a process that can be revised, reshaped, and resounded. As Gearóid Ó hAllmhuráin has noted, the music of regions has changed as it has moved into new spaces, such as the pub, and more recently become part of the tourism industry (2016: 228). The development of academic communities engaged in the research and performance of Irish traditional music involving selectivity, creativity, and ongoing reconstruction has also created new spaces and soundscapes that are part of the evolution of regional identities in the tradition. While the community of musical practice may itself be divided, sometimes by oversimplified binaries of 'tradition' and 'innovation', the academic institution becomes a space for research, dialogue, and experimentation (see Hill, 2009b). The institution can instigate change but is more likely to reflect changes, attitudes, and practices in the wider traditional music communities. Reflective practice, and increasing global interaction amongst academics, brings new perspectives on and to the local.

Understanding changing contexts for the transmission, performance, and commercialization of Irish traditional music is integral to its academic study. A central focus of this is the music's globalization, which can be examined historically (Motherway, 2013) but is particularly prominent in the 1990s. Even after the significant influence of *Riverdance* (1994) on the commercial market for Irish traditional music, Seán Laffey, editor of *Irish Music Magazine*, stated

> Riverdance was a phenomena that raised many boats on the tide of its popular commercial success, and yet running counter to its jazzy glamour has been a strong re-awakening of the local traditions, the rise of a new generation of solo and duet players re-interpreting the best of the past in a faithful and diligent fashion.
>
> (2007: 1)

Laffey's statement suggests a 'revival' but, alongside a rediscovery of the past through music, there is also a process of constructing new local soundscapes within the commercial music industry. Connell and Gibson critique the association between music and place in the commercial music industry and examine the process of deterritorialization in music, acknowledging that

> ... musicians are situated in multiple cultural and economic networks — some seeking to reinvent or revive traditions, others creating opportunities in musical production to stir national political consciousness or contribute to transnational political movements, and some merely seeking to achieve commercial success.
>
> (Connell and Gibson, 2004: 343)

The attachment of a commercial value to musics that are accompanied by a local or regional narrative is examined by Vallely (1997) and Keegan (2011), but these narratives do not necessitate a distinctive local sound or an adherence to older musical styles.

5.4 Irish Traditional Music in Academic Spaces

Academic institutions have a role in changing and shaping local musical processes through interaction with students who travel from and study at a distance from the institution's location and through engagement with their local communities. These local communities can and do become the subject of research. Within Irish traditional music, Talty has noted that 'young practitioner-researchers are engaging in "ethnomusicology at home" at an unprecedented rate. Their research projects explore cultural and musical aspects of Irish traditional repertoire in great detail and in the process diversify students' understanding of it within and beyond academic institutions' (2017: 105). Moves towards applied ethnomusicology include facilitating workshops for, and sharing research outputs with, schools and community groups, leading to the reintroduction of forgotten repertoires and enhancement of local festivals (see Nettl, 2005; Pettan and Titon, 2015). In this chapter, I critically examine these processes from my own experience in Co. Louth and the surrounding Oriel cultural region.

The Irish traditional music community has conventionally existed outside of academia, but through the twentieth century it gained a greater presence at Irish institutions. This is not unproblematic. As Talty notes, 'Traditional musics are communal, extra-institutional forms of expression associated with unique processes of transmission, enculturation, and social interaction' raising questions as to how these cultural processes are represented in higher education and 'what aspects of community music making... traditional music curricula hope to impart to their students' (2017: 102). Talty challenges academics to consider their role not only in the transmission of knowledge, but also as factors in the evolution of a musical community. As members of the Irish traditional music community increasingly engage in academic studies, and as academic institutions increasingly include the study of folk music traditions in their curricula, the role of the institutions in shaping the geography of Irish traditional music becomes more apparent. Nevertheless, an academic institution is only one factor in the development of music and regional identities, and its role in a musical ecosystem should not be overplayed.

Academic institutions are spaces in which social relations are both constructed and analyzed, but as they draw upon international literature and research and teach students from a wide geographical area, they facilitate the opening up of their regions intellectually, creatively, and geographically. Academic institutions can assert a local identity and highlight their role within their region — 'DkIT has earned a reputation as the leading higher education provider in the North East of Ireland... [and] we have contributed to the transformation of our region' (Dundalk Institute of Technology, n.d.) — while simultaneously promoting a strategy of internationalization. The music of the Institute's community reflects this Janus-like vision, but this does not necessitate a fixed musical sound or style.

The study of traditional music is a component of both undergraduate and postgraduate programmes at DkIT and is an integral area of research in the Creative Arts Research Centre. Students engage with the cultural, social, and historical studies of traditional musics while also developing their performance skills. The *Ceol Oirghialla* Traditional Music Ensemble, which consists of staff and students from the Institute, and, latterly, the OTO, an intergenerational community orchestra that is an ensemble-in-residence at DkIT, have been to the fore in the performance of traditional music in the Institute. The ensembles enrich the cultural life of the Institute and the region, performing at a number of events throughout the year.

Keegan (2012) highlights a focus on solo performers in the tradition but also acknowledges a growing relevance in ensemble performance. Academic institutions provide a space in which musicians come together and explore their practice in groups, often with credits assigned to ensemble playing. Thus, it is useful to think of tradition as a process (Glassie, 1995) or a 'work in progress' (Spalding and Woodside, 1995: 249). New sounds that shape regional music identities can evolve and develop, but with institutions attracting new students each year, the sounds produced by these musicians might constantly change.

5.5 Academic and Community Engagement

Traditional music in Dundalk and the wider Oriel region was well established prior to the development of music programmes at DkIT. A branch of CCÉ was first founded in Dundalk in 1958 and the Dundalk-based Siamsa Céilí Band won the All-Ireland senior title three years in a row from 1967 to 1969. The music of the area was significantly influenced by outsiders, including Sligo musician John Joe Gardiner (1893–1979). The emulation of a Sligo style and focus on a Sligo repertoire significantly influenced music-making in the area in the mid-twentieth century and may have slowed the emergence of a distinctive regional musical identity centred on Dundalk. Later artists including Gerry O'Connor (b. 1958), who was a faculty member at DkIT for a time, and his wife Eithne Ní Uallacháin (1957–1999), evoked a regional identity in their recordings (Lá

Lugh, 1995; Ní Uallacháin, 2014) through reference to local folklore and themes and the performance of repertoire from local sources. This sense of regional identity was further promoted through the publication of *A Hidden Ulster* (Ní Uallacháin, 2003), recordings by Pádraigín Ní Uallacháin (b. 1950) including *A Hidden Ulster* (2007) and *Ceoltaí Oiriala* (2017), and by the group Oirialla (2013). These local traditions inform the development of Irish traditional music at DkIT, which engages with local cultures alongside national and international ethnomusicologies.

Commenting on the institutionalization of numerous oral folk and traditional musics in formal education programmes in Western-style conservatories and music academies, Juniper Hill notes: 'These programmes can have huge impacts not only on musical transmission methods, but also on aesthetics, repertoire, style, performance practices, creative opportunities, hierarchies, political manipulation, economic considerations, valuation, status, and public perception' (2009a: 207–208). Focusing on the goals of the Sibelius Academy in Finland, she outlines aspects that are shared with DkIT: '... to resuscitate moribund traditions, to diversify the field of folk music, to increase the status and image of folk music, to produce highly skilled and knowledgeable folk musicians, and to turn folk musicians into artists and folk music into a respected art form' (Hill, 2009b: 88–89). Through engagement in traditional music in the form of academic research and recognition, academic institutions inform, support, and even advocate for the music in the region, engaging with local musicians, attracting musicians from outside, and facilitating rehearsals and performances.

Linking between regional and academic communities of musical practice can be mutually beneficial and reciprocal but will not, for various reasons, include everybody. The OTO was established in 2017 and, as well as rehearsing in DkIT, it includes members of faculty amongst its directors and membership. The OTO received two EPIC Awards from the organization Voluntary Arts in 2019. The Institute is also a partner in Music Generation Louth and hosts some of their activities. Music Generation Louth have also recently partnered with CCÉ to establish a youth orchestra engaging with Irish traditional music. Researcher-practitioners at DkIT facilitated a variety of activities for Fleadh Cheoil na hÉireann in 2018, the festival that focused on the musical traditions of Louth and Oriel,[2] and this follows a series of concerts since 2012 that have presented and highlighted local music and musicians. Féile na Tána (est. 2015), organized by local musicians Zoe Conway and John McIntyre — who have also contributed to instrumental tuition at DkIT — has also promoted some regionally focused musical projects. Staff and students from DkIT have also performed in An Táin Arts Centre and the Oriel Centre in Dundalk,[3] both venues that regularly programme Irish traditional music. Thus, as in other music scenes, it is the emergence of several actors that underpins musical development; the academic institution is one, and the incorporation of traditional music into academic programmes is only one aspect of the role of the institution.

Shortly after moving to Dundalk in 2011, I joined the local branch of CCÉ and began participating in local music sessions. I became more aware of the sense of tradition and, in some ways, of a lack of this sense amongst a greater part of the community, particularly in terms of a regional identity.[4] And yet a sense of regional identity is explicitly present in the work of a variety of musicians and ensembles in the region, reinforced by promotional rhetoric and reviews. The music department with which I took up a lecturing position was referred to as *Ceol Oirghialla*, directly translated as 'the music of Oriel'. This potentially sought to create a status for the department and connect its identity to the region. Building upon the work of my colleagues and predecessors, I had an opportunity to 'create a new music (sub)culture… through a combination of ideas with education and institutional power' (Hill, 2009b: 86). Over the past nine years, I have endeavoured to develop a focus in my teaching, particularly through the DkIT *Ceol Oirghialla* Traditional Music Ensemble, on the

[2]Organized by Comhaltas Ceoltóirí Éireann, Fleadh Cheoil na hÉireann, translated as the 'Festival of the Music of Ireland', is the largest annual festival of Irish traditional music in the country and places an emphasis on competitions.

[3]The Oriel Centre at Dundalk Gaol is the Regional Resource Centre for Comhaltas Ceoltóirí Éireann and, in addition to performances, it hosts classes by the Dundalk branch of the organization.

[4]My reflections as a 'blow-in' engaging in a local Irish traditional music scene are informed by the work of anthropologist Adam Kaul (2013) and his study of Irish traditional music in Doolin. 'Blow-in' is usually used to refer to a person who becomes resident in a community other than their place of birth and can often be applied for a long and indefinite period of time.

traditions of the region, but placing these in a broader soundscape of Irish traditional music. Concerts have included Bearna Uladh (2017), which focused on the musical traditions around the Irish border, and Oirghialla Oscailte (2018), when the Ensemble was joined by the OTO. Other concerts, such as Imirce an Cheoil (2012) and Ó Chladach go Cladach (2015) reflected the region's links with musicians from Sligo and the influence of Irish-American musical traditions on the development of traditional music in Ireland. Performers in all of the concerts included many students encountering Irish traditional music from a background in other musical genres, and their perceptions, creativity, and interpretation shaped the music heard.

Two aspects of academic engagement with Irish traditional music at DkIT worth considering are how development of historical research informs current practice and leads to a revival in aspects of a musical tradition (see Rosenberg, 2014) and how engagement in creative arts practice and teaching impacts on changing musical aesthetics within the tradition. In addition to research undertaken by academics in the department, a number of undergraduate and postgraduate research dissertations have focused on local or regional issues, including through creating critical editions from manuscript sources, developing biographical studies of local musical figures, and exploring performance practice and the musical styles of influential performers. Performing groups and teachers incorporate knowledge and skills from this research into their practice, changing pedagogy and using new or rediscovered repertoire, which may also be arranged to fit various instrumentations or aesthetic concerns. The development of larger scale ensembles or 'traditional orchestras' reflects a changing aesthetic and performance practice in the tradition (see also Keegan, 2012).

The DkIT *Ceol Oirghialla* Traditional Music Ensemble draws inspiration from a variety of Irish traditional music groups, exploring inspirations and arrangement possibilities with respect for both tradition and new tastes. The ensemble also draws upon research into the musical traditions of Louth and Oriel by staff and postgraduates in the Institute. Concerts in recent years have celebrated famous local musical figures, including fiddle players John Joe Gardiner (2012) and Josephine Keegan (2013), dancer Mona Roddy (2014), and piano accordion player Brian O'Kane (2017). Students had the opportunity to meet, speak, and perform with Keegan and O'Kane, the central musical figures in the concerts Ómós Josephine Keegan (2013) and Marching in Tradition (2017), respectively. For Marching in Tradition, the compositions of O'Kane were transcribed from both manuscript and audio sources as part of a research project. Some local performers have participated in performances at the Institute with the students. These have included students of Mona Roddy and the Walsh School of Music. All of the concerts reflect students' learning from both aural and written sources within the classroom, and the students are involved in the development of arrangements and the final selection of repertoire. Some students who teach in the community and for local branches of CCÉ have incorporated this repertoire into their own teaching.

All the traditional music activity referred to in this chapter reflects a move towards ethnomusicology at home, and this brings with it several challenges. Writing in the second edition of the seminal ethnomusicological book on fieldwork, *Shadows in the Field*, Bruno Nettl contends:

> We came to realize that we should do field research in our own communities, something that was both easier (it's our turf) and harder (be 'objective' about one's own family and friends?) than working abroad. We began to question the role we were playing in the 'field' communities, whether we were doing harm or good, and about our relationship to ethnomusicologists from those host communities. We worried that our very presence would result in significant culture change (and sometimes it did).
>
> (2008: vi)

Similarly acknowledging the potential impact of academia on communities of practice, Scottish geographer and flute player Frances Morton argued:

> There is currently concern within Geography, surrounding the intrusion of academic research performances on lay social practices and performances. There is a worry that the lay practices may change due to the influence of academic research. However, recognizing that research is a performance in its own right allows better critique of how we undertake our research, accumulate and understand our geographic knowledge, and relate to our research participants.
>
> (2001: 67)

As outlined in this chapter, it is clear that there is a close connection between academic research and practice in Irish traditional music, with many academics identifying themselves and being identified as practitioners.

There are two competing geographies evident in the discourse created by my reflections and communication with stakeholders. One, a historical geography that focuses on the past and the creation of heritage (Ronström, 2014; Kirshenblatt-Gimblett, 1998). The second, a musicking geography (Kavanagh, 2020) that embraces the lived musical experience in the present. The former is potentially easier for higher education to engage with in the context of curriculum and the logistics of scheduling, but the latter may create more opportunities for engagement with the community of musical practice. Each connects people, music, and place, but with different learning outcomes and understandings of the role, value, and nature of traditional music. The emphasis defines the experience of learning Irish traditional music in higher education and the engagement with regional music identities.

5.6 Academic Obligations and Community Expectations

In contrast with the programmes to the fore in Gayraud (2016) and Talty (2019), the BA (Hons) Music is not a programme that specifically focuses on traditional or folk music. For many of the former members of the traditional music ensemble, the study of Irish traditional music performance amounted to less than half of a module each semester, with some additional opportunities to engage in theoretical or historical studies of Irish traditional music. It is also important to note that the programme is not a performance degree but does provide opportunities for solo performers to specialize in performance. This manifested itself in the context of the traditional music ensemble through opportunities for solos or small group performances at concerts.

In contrast with the concept of enculturation employed by Cawley (2013) in relation to the learning of Irish traditional music, the study of Irish traditional music at DkIT introduces many students to the tradition without facilitating their development to proficient practitioners of the culture. Nevertheless, some former students have transitioned from beginners to professional performers, while other students with extensive prior learning in the genre benefitted from an academic approach to the study of the culture. The positive aspect of including Irish traditional music on the programme at DkIT was identified by local musician and former fiddle tutor Zoe Conway:

> Having a traditional music element to the courses at DkIT has given a focus, drive and outlet to the traditional musicians taking part in the courses. These musicians have gone on to become traditional music teachers and practitioners in our area which helps to promote traditional music here.
>
> (personal communication, 25 May 2020)

A challenge for any academic programme is ensuring that graduates achieve a sufficiently high standard to meet the requirements of their industry or sector. Writing about world music ensembles in US universities, Solís notes the difficulty of developing an ensemble to 'a level of achievement commensurate with the director's hopeful expectation' (2004: 14). He highlights turnover in membership amongst the factors that impact on the ability to progress. Understanding the pedagogical goals of the ensemble is critical, and the purpose of the traditional music ensemble at DkIT may align more directly with the rationale provided by John Baily to Simone Krüger (2009), whereby learning to perform is understood as a research technique and it is not expected that all students should achieve a high standard. Nevertheless, all members of the traditional music ensemble at DkIT participate in public performance, and some members have pursued subsequent professional engagements as performers of Irish traditional music.

Under my direction from 2011 to 2019, the traditional music ensemble at DkIT developed a particular focus on local musical traditions, albeit with explorations of musical links between Ireland and Scotland and the USA. The local focus of the ensemble was complemented by a number of postgraduate studies that also focused on the surrounding region, many of which were under my supervision. This approach provided a specific focus and identity for the ensemble, placed a particular attention on aspects of Irish traditional music that had not previously received significant attention, and attempted to make connections with the local community. There was an inclusive classroom approach that sought to be student-centred, and there were no prerequisites or auditions for joining the traditional music ensemble. This resulted in a membership that presented a wide range of abilities and experience in traditional music, including international and Erasmus exchange programme students. This was noted, positively and otherwise, by former students in their responses to a survey I carried out in May 2020. Forty-five respondents, representing over half the membership of the Ensemble during this time period, provided often similar reflections that, in the following discussion, are complemented by interviews and personal communication with a variety of other local stakeholders in Irish traditional music.

The make-up of the ensemble has a significant impact on the pedagogical approach, learning outcomes, and performances. The ensemble did not train all members to be proficient performers of Irish traditional music, but the experience of musicking supported by lectures in history and critical theory developed the appreciation amongst members for traditional music, their sense of place, and the role of music in society. Some former students indicated a desire for a higher level of engagement in technical aspects of performance, but the overwhelming majority of respondents identified the approach as creating an opportunity to engage with a musical tradition that they had limited or no prior experience with.

Creating and sustaining a connection with the local community is a challenge for the academic institution, and this can be affected by the approach and ethos of the academy and the awareness and perception of the academy's activities amongst the community (see also Talty, 2019). Writing about the BMus Honours in Folk and Traditional Music degree at Newcastle University, UK, Gayraud notes:

> This Folk Degree initiative has not always been well received by the folk community. In discussion and interview, many of those whose musical lives are firmly rooted in local sessions expressed profound scepticism regarding the feasibility of communicating folk musical culture effectively within the context of a formalised higher educational programme that is characterised by the standard models of lectures and tutorials, graded learning, examination and assessment.
>
> (2016: 128)

In relation to the programme at DkIT, local musician and music teacher Olive Murphy noted that although she did participate in some sessions involving the students from the Institute, the concerts each semester were the only occasions that she was aware of (personal communication, 18 May 2020), while one of the graduates commented that 'performances outside of the college made us more present in the Dundalk arts community as a whole'. However, during the time period in question, a small number of the students participated regularly in local sessions, attended traditional music concerts, or engaged in teaching Irish traditional music locally. This, combined with a low presence of former members engaged in these activities in the region post graduation, limits the long-term impact of the traditional music ensemble on the local music scene. Local music teacher and DkIT fiddle tutor Noreen McManus also linked participation in local sessions to the learning experience, recognizing that the concerts gave the students confidence (interview, 22 May 2020), a point also made by many former ensemble members, but highlighting the value of the sessions for learning about the culture and enhancing their playing. David Hughes (2004) of SOAS, University of London, is one of a number of authors who point to the challenge of attempting to create a learning experience that would parallel a home culture. Even at DkIT, where the ensemble focuses on the culture of the surrounding region, a challenge remains to create links between the classroom and the wider community. Nevertheless, McManus pointed to a number of students and former students who did engage in teaching, noting that they were 'teaching what they were learning', thus bringing the ensemble repertoire into the wider community.

Musician Tommy Fegan, who initiated the opportunity to develop a concert on the music of Josephine Keegan in 2013, noted that activities at DkIT provided 'a deeper appreciation of our local repertoire, performers and traditions' (personal communication, 25 May 2020). While Fegan highlighted the potential for further concerts focusing on older generations of musicians, one critique of the approach taken to date was that there was an emphasis on historical aspects of traditional music in the area rather than a representation of the musicians currently active on the scene. Noting a focus on the 'big names', McManus highlighted the potential to include other musicians from the region who might not normally have such a platform. Acknowledging previous collaborations with music and dance schools, she stressed the potential for further collaboration and the impact that this might have. Other respondents also noted the potential of the concerts to serve as a public platform for local artists. McManus also highlighted the impact on her younger students of experiencing the traditional music concerts at DkIT. Teenagers who see the young adults in the DkIT Ensemble aspire to be like them: 'it's cool'. Paul Hayes, Director of An Táin Arts Centre in Dundalk, also recognized the potential for concerts and other activities to provide inspiration for young people in the region (interview, 25 May 2020).

Many of the traditional music concerts by the ensemble at DkIT would not have been possible without the support of external musicians such as Brian O'Kane, Josephine Keegan, and Mona Roddy. One graduate noted: 'I was introduced to a number [of] renowned Irish musicians around Dundalk'. This was echoed in the response of another graduate who wrote that participating in the ensemble activities '[i]ntroduced me to other musicians both local and attending DkIT'. On some occasions, when a particular musician was featured prominently, members of their family were often present in the audience and this impacted on the atmosphere and occasion. O'Kane noted that the concerts provided young musicians studying music the opportunity to hear an older generation and use this as a yardstick and, through hearing their stories, to understand the social aspect of music historically, learning where it came from and what it meant to people (interview, 24 May 2020). The concerts were valued by many of the local musicians as a public platform and, for O'Kane, an occasion to relive memories and create new ones.

In developing concerts on specific themes and, in some instances, honouring individuals, it is necessary to make choices. A number of musicians from the region have been identified, with some suggestions coming from the community but, with the norm of one concert per semester until 2019, it would take some time to feature all aspects. While some musicians have engaged and supported the concerts and the ensemble, the decision of some local musicians not to participate in activities, events, or reviews of traditional music activities in the Institute hampers the sharing of information and ideas that may work towards greater collaboration and integration. Some stakeholders imagine a public service role for the Institute, whereby it documents and features local musicians and traditions or acts as a patron for local artists. Many have considered the Institute a beneficial collaborator, but it is evident that some may view the

Institute as a competitor. Given the small number of public events at DkIT, it is unlikely that the Institute would have a negative impact on audiences in other venues, and efforts have been made to support, participate in, and promote events hosted and promoted by others in the region.

The opportunity to meet the students was welcomed by some of the musicians, such as O'Kane, who visited the campus in preparation for a concert of his music and returned on other occasions to help with research projects and attend other performances. While some graduates indicated that they had been involved in local sessions, many did not engage in this scene, which is critical to the local Irish traditional music community. However, another graduate noted how participating in a performance by the DkIT Traditional Music Ensemble in an external venue introduced them to other artists and a network to which they still belong. It is also evident from the responses that international students valued participation in the traditional music ensemble not only as an opportunity to engage with Irish culture and gain performance experiences, but also to integrate more into the Institute community. The opportunity to experience and learn about social contexts for music is important, as indicated by many respondents.

While the impact of the ensemble on the wider community is difficult to assess, there is a clear sense from former members that their awareness of regional diversity and local traditions was enhanced. One respondent stated: 'My participation in DkIT Traditional Irish Music Ensemble has facilitated a greater appreciation for local and regional collections of Irish Traditional Music' and '[b]rought to my attention the wealth of traditional music and musicians in the Oriel region'. Another graduate wrote: 'Each performance at the end of a semester was made all the more interesting and educational for its content and how it was focused on a particular area. It highlighted selected tunes composed or played in a specific region'. Another stated: 'Through the trad ensemble, I learned more about the music of the Oriel region that I may not have otherwise examined. For example, examining some of the works of Josephine Keegan, working with local musicians such as Gerry O'Connor, learning tunes with roots in the area'. Audiences and other stakeholders have also noted this impact. Mary Capplis, the acting

Arts Officer with Louth County Council, commented that the DkIT Traditional Music Ensemble:

> ... have provided a spotlight on their music, for example by focusing on the 'Gap of the North' and the music, song and dance traditions of the 'Oriel' region, they have substantially raised the profile of this rich heritage. For those previously unaware of Oriel music and traditions, the themed concerts provide an opportunity to highlight the cultural legacy both here in the region and further afield.
>
> (personal communication, 25 May 2020)

For Capplis,

> Promoting the research and performances of local musicians has in turn informed audiences of the dedication of local musicians Pádraigín Ni Uallacháin and Gerry O'Connor who strive to promote the Oriel region's rich cultural musical heritage. This consequentially has inspired young musicians to delve into the music, learn it and perform it, thus keeping the legacy alive for future generations.
>
> (*ibid.*)

Similarly, local Irish dancing teacher Dearbhla Lennon noted:

> The opportunities for platforms of this nature are somewhat limited in the area, particularly when it comes to more niche aspects of local culture — this provides one. Exposure on social media platforms ahead of events is also beneficial in raising awareness, not just of the upcoming event but of the artist/art in general.
>
> (personal communication, 26 May 2020)

Communicating to and attracting an audience beyond an established cohort for a genre or artform is a challenge not only for the academic institution, in terms of sharing research in the community, but also for local arts centres and organizations.

Another challenge shared with the local arts centre is striking a balance between programming local artists and visiting artists with a national or international reputation (Paul Hayes, interview, 25 May 2020). The concerts at DkIT may have focused on local themes, but there was space for individual expression, and guest artists from Ireland and America also featured. As Capplis reminded me, 'the membership spans from all over Ireland, they bring different styles of playing as well as different music (tunes) to the audience. It therefore broadens the players' repertoire significantly and brings previously unheard music to a wider audience'. While the concerts did include a number of well-known tunes in Irish traditional music, the themes and approaches distinguished the experience of studying music at DkIT from at other institutions. This is critical, as Talty notes: 'Challenging the processes and consequences of canonicity in music education is absolutely necessary in order to optimize the diversity, inclusivity, and relevance of third-level music curricula' (2019: 266). Writing about folk music in England, Gayraud notes a debate in folk music circles regarding its institutionalization and the development of new opportunities in higher education for the study of folk music, stating: 'It is clear that the institutionalization of resources and transmission processes is having a profound impact both on musical canons, determining which tunes become central and which become peripheral, and on the styles of performance that are accepted as normative' (2016: 26). It is not clear whether the activities at DkIT had a direct impact on the canon, but it is noteworthy that recent awardees of the broadcaster TG4's Gradam Ceoil prizes for Irish traditional music include Seán McElwain for a project related to his doctoral studies on Sliabh Beagh completed at DkIT (McElwain, 2014a, 2014b). In 2018, Gerry O'Connor published a book based on the MA that he completed in 2008 and has both recorded and taught many tunes from it. Padraigín Ní Uallacháin has developed a website on the traditions of the Oriel region, which also features material from Sylvia Crawford related to her MA research at DkIT (Crawford, 2019). Arrangements originally made by me for the traditional music ensemble at DkIT also became the basis for the early arrangements and repertoire of the Oriel Traditional Orchestra. Local bands involving graduates from the undergraduate music programme include Kern and Alfi, who have included reference to traditions from the region on their recordings including *False Deceiver* (Kern, 2016), *The Left and the Leaving* (Kern, 2019), and *Wolves in the Woods* (Alfi, 2019).

While there is an emphasis on historical aspects of the musical culture of the region evident in the programming of concerts involving the DkIT Traditional Music Ensemble, there is a need to challenge the perceived duality of tradition and innovation, a prominent theme in Talty's (2019) dissertation. Creativity is embedded in the ethos of the School of Informatics and Creative Media at DkIT, and the Carroll Building provides a space for a number of creative disciplines, including music, television, film, theatre, and creative media. The situating of the traditional music ensemble in this milieu mirrors Doherty's identification of the potential of higher education institutions to develop a creative environment that encourages experimentation and innovation in traditional and folk music performance (2002: 10). Creativity and experimentation were a critical aspect of the traditional music ensemble at DkIT, and the musicians within the ensemble created opportunity for this. One graduate noted: 'I found it very interesting and exciting at times, when it would all come together, but even more so when we would experiment and amalgamated styles, even if it didn't work out'. Another graduate wrote: 'I always found that the traditional music ensemble allowed for more creative expression, with musicians who were particularly passionate about Irish music', and another commented how they 'enjoyed exploring more contemporary ideas in the context of a traditional ensemble'. While the ensemble did not directly engage with popular music, it did draw influence from it. This echoes observations by Simone Krüger (2009), drawing in particular on interviews with David Hughes at SOAS.

A challenge for an institution can be related to its size and budgets, and this can impact on its ability to support a multitude of workshops and masterclasses involving visiting artists or other research-related activities such as documentation and archiving. Conway noted the involvement of high-profile practitioners as part of the faculty at DkIT previously, stating:

> I think promotion of traditional arts starts with the expertise of practitioners and the key to it is including and building on a strong base of people who are knowledgeable about traditional arts in Louth, and giving them opportunities to share and inspire students and the community in general.
>
> (personal communication, 25 May 2020)

While instrumental tutors are often drawn from the locality and have included a number of prominent local performers, it is not feasible for a small institution to host a large number of masterclasses or offer regular remuneration to local artists to be more involved in campus life. Fluctuating student numbers make it difficult to guarantee sustained employment, particularly for instrumental tutors of Irish traditional music. While some institutions engage and have developed a network that extends into its community, sometimes engaging this network in curriculum design, this strategy has been developed to only a limited extent at DkIT. This leads to a potential for pedagogues such as myself to 'impose personal biases and perspectives that may be at odds with the expectations and values of the wider community for whom ownership of this music remains a considerable concern' (Talty, 2019: 306). There is a celebration of local figures at DkIT, and the curriculum draws on the work of local researchers and practitioners and on research about the local area, but there is potential for stronger connection with the community of Irish traditional music in the region. Facilitating the OTO at DkIT has been one step in this process. The establishment of the OTO was noted by some respondents as something that, although not led by the institution, opened up the space to the wider community. While one local musician indicated that there was a large number of adults interested in taking courses in Irish traditional music, when the institution did offer a part-time programme, the number of applications was limited.

A further challenge is the sustainability of programmes and the curriculum. As part of Programmatic Review in 2019, the Traditional Music Ensemble was removed from the curriculum in spite of strong stakeholder support. Nevertheless, a traditional music ensemble continued at the Institute in a voluntary and extracurricular capacity during the 2019–20 academic year, but activities and events, including a planned concert, were cancelled due to COVID-19. It is apparent from meeting former attendees, and from the feedback of former students, that the ensemble was significant both in the context of adding to the learning experience of students and in the cultural life of the region.

5.7 Conclusion

Performances and discourses of Irish traditional music often express or refer to regional identities. A trend towards regionalization and regionality in the tradition is influenced by local politics, commercial endeavour, and academic study. In many instances, there is an emphasis on story over musical style through the processes of naming tunes and presenting narratives that associate repertoire with people and places, rather than on performing in a particular musical style or creating an identifiable regional sound. Audiences can relate to and interpret the musical performance based on their own prior experiences, knowledge of music and culture, and ability to connect with extra-musical geographical narratives. The authenticity of performances may be judged differently by local and global audiences, and understanding differences in the interpretation of authenticity in performance practice is central to identifying regional differences in aesthetics and musical identities.

Regionalization can challenge the established canon and narratives of the tradition, drawing attention to neglected places and highlighting alternative soundscapes and approaches. The academic institution is a space in which regional identities are constructed, deconstructed, and performed through research, learning, and teaching. Through a variety of research practices including archival research, performance practice, composition, and applied ethnomusicology, Dundalk Institute of Technology plays a role in the (re)construction and dissemination of a local regional identity for the Oriel region. Reflecting and contributing to the activities of a wider community of musical practice in the region, faculty and students are active agents in the processes of musical evolution and the expression of a regional identity locally and globally.

References

Cawley, J. 2013. *The Musical Enculturation of Irish Traditional Musicians: An Ethnographic Study of Learning Processes*. PhD thesis, University College Cork.

Corcoran, S. 1997. 'Concepts of Regionalism in Irish Traditional Music'. In *Selected Proceedings from Blas: The Local Accent in Irish Traditional Music*, edited by M. Ó Suilleabháin, and T. Smith, 25–30. Limerick: Irish World Music Centre, University of Limerick/Dublin: Folk Music Society of Ireland.

Connell, J., and C. Gibson. 2004. 'World Music: Deterritorializing Place and Identity', *Progress in Human Geography* 28(3): 342–361.

Crawford, S. 2019. *Towards the Potential Role of a Neglected Eighteenth-Century Harper in Cultural Tourism in the Oriel Region*. MA dissertation, Dundalk Institute of Technology.

Doherty, L. 2002. *A Needs Analysis of the Training and Transmission of Traditional Music in University and Professional Level Education throughout Europe* (unpublished manuscript, the European Network of Traditional Music and Dance Education Working Group).

Dorchak, G. 2008. 'The Formation of Authenticity within Folk Tradition: A Case Study of Cape Breton Fiddling'. In *Driving the Bow: Fiddle and Dance Studies from around the North Atlantic 2*, edited by I. Russell, and M. A. Alburger, 153–165. Aberdeen: The Elphinstone Institute.

Dowling, M. 1999. 'Communities, Place, and the Traditions of Irish Dance Music'. In *Crosbhealach an Cheoil — The Crossroads Conference 1996*, edited by F. Vallely, H. Hamilton, E. Vallely, and L. Doherty, 64–71. Dublin: Whinstone Music.

Gayraud, E. 2016. *Towards an Ethnography of a Culturally Eclectic Music Scene: Preserving and Transforming Folk Music in Twenty-First Century England*. PhD thesis, Durham University.

Glassie, H. 1995. 'Tradition', *The Journal of American Folklore* 108(430): 395–412.

Hill, J. 2009a. 'The Influence of Conservatory Folk Music Programmes: The Sibelius Academy in Comparative Context', *Ethnomusicology Forum* 18(2): 207–241.

_____. 2009b. 'Rebellious Pedagogy, Ideological Transformation, and Creative Freedom in Finnish Contemporary Folk Music', *Ethnomusicology* 53(1): 86–114.

Hughes, D. 2004. '"When Can We Improvise?": The Place of Creativity in Academic World Music Performance'. In *Performing Ethnomusicology: Teaching and Representation in World Music Ensembles*, edited by T. Solís, 261–282. Berkeley: University of California Press.

Kaul, A. 2013. *Turning the Tune: Traditional Music, Tourism, and Social Change in an Irish Village*. Vol. 3. New York: Berghahn Books.

Kavanagh, A. 2020. 'Researching Music- and Place-Making through Engaged Practice: Becoming a Musicking-Geographer', *Geographical Review* 110(1–2): 92–103.

Kearney, D. 2009a. *Towards a Regional Understanding of Irish Traditional Music*. PhD thesis, University College Cork.

_____. 2009b. 'Towards a Regional Understanding of Irish Traditional Music', *Musicology Review* 5: 19–44.

_____. 2012. 'Beyond Location: The Relevance of Regional Identities in Irish Traditional Music', *Sonus* 33(1): 1–20.

_____. 2013. 'Regions, Regionality and Regionalization in Irish Traditional Music: The Role of Comhaltas Ceoltóirí Éireann', *Ethnomusicology Ireland* 2(3): 72–94.

_____. 2020. 'Soundscapes'. In *International Encyclopedia of Human Geography*. Vol. 12, edited by A. Kobayashi, 297–304. Second Edition. Oxford: Elsevier.

Keegan, N. 2011. 'The Linguistic Turn at the Turn of the Tune: The Language of Contemporary Ensemble in Irish Traditional Music', *Ethnomusicology Ireland* 1: 37–48.

_____. 2012. The Art of Juncture–Transformations of Irish traditional Music. PhD thesis, University of Limerick.

Kenny, A. 2014. 'Practice through Partnership: Examining the Theoretical Framework and Development of a "Community of Musical Practice"', *International Journal of Music Education* 32(4): 396–408.

_____. 2016. *Communities of Musical Practice*. London: Routledge.

Kirshenblatt-Gimblett, B. 1998. *Destination Culture: Tourism, Museums, and Heritage*. Berkeley: University of California Press.

Kneafsey, M. 2003. '"If It Wasn't for the Tourists We Wouldn't Have an Audience": The Case of Tourism and Traditional Music in North Mayo'. In *Irish Tourism: Image, Culture and Identity*, edited by M. Cronin, and B. O'Connor, 21–41. Cork: Cork University Press.

Krüger, S. 2009. *Experiencing Ethnomusicology: Teaching and Learning in European Universities*. Abingdon: Ashgate.

McElwain, S. 2014a. *Opening up the Canon of Irish Traditional Music: The Music of the Sliabh Beagh Region of North Monaghan/East Fermanagh*. PhD thesis, Dundalk Institute of Technology.

Melin, M. 2012. 'Local, Global, and Diasporic Interaction in the Cape Breton Dance Tradition'. In *Routes and Roots: Fiddle and Dance Studies from around the North Atlantic 4*, edited by I. Russell, and C. Goertzen, 132–144. Aberdeen: The Elphinstone Institute.

Morton, F. 2001, *Performing the Session: Enacted Spaces of Irish Traditional Music*. MA dissertation, University of Strathclyde.

Motherway, S. 2013. *The Globalization of Irish Traditional Song Performance*. Abingdon: Ashgate.

Myers, H. 1993. *Ethnomusicology: Historical and Regional Studies*. London: Macmillan.

Nettl, B. 2005. *The Study of Ethnomusicology: Thirty-One Issues and Concepts.* New Edition. Urbana and Chicago: University of Illinois Press.

_____. 2008. 'Foreword'. In *Shadows in the Field: New Perspectives for Fieldwork in Ethnomusicology*, edited by G. Barz, and T. Cooley, v–vii. Second Edition. Oxford: Oxford University Press.

Ní Uallacháin, P. 2003. *A Hidden Ulster: People, Songs and Traditions of Oriel.* Dublin: Four Courts Press.

Ó Bróithe, É. 1999. 'Fiddle, Tyrone'. In *The Companion to Irish Traditional Music*, edited by F. Vallely, 128–129. Cork: Cork University Press.

O'Connor, G. 2008. *Luke Donnellan's Dance Music of Oriel: Volume 1 & 2.* MA dissertation, Dundalk Institute of Technology.

_____. 2018. *The Rose in the Gap.* Louth: Lughnasa Music.

Ó hAllmhuráin, G. 2016. *Flowing Tides: History & Memory in an Irish Soundscape.* New York: Oxford University Press.

Ó Riada, S. 1982. *Our Musical Heritage.* Fundúireacht an Riadaigh i gcomhar le. Mountrath, Ireland: The Dolmen Press.

O'Shea, H. 2008. *The Making of Irish Traditional Music.* Cork: Cork University Press.

Pettan, S., and J. T. Titon, eds. 2015. *The Oxford Handbook of Applied Ethnomusicology.* Oxford: Oxford University Press.

Ronström, O. 2014. 'Traditional Music, Heritage Music'. In *Oxford Handbook of Music Revival*, edited by C. Bithell, and J. Hill, 43–59. Oxford: Oxford University Press.

Rosenberg, N. 2014. 'A Folklorist's Exploration of the Revival Metaphor'. In *The Oxford Handbook of Music Revival*, edited by C. Bithell, and J. Hill, 94–115. Oxford: Oxford University Press.

Solís, T., ed. 2004. *Performing Ethnomusicology: Teaching and Representation in World Music Ensembles.* Berkeley: University of California Press.

Sommers Smith, S. K. 2001. 'Traditional Music: Ceol Traidisiúnta: Irish Traditional Music in a Modern World', *New Hibernia Review/Iris Éireannach Nua* 5(2): 111–25.

Spalding S., and J. Woodside. 1995. *Communities in Motion: Dance, Community and Tradition in America's Southeast and Beyond.* Westport, CT: Greenwood Press.

Talty, J. 2017. 'Noncanonical Pedagogies for Noncanonical Musics: Observations on Selected Programs in Folk, Traditional, World, and Popular Musics'. In *College Music Curricula for a New Century*, edited by R. Moore, 101–114. Oxford: Oxford University Press.

_____. 2019. *The Ivory Tower and the Commons: Exploring the Institutionalisation of Irish Traditional Music in Irish Higher Education (Discourse, Pedagogy and Practice).* PhD thesis, University of Limerick.

Thompson, D. 2006. 'Searching for Silenced Voices in Appalachian Music', *GeoJournal* 65: 67–78.

Vallely, F. 1997. 'The Migrant, the Tourist, the Voyeur, the Leprechaun...'. In *Selected Proceedings from BLAS: The Local Accent Conference*, edited by T. Smith, and M. Ó Súilleabháin, 107–115. Irish World Music Centre, University of Limerick.

_____. 2008. *Tuned Out: Traditional Music and Identity in Northern Ireland.* Cork: Cork University Press.

Other Media

Alfi. 2019. *Wolves in the Woods* (CD). Own label.

Dundalk Institute of Technology. (n.d.). *DkIT Strategic Plan 2017–2019.* Available online at https://www.dkit.ie/about-dkit/institute-reports-and publications/dkit-strategic-plan.html

Kern. 2016. *False Deceiver* (CD). Kern Music, AGACD01-16.

_____. 2019. *The Left and the Leaving* (CD). Kern Music, AGACD02-19.

Lá Lugh. 1995. *Brigid's Kiss* (CD). Lughnasa Music, LUGCD961.

Laffey, S. 2007. 'Feeling the Pulse' (magazine article, *Irish Music Magazine* 14(2)).

McElwain, S. 2014b. *Our Dear Dark Mountain with the Sky Over It* (CD). Dark Mountain Records, DMR01.

Ní Uallacháin, E. 2014. *Bilingua* (CD). Gael Linn, CEFCD206.

Ní Uallacháin, P. 2007. *A Hidden Ulster — The Gaelic Songs of Ulster Volume 1* (CD). Ceoltaí Eireann, CE5056

_____. 2017. *Ceoltaí Oiriala* (CD). Ceoltaí Eireann, CE6717.

Oirialla. 2013. *Oirialla* (CD). Lughnasa Music 965.

Oriel Arts. 2017. 'Oriel Arts' (webpage). Available online at www.orielarts.com.

Chapter 6

Preserving Cultural Identity: Learning Music and Performing Heritage in a Tibetan Refugee School

James Williams
Department of Therapeutic Arts, University of Derby,
Britannia Mill, Derby, DE22 3BL, UK
j.williams@derby.ac.uk

6.1 Introduction

I boarded a plane to Delhi, India in the summer of 2018, primarily as a representative of my university, with the aim of building international networking and communications with overseas charity providers and of developing work experience opportunities. I also travelled alongside a small group of university students who would be completing some volunteer work with school pupils, mainly teaching English. The opportunist in me also saw this as a chance to conduct ethnographic fieldwork, either with musicians or in music education — I knew we would be working in rural communities almost 350 km north of Delhi in the Himachal Pradesh region, and

Musical Spaces: Place, Performance, and Power
Edited by James Williams and Samuel Horlor

ISBN 978-981-4877-85-5 (Hardcover), 978-1-003-18041-8 (eBook)
www.jennystanford.com

so I travelled with recording equipment and ethics approval (from the College of Health and Social Care, University of Derby) in hand. Amongst traditional and veteran ethnomusicologists, this may seem quite normal. However, despite my self-proclaimed status as an anthropologist and ethnographer, I had never completed fieldwork outside of the UK — not in a physical sense, anyway. My prior experience with ethnomusicology had taken three forms: firstly, the study of interdisciplinary collaboration among composers and performers in South East London (see Williams, 2016, 2020a); secondly, Arts & Health music-based workshops for social well-being (which, although in Spain, was 'practice-led' and aligned with health humanities and medical ethnomusicology rather than ethnographic fieldwork; see Williams, forthcoming); and thirdly, the study of music-making online in media turned socio-political spaces (see Williams, 2020b). And although this latter area embraces cyber communities across the globe, 'fieldwork' had always been conducted from the sofa of my one-bedroom apartment in Derby, UK, with the nearest living mammal sat purring on my lap. And whilst the ideas of off-site ethnomusicology have been around for quite some time — see the seminal text *The New (Ethno)musicologies* (Stobart, 2008) — I began to feel free from both spatial and geographical shackles. For the first time, I felt as though my pseudo-fraud status had been voluntarily ditched, and I was, at last, a *true*, *travelling*, and *geographically orienteering* ethnomusicologist.

Delhi was 40-odd degrees Celsius and as lively as one would expect, but an overnight bus transported me and the group of undergraduate students to the quieter and cooler town of Palampur. Sitting at the foot of the Himalayas, it is around 35 km (of road travel) south-east of Dharamsala (or Dharamshala), the home of the Dalai Lama and Tibet's Government in Exile. Although Palampur would be our residence for the next two weeks, each day we would be travelling 15 km back in the direction of Dharamsala towards the settlement of Gopalpur to a school operated by the Tibetan Children's Village (TCV).

TCV Gopalpur is one of a collective of school sites (eight Residential Schools, in Dharamsala, Bylakuppee, Chauntra, Gopalpur, Ladakh, Lower Dharamsala, Selakui, and Suja, and three Day Schools, in Delhi, Ladakh, and McLeod Gang) with the 'mission' of ensuring that 'all Tibetan children under its care receive a sound education, a firm cultural identity, and become self-reliant and contributing members of the Tibetan community and the world at large' (TCV, 2017). With the pedagogical mantra of 'working for the care and education of Tibetan refugee children', TCV offer the following core goals:

> ... to provide parental care and love; to develop a sound understanding of Tibetan identity and culture; to develop character and moral values; to provide effective modern and Tibetan education; to provide [a] child-centered learning atmosphere in the schools; to provide [an] environment for physical and intellectual growth; and to provide suitable and effective life and career guidance for social and citizenship skills.
>
> (*ibid.*)

Alongside, they also offer the slogan: 'Come to Learn, Go to Serve!' and a quote from His Holiness the Dalai Lama: 'From the day we became refugees, out basic objective was to rise to the very place from where we have fallen down' (*ibid.*).

This chapter examines how music functions within the education provision and ethos of the TCV school in Gopalpur as a means of preserving and building cultural identity. More specifically, it aims to understand the significance of the educational and cultural 'site' in which music-making — singing, instrumental learning, and performance — occurs, with its simultaneous existence as a place of residence also making it a home and sanctuary for refugee children. My intention here is not so much to focus on how Tibetan heritage is celebrated through musicking, but rather to offer a deeper critical understanding of music education within the context of culture and, importantly, of *space*, in light of the geographic dispersion of Tibetan people. I report that although expressions of localness are rich within the area and within the school, evidenced through much of the music-making examples I offer, music education and performance remain localized (from my observations) and that perhaps music-making could act more extensively as a vehicle to expand 'cross-culturalness' for Tibetan school children. Thus, in this chapter I offer suggestions for the widening evolution of music-based projects based on my own perceptions of such localized musicking, in terms of both culture and space, and how future cohorts and delegates of the charity programme on which my students were enrolled may help support and lay foundations for such cross-cultural developments. Furthermore, I discuss my own positionality as a Western ethnomusicologist researching and studying music-making and education as a vehicle for cultural preservation and how I negotiate relationships between Westernized educational, musical, and cultural paradigms and the practices observed at TCV Gopalpur.

My approach to this work rests on a synthesis of external observations (documented through notes, photography, and video recording of classes and performances) with internal involvements and practical music-making (including participating in instrumental lessons, playing instruments, etc. and discussions and interactions with staff members and school pupils) at TCV Gopalpur. These are combined with personal reflections on my own presence within these settings, both in classes and at performances, as a participant and audience member, respectively. Much of my account is situated against, and compared with, the fieldwork and findings of Keila Diehl, as captured in her seminal text on Tibetan refugee music, *Echoes from Dharamsala: Music in the Life of a Tibetan Refugee Community* (2002). Diehl's text is perhaps one of the only academic sources of recent times to present a both thorough and overarching account of Tibetan refugee music, in alignment with approaches to cultural preservation through the arts. And whilst Diehl's fieldwork was much more substantial (in terms of both duration and depth of contact with participants), her accounts acted as an inspirational and influential template for my visit.

With this work in mind, dating back to fifteen years before my visit, the crux of my approach in this project is to identify *change* and *growth*, situating and comparing (where appropriate) my experiences of musicking, culture, and education in Himachal Pradesh with those of Diehl. In addition, I reflect on my role as a 'travelling ethnomusicologist' to inform better my understanding of culture and education and their relationship with site and space. A blend of specific angles approaching the overlapping space between ethnography, autoethnography, ethnomusicology, and geography (both as methods and subject areas) offered a multifaceted design and outlook for understanding the music education at TCV Gopalpur, and whilst being aware that each of these disciplines includes a wide array of lenses through which to observe, the areas outlined in Figure 6.1 naturally assumed particular importance in my visit to the school (see below).

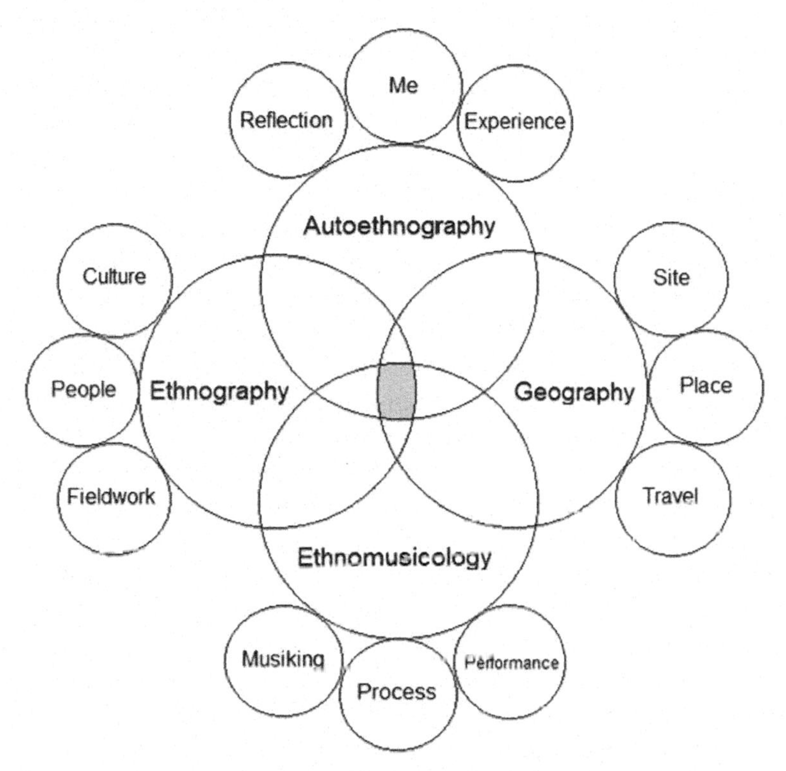

Figure 6.1 Diagram showing angles towards, and intersections between, ethnography, autoethnography, ethnomusicology, and geography.

6.2 'Preserving' Tibet Through the Arts

Sense of place sits close to the heart of many Tibetan communities, and these sentiments have a long history following the Chinese Annexation of Tibet (1950–1951). Diehl sees the notion of pilgrimage as central to Dharamsala's identity. For Tibetans, journey to the city is part of a roundtrip; to arrive there is an 'incomplete' experience (2002: 62) because of its 'dual character, as a holy destination for many Tibetan refugees and as a place they all hope to eventually leave' (*ibid.*: 29), to return to Tibet itself. Diehl suggests that the tension of 'simultaneously feeling at home and out of place informs the emotional power songs and other vocal music have as icons of places for Tibetan refugee listeners' (*ibid.*). Her account captures how music in Dharamsala is a vehicle for cultural preservation, and how such traditions offer rather complex cultural clashes with more modern and emerging musics, especially for younger generations — particularly pop, rock, and Hindi popular music (see Morcom, 2015). This is not only identifiable in Northern India; Mason Brown's fieldwork in Nepal finds modernizations of traditional Tibetan songs that are 'tied to the region's landscape and sacred sites' alongside other styles, such as Nepalese popular music (2018: iii). For young people, negotiating combinations of 'contemporary and global strains' with 'local tradition' is a way of 'recentering what it means to be Tibetan'; in other words, 'youth are connecting to the roots of their culture by learning a modern exile presentation of traditional Tibetan performing arts' (*ibid.*).

My initial intention was to explore, ethnographically, some of the tensions between the traditional and the modern within TCV Gopalpur's educational paradigm,[1] in particular the music classes that I discuss in more detail later. But importantly, as I outline in the next section, the nature of a *changing* place (i.e. a geographic location) or a changing *site* (of musical and cultural activity) was central to my fieldwork. I travelled to Himachal Pradesh in Diehl's footsteps in anticipation of seeing how music and its sites had evolved, particularly for the younger generations. How are today's young people living in exile negotiating the fine balance between cultural preservation and cultural development? Or, perhaps, how are they dealing with cross-cultural expansion? And to what extent are educational systems supporting these negotiations, allowing music-making to be a vehicle for *growth*?

These questions are, of course, partly political, especially in the politics shaping how arts (particularly music) play roles in heritage. In her book, Diehl

> ... present[s] some of the imagined and tangible ways in which the roots of the Tibetan homeland penetrate and are selectively cultivated in Indian soil... highlighting the historicity of popular memory and

[1]The contested terms 'traditional' and 'modern' are used here specifically to distinguish the 'conventional' and the 'contemporary'. I use 'traditional' to mean historically linked to Tibetan culture and 'modern' to mean recent and/or developed.

> what comes to be called 'tradition'. Through [her] descriptive accounts of traditional opera performances, school plays, and other musical events, [the] efforts by the Dalai Lama's government-in-exile to preserve what is commonly referred to as the 'rich cultural heritage of Tibet' [reveal] benefits and risks... including concerns about the long-term effects of an official cultural paradigm that authenticates the past and largely discredits the present.
>
> (2002: 29)

Some of these accounts that Diehl accesses are echoed in my own fieldwork captured in this chapter — notably in school plays/performances I witnessed. The Tibetan Government in Exile persists in a geographically complex state of being, as a home-from-home. But as time passes, the Tibetan diasporic community, living not only in Dharamsala but across Himachal Pradesh, continues to evolve as part of a re-localized culture. At what point do developments inspired by historical events and geographic movements alter perceptions of cultural authenticity held both by outsiders and by insiders, by members of the diaspora itself? What does Tibetan-ness mean today, and how does this differ from the Tibetan-ness of the past, if at all? And what roles do (and should) governments play in this? Preserving the 'rich cultural heritage of Tibet' is clearly an aim that persists to this day for Tibet's Government in Exile (seen in Diehl's fieldwork and my own), and I show evidence below of its influence not only in education, but also through the 'Culture Shows' given by TCV school pupils to celebrate the Dalai Lama's eighty-third birthday in July 2018 — events not dissimilar to the 'patriotic school pageants' that Kiehl draws attention to (2002: 63).

The identity of Tibetan people rests so heavily on geographic ideals, with the government's seemingly never-ending endeavours towards *returning* to Tibet continuing. An education so heavily dominated by the preservation of cultural ideals hints that there may be a sense of fear of Tibetan traditions and culture disappearing from people's lives, just as those people have been separated from their homeland. Fundamentally, I do not aim to criticize these attitudes, but in the examples I discuss below, I do seek to shed light on tensions between *growing* and, as Diehl puts it, 'staying the same' (2002: 65), particularly in music education, navigating this complex terrain alongside reflections on my own position as an educationalist living and working in the UK.

6.3 Our Mission in Tibetan Children's Village

The TCV slogan of 'Come to Learn, Go to Serve!' points, perhaps, to an education philosophy that is not particularly liberal — one guided by notions of service over notions of opportunity. This observation is not to impose any immediate value judgement on the system, but rather to recognize a point of difference (from my perspective, as used to Western educational paradigms). As outlined above, TCV lay out a clear and concise mission, and the ethos of the school in Gopalpur was communicated to us by the principal on arrival over local butter tea. But as a group primarily made of student artists, dancers, and musicians coming from the UK system of higher education institutions (HEIs), we had our own mission for the students' trip; the nature of our proposed contributions and how they would align with the educational practices of the school were up for negotiation.

Our goal was to introduce Tibetan pupils to the creative and performing arts as they existed in our worlds, inevitably contrasting with some elements of Tibetan doctrine and ways of life — a fact that led us to reflect critically on both our own and Tibetan education norms. This rendered strange power relations between us, as visitors, and the Tibetan staff and student residents, with a mutual understanding in place that we were all learning from each other, while nonetheless we (the visitors) were providing a service to the school. I certainly felt that whilst in the principal's office, we belonged to the school — not only as visitors, but also as cultural proponents of it. The power dynamics seemed instantly directive, as opposed to collaborative, and whilst I certainly felt welcome and enthused about our roles, I was also uncertain about the delivery of our Western practices. This feeling was made particularly acute given that we were visiting Gopalpur as a group of volunteers from the Challenges Abroad scheme, supporting their charity The FutureSense Foundation, which work with partner communities around the globe towards 'long term' and 'sustainable' impacts; Challenges Abroad offer UK students opportunities to 'develop a greater awareness of global development issues' (Challenges Abroad, 2020). As an international programme operating in six locations across the world, Challenges Abroad has its own *mission* and set of values, as laid out in its promotional material:

... to ignite compassion and create a generation of global citizens... Our partnership [with FutureSense Foundation] ensures that all of our volunteers work on ethical projects and can make a long-lasting and positive impact in the community. Challenges Abroad is much more than travel and volunteering. It's an opportunity for you to become a Global Citizen with a wider world view and an understanding of other cultures, all the while helping us tackle the challenges currently facing marginalised communities around the world.

Be the change you wish to see in the world.

— Mahatma Gandhi

(*ibid.*)

Of course, for the charity, this marketing rhetoric is designed to capture the attention of prospective volunteers, a rhetoric complete with hints at the quasi-colonialist assumptions its work may be accused of displaying. Even during the initial stages of the trip, it was obvious that there were both similarities and differences in the goals of TCV and Challenges Abroad. What stood out to me the most, though, was the simultaneous 'similar but different' approach to multiculturalism shown in the two; on the one hand, there was an educational institution whose primary focus was *preservation* of *local* culture, whilst, on the other hand, there was a Western organization pushing for *globalized* culture. Indeed, the TCV mission is for its pupils to succeed within communities across the wider world. But while *culturalism* always seemed paramount, there was a stark contrast when compared with the *cross*-culturalist ideological promotions of Challenges Abroad. There felt to me a slight tension, or at the very least a difference in perspective, on the related understanding of *growth* — whether personal, educational, or (cross-)cultural. And whilst it may be criticized as Western-centric to scrutinize the Tibetan *culturalist* perspective, there was certainly a quagmire of potentially clashing styles developing from early on in our visit, one that we as students and staff were in the middle of.

Further to this, the group of students with whom I travelled was part of a cohort from an Arts and Health degree I run at the University of Derby. Here our students engage in *applied arts*, using interdisciplinary and integrated creative practices to work to support the day-to-day well-being of a diverse range of people and groups. As their degrees progress, they are taught not only to be community artists, but also to be arts anthropologists. This is crucial within their experiential practice, for example, when on placement and indeed in their graduate vocations; it is essential that their approach adopts and adapts to principles of inclusivity, widening participation, and diversity — including across cultures, identities, races, religions, and genders. Before travelling to India, we agreed between us on a core aim of using the arts to promote and to enable well-being, working fluidly and inclusively. The 'de-colonization of curriculum' has been a prominent movement in HEIs over the past few years (Charles, 2019), and whilst the disciplines of therapeutic arts and arts and health (the courses which the students were studying) are typically thought of as based on Westernized pedagogic practices, there has been a growing interest in diversity and change at all levels of study — from studies in arts and health diversity at the undergraduate level, moving away from Eurocentric paradigms, to doctoral projects like the one I currently supervise looking at the decolonization of Dramatherapy master's degrees. Importantly, the respective approaches of Challenges Abroad and TCV, and our positionality *in between*, all had impacts on our roles and observations, and this is something I refer back to later when exploring examples of music education practices.

On arrival at the school, my students opted to work across several projects that were ongoing within its grounds. Mainly, though, they were tasked with teaching English to school children of different ages. It was immediately apparent that these children were used to visiting teachers, and they showed impressive evidence of their experience in participating in this kind of lesson. I audited a handful of classes and was surprised to find that many of the pupils, especially younger ones, had vast vocabularies for their age — they had evidently received an extensive education in English language. But despite the principal mainly allocating my students the work of teaching English, I was particularly interested in the role of music within the school: what music-making was happening, and what goals did it serve? The principal kindly provided me with a timetable of daily music lessons, and so for the next two weeks I spent some time between my other commitments participating in music classes. Much to my interest (coming naïve to the context), these lessons solely featured Tibetan music. The significance of Tibetan heritage throughout the school was becoming increasingly clear, and

I was determined to understand how the music education at TCV functioned as a vehicle for preserving Tibetan culture and, beyond that, for *growth*.

6.4 Music as Performing Arts in School Education

6.4.1 Music Classes

At TCV Gopalpur, I was fortunate to have sight of a number of music classes. However, I was able to engage only in ways permitted to me or deemed appropriate by our hosts. I sat in on classes offered by two of the music tutors. In one kind of lesson, of roughly 45 minutes long, pupils were given vocal music (both traditional and popular) to listen to, and then they were taught to perform the pieces together. Learning was achieved through a combination of ear and demonstration. After group-based listening, the music tutor spent some time listening back to the track (which was played through a portable CD player at the front of class) and transcribing the words out on the board. Meanwhile, some of the pupils collected instruments from around the room and began trying to figure out how to play the accompaniment, partially guided by the tutor while he transcribed the words.

Unsurprisingly, neither tutor nor pupils spoke any English with one another, and so I was reduced to interpreting physical and gestural cues. I gradually began to discern some order in what initially seemed like chaos, and I learned that they were rehearsing for a Culture Show that was on the horizon (see Section 6.5), where they would all be performing. After transcribing the lyrics, the tutor guided the pupils through the words — in some cases through the pupils' own handwritten copies of the transcriptions (Figure 6.2). After teaching the words, the tutor sat at the front of the class with the group gathered around; they performed the song together, sometimes with the CD playing as a backing track and at other times without. Figure 6.3 shows one instance where a boy had worked through a complete transcription of a song (learning it seemingly by ear) on the dulcimer, providing much of the harmonic accompaniment for the group. Although pupils performed along with accompanying music, it was apparent that the learning of the lyrics was paramount. I had the sense throughout the school that Tibetan culture was considered special in a way that it was to be *protected* and not just preserved.

Figure 6.2 Transcribing words in a Tibetan music class, Palampur, 4 July 2018. Photo by James Williams.

Lyrics are a central component of Tibetan music, and learning them sat at the heart of these music classes. As Diehl notes, '*speaking* Tibetan, and being able to *read* and *write* Tibetan are distinct skills' (2002: 212), associated with 'colloquial Tibetan' (spoken) and 'literary Tibetan' (written and chanted texts), respectively. Learning in the classes I observed straddled Tibetan music and Tibetan reading and writing.

Most fascinating to me about these classes was seeing Tibetan contemporaneity expressed through current popular music. This element of modernity did not so much *clash* with tradition (as I had expected), but instead it seemed to build on it. Anna Morcom writes:

> In their lyrics, Tibetan pop music (whether in Tibetan or Chinese) and *dunglen* have been recognized as a powerful expression of Tibetan identity, in what Brubaker and Cooper would term 'hard' form — bold and proud delineations of Tibetanness in terms of land, people, culture, religion and so on.
>
> (Morcom, 2018: 137)

Figure 6.3 Singing and performing popular Tibetan music in a school class, Palampur, 4 July 2018. Photo by James Williams.

Tibetan popular music carries the same focus on 'abiding themes of identity in lyrics and their expression' as found in traditional music (Morcom, 2018: 133), but set in modern musical forms; as Morcom notes, 'it is lyrics more than musical style where political expression takes place in today's Tibet' (*ibid.*). In TCV's music education, pupils were engaged simultaneously in conventionalism and modernism. This openness to modern-day creativity in music classes in 'Tibet in Exile' represents a notable departure from the contents of Diehl's accounts:

> The story behind 'modern Tibetan music' is largely a story about the social and artistic challenges young Tibetan refugee musicians face in their efforts to convince their community-in-exile that there is room for a new musical genre alongside, or even within, the politically charged paradigm of cultural preservation presiding in Dharamsala and elsewhere in the Tibetan diaspora. It is, put more generally, the story of determining what it means to be 'modern'... and further, what the role of contemporary music might be in answering this question.
>
> (2002: 176–177)

With this in mind, knowing that schools such as TCV Gopalpur not only show openness to modernity in music, but that they actually embed it into the music curriculum of the school, demonstrates the *change* and *growth* that they outline in their *mission*. These music classes are sites for new *and* old, with both representing contemporary Tibetan culture.

6.4.2 Instrumental Tuition

The other type of music class that I was permitted to take part in revolved around instrumental tuition. These lessons involved groups of around twenty pupils who were notably younger than those in the other classes I sat in on. Pupils were learning Tibetan instruments, mainly the traditional lute (sometimes referred to as *tungna* or *tugna*) (see Figure 6.4, left). These sessions taught traditional performance practices for Tibetan instruments in a blended format, with group instruction from the tutor followed by pupils taking turns to receive one-to-one instruction from the tutor (again, the session lasted 45 minutes to an hour) (see Figure 6.4, right).

Western models of instrumental tuition typically rely on a one-to-one format (Gaunt, 2011), the limitations of which (notably to do with well-being) have not gone unnoticed in recent times (Norton, 2016a, 2016b). But here, I observed that the children did not so much learn together (i.e. side-by-side), but rather they learned *collaboratively*. The tutor offered opening instructions and then circulated to check each pupil as they practised, but it was clear that the pupils also helped each other, in the placement of fingers, posture and holding the lute, tuning, and other aspects. Those less able were supported by those more able; hence, more advanced learners began developing coaching skills at a remarkably young age. Pupils were of varying abilities, with the tutor explaining to me that those at earlier stages had fewer strings on their lute (beginning with one), whilst those at a more advanced level had a greater number (up to three).

Figure 6.4 One-to-one instrumental tuition between teacher and pupil within a group class, Palampur, 3 July 2018. Photos by James Williams.

As a violinist and guitarist, I was keen to move from my position on the sidelines and 'have a go', while also being apprehensive about 'doing it correctly'. I soon found myself and the school pupils in role reversal — I was meant to be watching them learn, not the other way around. All of a sudden, I was learning and playing, while the pupils observed, helped, and taught me. There were unforgettable moments as a young boy took the hand of this reasonably accomplished violinist and daily guitarist, readjusting the posture of my fingers for accuracy. These pupils were collaborative, supportive, and confidently kind-hearted. But I quickly became aware of the impact I was having on the class; students were clearly more interested in me, the way I was playing, and my style of improvising on the lute than in their instrumental practice. Even though the tutor smiled at me as she looked over, I could see I was becoming a distraction as she frequently checked that the pupils around me were not misbehaving, neglecting their practice. I had entered with the full intention of respecting the rules and boundaries of the classroom, but now I felt as though I was breaking them.

The curiosity of the pupils in that particular class outweighed their commitment to instrumental practice; they were seeing something new, and they wanted to watch the musical creativities that I was bringing into the space. But whilst they showed a collaborative spirit of supporting each other with their instrumental learning, they seemed adversely unaware of their own creativity and creative potential. Indeed, their learning took a specified shape, unfolded through a specified routine, and was for a specified reason. This level of prescription may lend itself to the valuing of Tibetan music performance as a means of cultural preservation, but it seems less likely to inspire creativity in performance. And it is this point which leads me to the overarching conclusion I arrived at through studying and being part of the music sessions at TCV Gopalpur, and it also explains my emphasis on the word *performing* in this section's title.

In all music sessions I visited, I saw playing, singing, listening, demonstrations, collaboration, teamwork, and peer support — all activities directed towards performance. But the concept of performance itself, within the ethos of the institution, seemed only geared towards preserving and celebrating the Tibetan-ness (even if it was a *modern* Tibetan-ness) I discussed earlier in the chapter. I saw very little making or creating of *new* music, whether overtly Tibetan or otherwise. Although, of course, I was limited to my own situated observations, I struggled to identify much *creativity* in performance (see Clarke, 2005; Sawyer, 2006) — all these observations made by someone brought up in the Western classical idiom of music learning. I was reminded of the limitations of the Western classical model of music performance and aesthetics (see Hamilton, 2000) — but the goal here in this class was not so much optimization or achievement of perfection, but rather cultural preservation. There is perhaps an argument to be made for more *creative* musical structures to be taught and the *creating* of *new* music to have a place within the educational setting. Diehl (2002) also found composing or songwriting projects — or simply encouragement to deviate from conventions in performance — limited in her field experience. On the other hand, it is worth noting that the real benefits of the pedagogic example I observed involved being *collaborative* — in itself a rich form of creativity (see John-Steiner, 2006).

6.5 A Show of Culture

Earlier in this chapter, I noted Diehl's mention of 'traditional school plays' or 'patriotic school pageants'. I was fortunate enough to experience something similar at the culmination of my two weeks in TCV Gopalpur. Diehl writes,

> Children born in exile — second- and third-generation refugees who have never actually had to leave anywhere — must be taught to remember the experiences of others... In exile, this sense of history, identity, and responsibility is conveyed to young Tibetans and is reinforced in others largely through narratives and performances in schools or other institutional settings, events akin to what Connerton describes as 'acts of transfer that make remembering in common possible'.
>
> (2002: 66)

Whilst I was not present to see school performances explicitly concerned with cultural preservation, the invitation towards the end of my visit was to an evening of performances by school classes in celebration of the Dalai Lama's birthday. This event was an exhibition of Tibetan culture and thus, in turn, *implicit* of preservation (see Figure 6.5). Much of the classroom-based music learning discussed in the previous section was in preparation for this celebration. It featured a literal stage, set in front of audience attendees including pupils, teachers, senior school officials, and guests. It was a platform for Tibetan music and theatre performed by each school class, who moved in turn from their places in the audience to places on the stage. Lasting well into the night, the programme of this culture show mostly consisted of group music and singing, with some drama-based performances. It took place in the school's gymnasium, decorated with Tibetan art and, most noticeably, photographs of the Dalai Lama mounted on the walls all around the space.

As an offering to the Dalai Lama as part of a birthday celebration, only Tibetan music was performed — some traditional, some modern (as discussed previously), representing patriotic and deeply rooted Tibetan-ness, whilst also exhibiting some of the more forward-looking Tibetan identity. Whilst I enjoyed the programme, I found myself becoming intrigued about the impact of such cultural ceremonies on the younger generations studying and living at TCV; the breadth and modernity of Tibetan-ness was certainly identifiable and celebrated, but was there 'enough' *cross*-cultural breadth?

Figure 6.5 The Culture Show at TCV Gopalpur, 6 July 2018. Photo by James Williams.

During my time in Himachal Pradesh, I also visited some of the other partner schools of The FutureSense Foundation, where other arts and music projects were taking place. Neel Kamal Public High School, Jamula (one of the Indian schools not for Tibetan refugees that I visited just for one day), had arranged a programme of musical performances on the day of my visit, showcasing some of the individual musical abilities of its pupils. Although much smaller, this school event offered pupils the opportunity to perform genres of their choice, ranging from classical Indian music to popular styles. Here, I was even asked to play some of my own music for everyone to enjoy and sing along with. Whilst I spent very little time at this school, the different approach to performing arts was striking to me. The preservationist traits of the Tibetan Culture Show were very different from the programme at the Indian school, which seemed to adopt a comparatively open feel. Being invited to perform on my only day visit was, to me, a gesture of welcome, and whilst I felt incredibly welcome at TCV, neither myself nor my students were invited to perform at the Culture Show. The two schools were geographically quite close (around 45 minutes travel by car), but their socio-political culture, in terms of *performance*, seemed very different.

The contrasts in the school visits encouraged me to think about the localities and about the neighbouring cultures. Whilst I enjoyed the experiences, for me the Tibetan Culture Show stood out as uniform and as almost regimented in its conception of birthday festivities. Staying in Palampur, I found myself drawn into pockets of multiculturalism — be it through food, fashion or even ceremonies (at one moment I was spontaneously invited into a wedding party, where there was a range of popular musics and dance; the next I was in a small local house listening to a married couple play and sing traditional songs together). But what struck me most about these neighbouring cultures was how I found myself interpreting their differences, based on my own cultural positions — as an educationist and as a musician myself carrying my own Western schooling in performance. I discuss this more below.

6.6 Conclusion

6.6.1 The (Un)changing Sound of Tibet in Education

My visit sheds light on the roles of learning and performing in enabling both understanding and preservation of Tibetan culture for refugee school children in Himachal Pradesh. It shows some of the complexities of how music education supports the contemporaneity of Tibetan-ness in the diaspora. This project has shed light on the educational developments for Tibetan refugee school children happening in Himachal Pradesh — specifically at TCV, Gopalpur — and not only how the role of music and performing arts enables both understanding(s) and preservation(s) of Tibetan culture, but also how the music curriculum supports the contemporaneity of the Tibetan diaspora. Education acts as a core component of Tibet's cultural heritage, and what it means to practise modern Tibetan-ness through the arts clearly continues to have relevance as a part of children's lives. Some of my observations in the field, however, lead me to consider the possibility that modern-day Tibetan children may not be offered the same kinds of cross-cultural opportunities in their music learning as children in neighbouring educational communities. Music-making in groups lends itself to communication, collaboration, and interaction, and whilst this was especially evident in some of the instrumental tuition classes at TCV Gopalpur, I observed that much of the music tuition formed a basis to preserve Tibet's rich cultural heritage. *Creative* musical practices and education in *other* (non-Tibetan) musics seem limited, even though the curiosity of school pupils in music classes, as they took an interest in me joining in, was very clear.

Within the school classes, I identified no tensions whatsoever between the traditional and the modern: music classes exhibited an education involving both conventional song and modern popular music — but my concern that it was *all* Tibetan (just as the instrumental classes were all on Tibetan instruments) is something that I fear will not change soon. Music pedagogy examples such as these demonstrate how detached communities can be from wider arts cultures, even if their institutions are afforded opportunity and access to diversity through charity and volunteering programmes such as Challenges Abroad and FutureSense. My recommendation, as an outcome of this study, is that future cohorts of volunteers — comprising musicians, artists, and dancers — should be engaged both in supporting the culture and communities under ideas of growth and preserving Tibetan-ness, but additionally that they are involved in transmitting a wider and more diverse *cross*-cultural growth, where performing arts (such as music) can be used as a platform to mediate these explorations (and tensions, should they arise). I would not only encourage the volunteering individuals themselves to be leaders in this endeavour, but also call for the charities to promote such projects. Diehl writes,

> The popular music being made by Tibetans living outside of their homeland provides an interesting case study of what happens to

> culture 'on the road', of musical creativity unhooked from a particular geographical place. In fact, the boundaries of cultural practices rarely coincide, of course, with the lines drawn on maps. So, refugee culture is, then, just an extreme example of the very common 21st-century experience of living multi-local, multi-cultural, hybrid lives. Paradoxically, due to the pressure to preserve Tibetan culture in exile, Tibetan refugee musicians may be less free to borrow from foreign cultures and genres than others whose communities are not threatened by the dual forces of colonization in the homeland and assimilation in the diaspora. Underlying these dynamics is a call to reexamine the meaning and usefulness of the concept of 'cultural preservation', including the risks of canonizing particular traditional practices as 'authentic' and dismissing contemporary or popular innovations as 'inauthentic' or even threatening to the community's cultural and ethnic integrity.
>
> (2002: 12)

A consideration I made on my return flight home is that over fifteen years later, and despite my relatively short time in Himachal Pradesh, Diehl's remarks feel very much true today. Both TCV and Challenges Abroad offer fantastic support for communities, and this is evidenced at the site in Gopalpur. I worry, however, that TCV is not harnessing fully the opportunities brought to the school and its staff and pupils by volunteering arts students — the school could be a culturally significant *place* for refugee children and a *site* for growing and expanding musical creativities. I hope to return again and to see more closely the growth of Tibetan culture through musicking and education.

6.6.2 Positionality and Reflection

My final concluding remarks refer back to my role as a travelling ethnomusicologist and my reflections on positionality in the evaluation and discussion of this fieldwork. My privilege to be able to complete the fieldwork originates in a Western educational idiom, accompanied by a similarly open and reflexive musical viewpoint. For me, this fieldwork was about developing an understanding of the education, performance, and heritage of Tibetan refugee children's context, but just as much about examining those of my own —

how does my own position in all of this influence how I see and explore the worlds of others? How does my own musical education constrain or free me to explore the musical education of others? And when travelling to a site of heritage, to what extent does my own cultural background influence my research methodologies? These are all questions relevant to musicologists, ethnomusicologists, geographers, and ethnographers in the context of decolonizing the work of HEIs.

References

Brown, M. 2018. *From Pungyen To Palyul: Recentering Identities through Alliance and Music in Trans-Himalayan Nepal*. PhD thesis, University of Colorado.

Brubaker, R., and F. Cooper, 2000. 'Beyond "Identity"', *Theory and Society* 29(1): 1–47.

Charles, E. 2019. 'Decolonizing the Curriculum', *Insights* 32(1): 24, 1–7.

Clarke, E. 2005. 'Creativity in Performance', *Musicae Scientiae* 9(1): 157–182.

Diehl, K. 2002. *Echoes from Dharamsala: Music in the Life of a Tibetan Refugee Community*. Berkeley: University of California Press.

Gaunt, H. 2011. 'Understanding the One-to-One Relationship in Instrumental/Vocal Tuition in Higher Education: Comparing Student and Teacher Perceptions', *British Journal of Music Education* 28(2): 159–179.

Hamilton, A. 2000. 'The Art of Improvisation and the Aesthetics of Imperfection', *The British Journal of Aesthetics* 40(1): 168–185.

John-Steiner, V. 2006. *Creative Collaboration*. Oxford: Oxford University Press.

Morcom, A. 2015. 'Locating Music in Capitalism: A View from Exile Tibet', *Popular Music* 34(2): 274–295.

_____. 2018. 'The Political Potency of Tibetan Identity in Pop Music and Dunglen', *HIMALAYA: The Journal of the Association for Nepal and Himalayan Studies* 38(1): 127–144.

Norton, N. 2016a. *Health Promotion in Instrumental and Vocal Music Lessons: The Teacher's Perspective*. PhD thesis, Manchester Metropolitan University.

_____. 2016b. 'Health Promotion for Musicians: Engaging with Instrumental and Vocal Teachers', *Arts and Humanities as Higher Education* 15.

Sawyer, K. 2006. *Explaining Creativity: The Science of Human Innovation*. Oxford: Oxford University Press.

Stobart, H., ed. 2008. *The New (Ethno)musicologies*. Lanham, MD: Scarecrow Press.

Williams, J. 2016. 'Creative Departures from Compositional Principles: A Case Study of Contemporary, Theatrical Minimalism with Live Electronics', *Principles of Music Composing* 16: 116–123.

_____. 2020a. 'Antiphonal Authorities: Exchanges of Control and Creativity in Collaborative Electroacoustic Performance'. In *Sound Art and Music: Philosophy, Composition, and Performance*, edited by J. Dack, T. Spinks, and A. Stanović. Newcastle upon Tyne: Cambridge Scholars Publishing.

_____. 2020b. 'Cassetteboy: Music, Social Media, and the Political Comedy Mash-Up'. In *Popular Music, Technology, and the Changing Media Ecosystem: From Cassettes to Stream*, edited by T. Tofalvy, and E. Barna, 233–251. Cham, Switzerland: Palgrave Macmillan.

_____. (forthcoming). 'Collaborative Notation and Performance in Arts & Health Workshops: How Musical Scoring in Groups Enables Creative Interaction and Communication for Social Wellbeing'. In *Together in Music: Participation, Coordination, and Creativity in Ensembles*, edited by R. Timmers, F. Bailes, and H. Daffern. Oxford: Oxford University Press.

Other Media

Challenges Abroad. 2020. 'About Us' (webpage, *Challenges Abroad*). Available online at https://challengesabroad.co.uk/about-us/about-us/

TCV. 2017. 'TCV's Mission' (webpage, *Tibetan Children's Village*). Available online at https://tcv.org.in/tcvs-mission/

Chapter 7

Claiming Back the Arctic: Evaluating the Effectiveness of Music as a Voice for the Indigenous Subaltern

Kiara Wickremasinghe
Department of Anthropology and Sociology,
SOAS University of London, 10 Thornhaugh Street,
Russell Square, London WC1H 0XG, UK
644916@soas.ac.uk

7.1 'The Scramble for the Arctic': Introducing the Context

> ... often through the music which had sustained them through everything, the music allowed their human dignity to shine again.
>
> (Sara Wheeler, *Private Passions,* BBC Radio 3, 2015)

Drawing on her travels across the Arctic Circle, during an interview with Michael Berkeley on BBC Radio 3's *Private Passions,* the travel author Sara Wheeler reports on music being a voice for the

Musical Spaces: Place, Performance, and Power
Edited by James Williams and Samuel Horlor

ISBN 978-981-4877-85-5 (Hardcover), 978-1-003-18041-8 (eBook)
www.jennystanford.com

marginalized and a pathway to indigenous self-representation. Her reaction to the sublime beauty and brutal reality of the landscape is powerful in her description: 'overwhelming and uplifting and terrifying, all at the same time' (BBC Radio 3, 2015).

The Arctic is often imagined as a pristine, silent, and dreamy place that is surpassed by the power of nature. In reality, the Arctic Region is a fragile, politicized, and contested space, among the first to experience the impacts of climate change, posing consequences for the wider world (Leduc, 2010). While global warming increases the rapidity of polar ice melt, sea level rise, and other alarming forms of environmental change, the region also presents geopolitical opportunities where oil reserves are exposed and shipping routes opened, ultimately causing a 'scramble for the Arctic' (Craciun, 2009: 103). This 'scramble', as Craciun (2009) argues, reflects both twenty-first century struggles to access the Northwest Passage and North Pole and historical struggles for circumpolar control motivated by the Arctic being a transcontinental access point to several other regions. Often dominated by imperial projects, defeated by energy corporation agendas, and overshadowed by polar bears and other poster boys for climate change are the narratives of indigenous people living in the Arctic Region. While the Arctic has not experienced an emblematic version of colonialism, the framing of indigenous people as 'other' and the suppressing of their voices perpetuate colonial relations (Cameron, 2012: 103), echoing Spivak's (1988) concept of the *subaltern*. Adopting a postcolonial theoretical approach, this chapter explores the use of music as a medium for attaining *subaltern* self-representation through exercising indigenous advocacy and communicating indigenous knowledge and counter-narratives. Depending on the message carried, voices represented, audience reach, and response of listeners, music can be a vehicle for decolonization, promoting indigenous sovereignty in the Arctic Region. This chapter argues that sovereignty and 'claiming back the Arctic' manifest not merely in a physical territorial sense but also as cultural preservation. Along with promoting mediums for the *subaltern* to speak, fostering platforms for effective listening is equally important for purposes of generating response and action. Will we stop and listen?

7.2 'Oriental Undertones in the Arctic': Applying Postcolonial Theory to an Arctic Context

This chapter begins by examining why the Arctic context and the plight of its indigenous populations may be examined through a postcolonial theoretical lens. Postcolonial scholars direct attention to the position of a subject in relation to others, demonstrating the existence of hierarchies such as those visible in colonial relations. By problematizing dominant (often Western) knowledge systems, postcolonial scholars propose alternative views of the world and offer counter-narratives that privilege marginalized voices (Sharp, 2009). The cultural imagination of the Arctic Region as a pristine and dreamy wilderness can be viewed in parallel to Said's scholarship on *Orientalism* (1978). This seminal text ingrained the notion that patronizing and exoticized views of place and people in the *East* (*Orient*) legitimized *Western* (*Occidental*) dominance in this region. Said's claim that culturally imagined binary geographies shape real geographies practised in the *Orient* resonates with the Arctic context, where romanticized perceptions of its wilderness and populations may manifest in regional power dynamics. While the Arctic has not been subjected to formal colonial rule, it is reported that indigenous beliefs and practices were historically suppressed by nation-states and Christian missionaries and that this suppression continues into present times through the framing of indigenous populations as 'local' and 'traditional' (Cameron, 2012: 103). Indigenous identity, whether self-assigned or imposed by others, is a positioning that has emerged 'through particular patterns of engagement and struggle' (Li, 2000: 3) and one that poses both opportunities and risks. One risk is that indigenous populations are often equated with the image of the native, resulting in 'intellectual and spatial confinement' (Appadurai, 1988: 38), where the thinking and knowledge systems of these people are bounded by the geographical space they occupy. This experience of marginalization echoes Spivak's (1988) concept of the *subaltern*, which points to those at the margins of society who are rendered without agency under the hegemonic terms set by postcolonial culture. Claiming that the *subaltern* face 'epistemic violence' (*ibid.*: 25) where their knowledges are undermined while

Western knowledge systems are privileged, Spivak emphasizes the need for self-representation. She asserts that *subaltern* heterogeneity, which generates diversity in voice, must be recognized to avoid certain perspectives being privileged over others. In the Arctic context, this may involve unpicking the complex positionings of indigenous people in the Canadian Arctic (Inuit) versus those in the European Arctic (Sámi) and how these ambiguities may impact the counter-narratives that postcolonial scholars are eager to convey.

Inuit communities are particularly caught up in climate change debates as their physical territory and livelihoods are threatened by environmental change and political processes. While there are ongoing efforts to recognize their vulnerability as linked to climate change, Inuit rhetoric points to historical and contemporary colonial practices underlying climate change itself, where the dispossession of indigenous lands and resources has paved the way for industrial shipping and resource extraction in the region and increased greenhouse gas emissions (Cameron, 2012). Since the human dimensions are gaining traction in climate change debates, Cameron (2012) observes that Inuit knowledge is beginning to be valued and sought when designing mitigation and adaptation strategies for climate change. Research studies and mapping projects which incorporate the narratives of indigenous people are examples of non-musical initiatives for indigenous autonomy in the Arctic. Cameron (2012) cites scientific research documents such as the Arctic Climate Impact Assessments (ACIA), which include indigenous perspectives on climate history and current climate change. Another such endeavour is the Pan-Inuit Trails Atlas. A collaboration between Inuit and researchers, this interactive atlas draws on indigenous knowledge to digitally map Inuit trails over satellite imagery, thereby preserving indigenous knowledge and history while asserting historical Inuit sovereignty and mobility in this region (University of Cambridge, 2014). In the wake of sea ice fragmentation resulting from global warming, preserving knowledge of trails is important to Inuit, for whom sea ice represents home, connects communities, provides a space to travel, and contains fishing lakes and hunting grounds (Bravo, 2010).

On the theme of maps that subvert dominant narratives, Craciun (2009) cites a Sámi map maker named Davviálbmogat known for his anti-colonial maps of Sámiland depicting human and non-human habitation and inscribed with the names of indigenous peoples. In the European Arctic, Sámi inhabit the northern regions spanning the nation-states of Norway, Finland, Sweden, and Russia and have sought recognition as a distinctive group of people since the nineteenth century (Ramnarine, 2013). 'Positioned ambiguously within the borders of the western, the indigenous, the European and the colonised' (*ibid.*: 251), Sámi have historically been denied agency, including the right to map and settle in their own lands, practise shamanism, and sing *joik*, as their practices are perceived negatively by scientists, Christian missionaries, and Nordic governments (*ibid.*). The indigenous vocal style of *joik* comprises a multi-layered narrative with no beginning or end and portrays relationships between music, environment, and the sacred (*ibid.*). Thus, for them, music is not only a medium for communicating indigenous counter-narratives but also a direct vehicle for decolonization, as reviving Sámi *joik* and language practices promotes cultural sovereignty and helps 'claim back the Arctic' in a different sense. The heterogeneous nature of indigencity and complexities attached to historical and contemporary colonial practices can influence counter-narratives; so while counter-narratives represent a different vantage point and give agency to the marginalized group, they may not directly oppose the dominant narrative. Interestingly, as postcolonial scholar Bamberg (2004) notes, subjects can tack between dominant- and counter-narratives and position themselves in the process. Having now laid out the context for indigeneity in the Arctic Region, this chapter proceeds to critically explore music as a medium for projecting indigenous counter-narratives and promoting *subaltern* self-representation.

7.3 'Music to Direct Change': Indigenous Advocacy and Counter-Narratives

Through music, indigenous people can offer an alternative environmental critique to Western schools of thought, one that situates the human subject as part of the environment rather than in relation to it (Ramnarine, 2009). The dissemination of these indigenous counter-narratives enhances intercultural dialogue

(Leduc, 2010) as the earth enters a new epoch conceptualized by Crutzen and Stoermer (2000) as the *Anthropocene*. The earth's exit from its current geological epoch, namely, the Holocene, and into this new epoch, the *Anthropocene*, is largely attributed to adverse human activities which are shifting human and global environmental relations. At a time where climate scepticism is rising and human development is being challenged, many are unwilling to accept that humankind has become 'a global geological force in its own right' (Steffen *et al.*, 2011: 843).

Sámi composer and political activist Nils Aslak Valkeapää's (1943–2001) *Goašše Duse* (*Bird Symphony*) (1993) portrays a 'special bond' that indigenous people have with ecology and reminds listeners of human coexistence with nature. Featuring recorded birdsong and waterscapes as improvising agents from his home environment, with *joik* singing and reindeer bells signalling human presence, Valkeapää captures the integration between all subjects and calls for listening to the flow of one another (Ramnarine, 2009). While justifying coexistence and serving as a reminder that human musical creativity is often derived from sonic ecosystems, *Bird Symphony* is not past focused or myth driven. Rather, it is a political work that critiques Western ways of approaching nature as a resource for human ownership and exploitation (*ibid.*). Considering that the Arctic itself has been subject to resource extraction, forest logging, ozone layer depletion, and transboundary pollution, musical works such as *Bird Symphony* are a symbol of protest against political processes that threaten indigenous territory and natural ecosystems. Such musical works are also modes for self-representation, critical to processes of 'asserting land rights, histories, and the validity of indigenous philosophies, as well as to rejecting external (colonial) representations' (*ibid.*: 208).

Alongside conveying indigenous wisdom, revivals in *joik* symbolize decolonization, considering that this form of Sámi music-making was historically prohibited. Owing to their association with shamanism and the supernatural, *joiking* and drumming were negatively perceived by Christian missionaries (Hirvonen, 2008). Thus, in reviving these cultural and spiritual practices, the *subaltern* Sámi are speaking out, expressing their identity, and declaring cultural sovereignty. *Joik* has become an integral part of the indigenous political project as evidenced in another of Valkeapää's works, a musical collaboration with Seppo Paakkunainen (b. 1943) named *Juoigansinfoniija* (*Joik Symphony*) (1989). Composed for symphony orchestra, an improvising instrumental group, two *joik* singers, and solo saxophone, and formed of four movements as in Western classical music, *Joik Symphony* employs what Ramnarine terms 'symphonic activism' to protest against negative representations of Sámi and struggles over minority status (2009: 187). Based on the activist Valkeapää's melodies and poems, the titles of the four symphonic movements are striking in their communication of nature-based discourse and indigenous philosophy: 'Polar Night Resounding with Cold' (Gumadii galbmasit skábma), 'Drone, *Joik* of the Hills' (Humadii, duoddarat juige), 'Sisters, Brothers, the Wind in my Heart' (Oappát, vieljat váimmustan, biegga), and 'The Ocean of Life' (Eallima áhpi) (*ibid.*). Although *joik* can represent a mode of decolonization, Helander and Kailo (1998) warn of 'white shamanism', a contemporary form of colonial practice where indigenous songs, knowledge, beliefs, or healing methods are appropriated by non-indigenous people and sold for profit. This commercialization of Sámi spirituality moulds it into an exotic commodity, misinterprets and distorts shamanistic wisdom, and consequently diminishes the effect of the indigenous project (*ibid.*).

Returning to indigenous counter-narratives, Valkeapää's *Joik Symphony* ironically adopts Western musical structures when furthering the indigenous political project, illuminating the complexities of indigenous positionings. This supports postcolonial scholar Bamberg's (2004) observations about subjects tacking between master- and counter-narratives when stating their vantage point. A more recent musical initiative that demonstrates ambiguous indigenous positioning is the work of young Sámi rapper Amoc. Rapping since 2005, Amoc has pioneered the use of Inari Sámi language (a minority language with only around 300 remaining speakers), dispersing it across local and global platforms and ultimately helping to preserve this endangered language (Leppänen and Pietikäinen, 2010). Moreover, through music he has helped Inari Sámi regain their collective voice, strengthened ethnic identity within the community, and motivated younger Sámi to learn the language, enabling it to sustain into the future (Ridanpää and Pasanen, 2009). A striking contribution of Amoc's musicianship is his bridge-building between two stereotypically opposite worlds, that is, Western urban rap culture and rural, nature-centred Sámi culture. In this respect, his rap 'functions as an emancipatory tool deconstructing the stereotypical ways of approaching ethnic heritages' (*ibid.*: 213), and it overcomes the 'native' image of 'intellectual and spatial confinement' that Appadurai describes (1988: 38). Sámi tend to be imagined through colonial eyes as modest, submissive, communal, and primitive, but Amoc's rap features lyrics such as 'I am Amoc, and when I go mad / People are frightened to death' and 'I have been alone for my whole life, I'm alone everyday'. These examples of lyrics reveal aggression and individual loneliness, respectively, deconstructing stereotypes of Sámi being purely peace-loving and communal people (Ridanpää and Pasanen, 2009). Said's (1978) explorations in *Orientalism* show how art and other cultural practices contribute to 'othering' by fuelling stereotypes and imaginary opposition, and so Amoc's work can be credited for deconstructing stereotypes of Sámi people while conserving a minority language. Ironically, however, Ridanpää and Pasanen (2009) suggest that it is Amoc's exotic ethnic roots and related Sámi stereotypes that make his music hugely appealing and a topic of discussion in academic, media, and political circles.

This section has highlighted how indigenous music-making aids both physical and cultural aspects of 'claiming back' the Arctic. Works such as Valkeapää's *Bird Symphony* protest against the destruction of indigenous territory and natural ecosystems due to resource extraction and other political processes and alert listeners to land rights and issues relating to physical territory. Calling for indigenous knowledge to be taken seriously and reviving historically silenced methods of music-making such as *joiking* mark decolonization and declare cultural sovereignty. However, the misappropriation and commercialization of indigenous cultural assets threaten the power of the indigenous political project to effect meaningful change. At times, the ambiguous positionings of indigeneity as revealed through music can confuse or diminish the distinct message conveyed. On the other hand, cases such as Amoc prove that mixing styles and leaving ambiguities unresolved does appeal to listeners and deconstructs stereotypes in the process. This chapter now turns to imaginations of the Arctic wilderness and how anxiety about losing it fuels musical initiatives for environmental advocacy, while debating whether indigenous populations reap any benefits from these efforts.

7.4 'Ecotopian Arctic Soundscapes': Music for Environmental Advocacy

When sound and music are channelled positively to circulate environmental messages via social justice events, they create 'ecotopian soundscapes' according to Morris (1999: 129). Greenpeace's Save the Arctic campaign has fashioned ecotopian soundscapes in various ways to raise awareness about the fragile state of the Arctic and to stimulate environmental action. In 2015, for example, a group of string players set up outside the headquarters of global energy company Shell in London and performed a *Requiem for Arctic Ice* in an effort to halt Shell's plans for Arctic oil drilling expansion (Gayle, 2015). Employees and passers-by alike stopped to listen, reflecting ideas that music is 'not merely a meaningful or communicative medium' but that it has the power to influence 'how people compose their bodies, how they conduct themselves, how they experience the passage of time, how they feel — in terms of energy and emotion — about themselves, about others, and about situations' (DeNora, 2000: 16, 17). The ecotopian soundscape of *Requiem for Arctic Ice* captured the attention of people, composed their bodies by stopping them in their tracks, and turned their attention towards a specific situation and place, that is, global warming in the Arctic Region.

On another occasion in 2016, Greenpeace produced a music video featuring the popular Italian pianist Ludovico Einaudi performing his composition *Elegy for the Arctic* on an iceberg in Svalbard (Classic FM, 2018). With pristine Arctic wilderness and a crumbling glacier as a backdrop, this solemn and minimalist musical reflection accents what could be lost at the expense of human development and why urgent action is required. Attracting almost two million viewers on YouTube, this music video has an impressive reach owing to the fame of the pianist and alluring production techniques. Ecotopian soundscapes may have discernible emotional impacts on audiences but monitoring correlations between such events and any environmental action taken afterwards is challenging. Nevertheless, while artists such as Einaudi and organizers of environmental initiatives such as Greenpeace cannot guarantee activating change, they do provide 'aesthetic infrastructure and environmental behaviour' that viewers can incorporate into everyday life choices if the soundscapes move them to do so (Galloway, 2014: 71). This echoes ethnomusicologist Titon's (2016) comparison between music and ecosystems and his claim that cultural sustainability and ecosystem maintenance are inextricably linked.

Both of the Greenpeace initiatives above capitalize on the emotional affordance of music to effect change and carry a key message, reminding audiences of the view that humans are only one group in a larger interconnected ecosystem and should therefore maintain reciprocal relationships with other groups. Privileging the image of silent, pristine wilderness in its key message resonates with Schafer's (1977) pioneering work on soundscapes, which defines ideal soundscapes as those free of the mechanical, electrical, and industrial noises of increasing human development, often at the expense of nature. The key message also echoes indigenous counter-narratives on coexistence between humans and nature, and yet these populations are largely excluded. While silence and isolation are highlighted, people — especially indigenous populations of the Arctic — are missing in image and voice in the campaign, music video, and composition. Thus, these ecotopian spaces in the examples discussed can be contested in terms of whose gaze and narratives they privilege. Further, the effectiveness of these idealized spaces depends on how participants engage; that is, whether they choose to actively listen and proactively adjust their actions (Galloway, 2014). The following section addresses this notion of effective listening, which applies to any form of musical advocacy, whether indigenous or environmental.

7.5 'Hearing the Subaltern Speak': Fostering Platforms for Effective Listening and Action

Ecotopian soundscapes 'consolidate a listenership who is united by common ecological values' (Galloway, 2014: 71). Alongside creating mediums for the *subaltern* to speak, promoting active listening is equally vital to the process of advocacy. Elements such as audience reach and response to messages conveyed via music influence the level of action taken. Alaskan composer John Luther Adams' composition *Sila: The Breath of the World* is an example of music that facilitates different experiences of ecological listening depending on where it is performed and with what configuration of instruments. Through this work, Adams encourages the audience to search for music within surrounding sounds with the expectation that they will leave the performance transformed (Lincoln Center, 2014). Its outdoor premier in 2014 at Lincoln Center's Hearst Plaza, New York City featured five choirs of woodwind, brass, strings, percussion, and voice, superimposed over the city's soundscape. Though non-indigenous, Adams derives compositional inspiration from the Inuit concept of '*Sila*', a force that broadly signifies and is experienced in weather. Spiritual and ideological aspects of Inuit knowledge are often marginalized by Western interpreters due to their presumed inaccessibility, isolation in the past, and non-empirical nature (Leduc, 2010), yet Adams 'aligns with Inuit thinking and further shapes an ear for climate change in this era of global warming' (Chisholm, 2016: 174). When depicting '*Sila*', Adams is not focused on specific harmonic, melodic, and rhythmic patterns but rather on 'an ecology of music' which centres around 'the totality of the sound' (Adams, 2009: 1). He seeks to attune listeners to unsettling noise, which has now become 'the breath of the world' (*ibid.*: 4), and to reorient them so that they become outward focused and conscious of larger patterns of life on earth.

As well as in musical compositions, sonic images in both indigenous and non-indigenous cinema drawing on Arctic themes can aid advocacy efforts and possibly secure a wider reach than pieces of music, in number and demographic variety. Combining the sonic and visual can be effective in evoking emotion and crafting memorable themes and messages for audiences. *Ofelaš* (*Pathfinder*) (1987), an example from Sámi cinema, and Disney's *Frozen* (2013), a Western animation that reflects northern politics and people, both illuminate indigenous ideologies, though to varying degrees and targeting different audiences. The former is an action film based on a Sámi legend, filmed in Norway and broadcast in Sámi Lapp language (with English subtitles), while the latter is a recent addition to the Disney collection catering to younger, English speaking audiences. While *Ofelaš* is based on a Sámi legend, Ramnarine claims that it is not past focused but rather 'tells a story about conflict in the past to promote Sámi indigenous sensibilities in the present, as well as point to future political possibilities for global cooperation' (2013:

251). Considering Sámi oral tradition holds keys to records of past geophysical events, which can contribute to current climate change debates; *Ofelaš* illuminates indigenous sacrality and reminds viewers of the value in community and global interdependency. *Frozen* tells a fictional tale of sisterhood, mystical icy powers, and a quest to reverse a perpetual weather condition and save a kingdom.

Despite stark differences on the surface, both films begin by featuring Sámi *joik* in their opening credits sequence. *Ofelaš* portrays *joik* and shamanic drumming with Sámi composer and activist Valkeapää contributing to the film soundtrack, while the opening *joik* 'Vuelie' in *Frozen* is composed by Norwegian-Sámi musician Frode Fjellheim. 'Vuelie' (Song of the Earth) has no beginning or end, which is a characteristic of indigenous *joik*, but Fjellheim also marks the influence of Lutheran missionaries in Scandinavia by overlaying the *joik* with a church hymn line (Ramnarine, 2016). Both films continue to capture elements of indigeneity, though *Ofelaš* perhaps portrays links between the environment and the sacred, including animal symbolism, more accurately and through a Sámi lens. In one scene, a reindeer sighting foretells the shaman's death and in another a shamanic vision of a reindeer after Aigen (the boy protagonist) defeats and survives invaders symbolizes his appointment as the next shamanic leader/pathfinder (Ramnarine, 2013). Though much less solemn in its depiction, *Frozen* describes the closeness between reindeer herder Kristoff and his reindeer Sven, especially through the song 'Reindeers Are Better than People'. *Frozen* also refers to the sacred Sámi landscape in the song 'Fixer Upper' when rocks come to life as spirit beings. In citing links between the environment and the sacred, *Frozen* departs from master narratives and presents an alternative view of the world, in this case a Sámi one. Additionally, while its plot is fictional, *Frozen* raises awareness about climate change and the fragility of the northern landscape, though this is depicted in an inverted sense, as perpetual winter, as opposed to as global warming (Ramnarine, 2016). The animation also illuminates the power dynamics at play in claiming the kingdom, somewhat resonating with current political claims to the Arctic Region (*ibid.*). Combining sonic and visual elements can be an effective means of capturing the attention of audiences and a platform which promotes active listening.

7.6 'Claiming Back the Arctic': Conclusion

Imagining the Arctic as pristine and sublime wilderness masks its fragile reality as a contested space, scrambled over by political, indigenous, and environmental parties. Fear and nostalgia associated with losing the Arctic wilderness have sparked multiscale efforts to save this region. Where music has been incorporated in environmental advocacy, such as in Greenpeace's Save the Arctic campaign, it has afforded emotional uplift, united listeners, and encouraged environmental behaviour. However, as environmental advocacy efforts, inter-state dialogue, and regional development agendas progress, indigenous populations find themselves on the sidelines. From a postcolonial theoretical perspective, *subaltern* populations such as Inuit and Sámi face marginalization and often lose out in scrambles for the Arctic. As argued in this chapter, music can provide a voice for the indigenous *subaltern* and promote physical/territorial and cultural 'claiming back' of the Arctic. Indigenous musical works such as *Bird Symphony* can assert physical sovereignty by serving as icons of protest against land rights issues or destructive resource extraction. Through conveying indigenous knowledge, wisdom, and narratives, such musical works also express cultural sovereignty. Works such as *Joik Symphony*, which revive historically suppressed practices like *joiking*, or Inari rap by Amoc, which deconstructs ethnic stereotypes, also declare cultural sovereignty by breaking free from colonial barriers and representations. However, a critical examination of these musical initiatives, particularly how they assume musical forms such as 'symphony' and 'urban rap' that have originated in the Western world, unravels the more problematic side of these musical advocacy efforts. Further, projecting marginalized indigenous voices through music only represents one side of the coin. The other side demands platforms for effective listening that generate response and action. This entails reorienting listeners through musical initiatives such as Adams' *Sila* or conveying memorable sonic and visual narratives through films such as *Ofelaš*. Only once platforms for effective listening are fostered will we hear the *subaltern* speak.

References

Adams, J. L. 2009. *The Place Where You Go to Listen: In Search of an Ecology of Music*. Middletown, CT: Wesleyan University Press.

Appadurai, A. 1988. 'Putting Hierarchy in its Place', *Cultural Anthropology* 3(1): 36–49.

Bamberg. M. 2004. 'Considering Counter-Narratives'. In *Considering Counter-Narratives: Narrating, Resisting, Making Sense*, edited by M. Bamberg, and M. Andrews, 351–371. Amsterdam: John Benjamins Publishing Company.

Bravo. M. 2010. 'Epilogue: The Humanism of Sea Ice'. In *SIKU: Knowing Our Ice*, edited by I. Krupnik, C. Aporta, S. Gearheard, G. J. Laidler, and L. Kielsen Holm, 445–452. London: Springer.

Cameron. E. S. 2012. 'Securing Indigenous Politics: A Critique of the Vulnerability and Adaptation Approach to the Human Dimensions of Climate Change in the Canadian Arctic', *Global Environmental Change* 22(1): 103–114.

Chisholm, D. 2016. 'Shaping an Ear for Climate Change: The Silarjuapomorphizing Music of Alaskan Composer John Luther Adams', *Environmental Humanities* 8(2): 172–195.

Craciun, S. 2009. 'The Scramble for the Arctic', *Interventions: International Journal of Postcolonial Studies* 11(1): 103–114.

Crutzen, P., and E. Stoermer. 2000. 'The Anthropocene', *Global Change Newsletter* 41: 17–18.

DeNora, T. 2000. *Music in Everyday Life*. Cambridge: Cambridge University Press.

Galloway, K. 2014. 'Ecotopian Spaces: Soundscapes of Environmental Advocacy and Awareness', *Social Alternatives* 33(3): 71–79.

Helander, E., and K. Kailo, eds. 1998. *No Beginning, No End: The Sami (Lapps) Speak Up*. Edmonton: Canadian Circumpolar Institute/Nordic Sami Institute.

Hirvonen, V. 2008. *Voices from Sápmi: Sámi Women's Path to Authorship*. Guovdageaidnue (Kautokeino), Norway: DAT.

Leduc, T. 2010. *Climate, Culture, Change: Inuit and Western Dialogues with a Warming North*. Ottawa: University of Ottawa Press.

Leppänen, S., and S. Pietikäinen. 2010. 'Urban Rap Goes to Arctic Lapland: Breaking through and Saving the Endangered Inari Sámi Language'. In *Language and the Market*, edited by H. Kelly-Holmes, and G. Mautner, 148–158. London: Palgrave Macmillan.

Li, T. M. 2000. 'Articulating Indigenous Identity in Indonesia: Resource Politics and the Tribal Slot', *Comparative Studies in Society and History* 42(1): 149–179.

Morris, M. 1999. 'Ecotopian Sounds; or, the Music of John Luther Adams and Strong Environmentalism'. In *Crosscurrents and Counterpoints: Offerings in Honour of Bengt Hambraeus at 70*, edited by P. Brosman, N. Engebretsen, and B. Alphonce, 129–141. Göteborg: Göteborg University Press.

Ramnarine, T. K. 2009. 'Acoustemology, Indigeneity and Joik in Valkeapää's Symphonic Activism: Views from Europe's Arctic Fringes for Environmental Ethnomusicology', *Ethnomusicology* 53(2): 187–217.

_____. 2013. 'Sonic Images of the Sacred in Sámi Cinema: From Finno-Ugric Rituals to Fanon in an Interpretation of Ofelaš (Pathfinder)', *Interventions: International Journal of Postcolonial Studies* 15(2): 239–254.

_____. 2016. 'Frozen through Nordic Frames', *Puls: Swedish Journal of Ethnomusicology and Ethnochoreology* 1: 13–31.

Ridanpää, J., and A. Pasanen. 2009. 'From the Bronx to the Wilderness: Inari-Sami Rap, Language Revitalisation and Contested Ethnic Stereotypes', *Studies in Ethnicity and Nationalism* 9(2): 213–230.

Said, E. W. 1978. *Orientalism*. New York: Vintage Books.

Schafer, R. M. 1977. *The Soundscape: Our Sonic Environment and the Tuning of the World*. Vermont: Destiny Books.

Sharp, J. P. 2009. *Geographies of Postcolonialism*. London: Sage.

Spivak, G. C. 1988. 'Can the Subaltern Speak?' In *Marxism and the Interpretation of Culture*, edited by C. Nelson, and L. Grossberg, 21–78. Urbana, IL: University of Illinois Press.

Steffen, W., J. Grinevald, P. Crutzen, and J. McNeil. 2011. 'The Anthropocene: Conceptual and Historical Perspectives', *Philosophical Transactions of The Royal Society A: Mathematical, Physical and Engineering Sciences* 369(1938): 842–867.

Titon, J. T. 2016. 'Why Thoreau?' In *Current Directions in Ecomusicology: Music, Culture, Nature*, edited by A. S. Allen, and K. Dawe, 69–79. Abingdon, Oxon: Routledge.

Other Media

BBC Radio 3. 6 December 2015. 'Northern Lights: Sara Wheeler', *Private Passions* (radio broadcast, BBC Radio 3). Available online at http://www.bbc.co.uk/programmes/b06r44b4

Classic FM. 21 December 2018. 'Einaudi Plays Piano on an Iceberg as an Arctic Glacier Crumbles around Him' (online article, Classic FM). Available online at https://www.classicfm.com/composers/einaudi/news/iceberg-video-greenpeace/

Gayle, D. 3 August 2015. 'Greenpeace Performs Arctic Requiem in Effort to Touch Hearts over Shell Drilling' (newspaper article, The *Guardian*). Available online at https://www.theguardian.com/environment/2015/aug/03/greenpeace-adds-a-new-string-to-its-bow-with-musical-protest-at-shell

Lincoln Center. 2014. 'Sila: The Breath of the World' (online video). Available online at https://www.youtube.com/watch?v=rUDjOyacZoU

University of Cambridge. 10 June 2014. 'First Atlas of Inuit Arctic Trails Launched' (online article, University of Cambridge). Available online at http://www.cam.ac.uk/research/news/first-atlas-of-inuit-arctic-trails-launched

Music and Spatial Imaginaries

Chapter 8

'He Is a Piece of Granite...': Landscape and National Identity in Early Twentieth-Century Sweden

Anne Macgregor
Department of Music, University of Sheffield,
34 Leavygreave Road, Sheffield, S3 7RD,
United Kingdom
a.e.macgregor@sheffield.ac.uk

8.1 Introduction

One of the most consistent features of the reception of the Swedish composer Ture Rangström (1884–1947) is the use of landscape imagery to describe and explain his music. There is, of course, nothing particularly unusual about composers depicting landscapes in their music, or their music being interpreted in terms of landscape, and both were common practice in early twentieth-century Swedish culture. However, Rangström is noteworthy even within this context for several reasons: the frequency with which his music was linked to landscape, the consistency of the imagery used, and the extent to

Musical Spaces: Place, Performance, and Power
Edited by James Williams and Samuel Horlor

ISBN 978-981-4877-85-5 (Hardcover), 978-1-003-18041-8 (eBook)
www.jennystanford.com

which his identity — especially his national identity — was conflated with the character of his music. One of the few contemporary English language summaries of Rangström's music, an article published in the New York magazine *Musical Courier*, is a case in point:

> Some of the younger composers who have during the last years fought their way to the front are Ture Rangström, Nathanial Berg and Kurt Atterberg, forming a kind of trio by themselves... Rangström [has] a stern, honest and sincere nature of absolute originality. He is a piece of granite with a peculiar, charming melancholy, perhaps a little difficult to approach by one not familiar with his development and that of the Swedish music, but he has produced works of real, established value, as can easily be seen when his art is studied.
>
> (Westberg, 1920: 6)

After this commendation of Rangström, Westberg goes on to describe Berg as perhaps the most talented of the three and Atterberg as the most productive. However, there is an undercurrent of disapproval in his depiction of Atterberg's chameleon-like ability to absorb and replicate a variety of international styles — an ability that contrasts with Rangström's honesty, sincerity, and originality. Westberg's pithy conclusion is that 'Rangström is national, Berg is international, and Atterberg is whatever he chooses to be.' By implication, Westberg ties the idea of national identity to authentic self-expression: Rangström is, for him, thoroughly genuine and completely national. The image that he uses to convey this character is granite, which forms the bedrock of much of the Swedish landscape and thus has connotations of unyielding permanency.

Rangström played an active part in the discourse about the relationship between landscape and music, and he may even have been the first to emphasize granite as the fundamental descriptor of his (and Swedish) music. Early in his career as a critic-composer, he wrote an eloquent article about the need to cultivate a distinctively Swedish art music, and he promoted the Swedish landscape as the obvious source of inspiration for such a change in direction. As in Westberg's article, straightforward and personal expression is prioritized along with landscape imagery:

> Let it be that our temperament waits only for the simple, meagre, but strong art-music, which knows its innermost, highest individuality

> and melds it together with Swedish nature and the landscape's own essence. Much granite — fir and pine in thorny dark stretches — a light, melancholy, shimmering Mälar idyll — a red cottage with white trim — long winter nights, short sunny days and much, much granite...
>
> (Rangström, 1911)[1]

There is some evidence that 'much granite' became Rangström's mantra: he used it in 1918 when writing to the older composer Wilhelm Stenhammar about his intentions for his second symphony (named *Mitt Land* — My Country), and in 1934 the music writer Folke H. Törnblom used the phrase in quotation marks when describing Rangström's style. Most of the references to granite in Rangström's reception, though, appear without such apparent awareness of precedent, and so the extent to which Rangström helped to shape his own reputation is difficult to pin down. (Stockholm's musical life was driven by a small and interconnected circle of composer-critics; Westberg, for example, was a personal friend of Rangström, and so it is possible that the Courier article was vetted or even guided by the composer.) What Rangström and his contemporaries understood as a musical manifestation of granite is comparatively clear though hardly specific: when they use the term in writing, the immediate context consistently includes words such as abrupt, harsh, angular, or barren. Through this cluster of adjectives, granite-like-ness could be heard in the blocky structures of Rangström's music, in the starkness of his harmonies, in harsh timbres, or in persistent ostinatos.

Even more prevalent in Rangström's reception than the refrain of 'much granite' were references to the Gryt archipelago in Östergötland, south of Stockholm. Rangström spent his summers there throughout his adult life and obviously found it a creatively stimulating environment; holidays there afforded both the opportunity and the inspiration for many of his compositions. This artistic connection with the archipelago landscape was underlined most dramatically at Rangström's fiftieth birthday celebrations, when he was presented with the title deeds for an island that had been bought for him by public subscription. His devotion to this little island (he called it his kingdom) was such that, after his death, his children sought permission to lay him to rest there. When they were

[1]All translations are the author's own.

unsuccessful, their second choice was not Stockholm, as might be expected given that the capital city was also Rangström's birthplace and hometown, but the local village of Gryt. Rangström's personal commitment to the archipelago region was thus demonstrated even in death, and many of his obituaries drew on landscape imagery in order to eulogize both his music and his character: 'He was *ursvensk* [originally Swedish] in type and disposition, gnarled as a wild oak in his artistic temperament, heartwood through and through as person and musician' (Sundström, 1947). Here, the wild oak conveys much the same perception of Rangström and his music as granite does: it is tough and untamed, thoroughly natural and sound. Crucially, this statement binds together Rangström's musicality and personality in the same image, and the aspect of his personality that is highlighted is his national identity. *Ursvensk* is a term that defies direct translation, but the prefix *ur* can mean 'from', 'from within', or 'out of', and so the whole word describes (as either noun or adjective) someone who is essentially Swedish, a Swede of Swedes, or a Swede from of old. It is one of three Swedish terms that appear in Rangström's reception and that incorporate ideas of place and belonging; the others are *landskap* (landscape) and *hembygd* (loosely, home-place). By using these terms to unpick the ways in which Rangström's music was understood by his contemporaries as an expression of both landscape and personal identity, we can examine the long-standing assumption that his music communicates national character. We will see that Rangström's music does not simply portray a specific geographical region with accuracy but shows that Rangström embraced and embodied an understanding of landscape that was distinctively Swedish.

8.2 *Landskap*: 'He Is a Piece of Granite'

Musical depictions of landscape have been a regular feature of Western classical music ever since Beethoven's *Pastoral Symphony*. Such pieces are often expressions of patriotism and are sometimes marked by precise detail: Smetana's *Vltava* is a prime example of both, in its portrayal of a river's journey from its twin sources through the Bohemian countryside to Prague and beyond. Many other works and composers could be cited, but Rangström stands out from most of them due to his sense of landscape as an experiential and not merely a visual phenomenon.

In English, the term 'landscape' usually refers to the physical features of an area of land and is often also concerned with their aesthetic appeal. This understanding of landscape came into the language around 1600 thanks to a genre of paintings which depicted natural scenery, and it has displaced the older meaning of 'region' or 'province' that existed in Old English, Old Dutch, and Old Norse. The Nordic concept of landscape still incorporates aspects of the older meaning, namely, the idea of human interactivity with the physical landscape. In the interdisciplinary volume *Nordic Landscapes* (Olwig and Jones, 2008), several authors take time to explain the origin of the word *landskap* and the difference that it makes to modern-day mentalities. Gabriel Bladh, a professor in cultural geography, writes that the term *landskap* has been used since at least the Middle Ages to denote a province that is more or less 'culturally and geographically homogenous' (2008: 221). Such regions adhered to a common law (Olwig and Jones, 2008: xviii), meaning that where someone lived directly affected how they lived. When the idea of landscape as observed scenery reached the Nordic countries in the Romantic era, it did not supersede this concept of landscape as a lived territory (Bladh, 2008: 221–223).

Bladh is relatively cautious in dating *landskap* to the Middle Ages; Tomas Germundsson, a professor of human geography, asserts that the regions 'often have their roots in prehistoric times and thus predate the Nordic nations' (2008: 157). In either case, there is some chronological justification for giving regional identity precedence over national identity, or at least seeing the first as fundamental to the second. These provinces, as recognized entities, certainly predate the modern form of the Nordic countries; as the balance of Baltic power shifted over the centuries, they were sometimes traded as part of peace treaties, and regional identity was thus more stable than national identity. The retentive power of local tradition is evidenced by the tendency of modern-day Swedes to describe their regional identity in terms of the ancient *landskap* rather than by the current administrative regions (Germundsson, 2008: 158).

The Nordic understanding of landscape as a lived territory immediately narrows the conceptual gap between a physical landscape and a piece of music. It means that we are not dealing with a translation of something visual into something audible; music is not merely sonic scenery. Rather, both landscape and music can be understood as an experience in which internal sensation and physical activity are as important as external perception. For Rangström, there was a broad overlap between these two kinds of experience: he titled his Third Symphony 'Song Under the Stars' and identified his inspiration for it as 'lonely, night-time sailing trips' (Helmer, 1998: 307); he had an epiphany about his opera *Gilgamesj* while stopping his boat from sinking at its moorings in a 'steaming southwester' (N., 1946); he even described a particular stretch of water sparkling under the sunset as 'an F sharp major triad for large orchestra with six trumpets!' (Rangström, 1943). Rangström clearly experienced landscape in musical terms at least as powerfully as his listeners then experienced his music in landscape terms. As he put it himself, 'music becomes our experience. I cannot sing a forest or spring or sea without my own impression — conscious or unconscious — of what a forest, spring or sea is becoming crucial for the music' (Rangström, 1942). By claiming that he could not help but incorporate his own experience of landscape into his music, Rangström emphasized the authenticity of his compositions. They could be intended and heard as both deeply personal and thoroughly national because the portrayal of experienced landscape was, in itself, Swedish — or at least Nordic, for the concept of lived landscape is common to the Nordic countries.

There is good reason to single out both Sweden and Rangström from the broader context of Nordic landscape music at the time. Early twentieth-century Sweden differed from the neighbouring countries in that distinct regional identities were celebrated rather than a single, nationwide ideal. Sweden's relative security and stability help to explain this cultural diversity; by way of contrast, the external threats and internal divisions faced by the other Nordic countries encouraged the promotion of a clear, core national identity. (It is worth noting that Norway, Finland, and Denmark all possess one iconic composer who was seen as representative of national spirit, whereas Sweden produced a cluster of contemporary composers, many of whom were associated with different regions of the country.) Rangström was not considered a national composer because his music portrayed a specific heartland region that was historically significant or geographically typical; it was enough that his music belonged somewhere. In fact, one of the ways that Rangström was unusual even among his Swedish contemporaries was that Östergötland was not his native region, and this leads us to our second key concept: *hembygd.*

8.3 *Hembygd*: A 'Landscape for Soul and Heart'

Hembygd loosely translates as 'home-place', from *hem* meaning 'home' and *bygd* meaning 'district'; it denotes a place in which one feels a sense of belonging. *Bygd* has an experiential aspect rather like *landskap* though on a smaller scale: a *bygd* is 'a small district that is first and foremost a cultural unit, but secondly it is a unit bound together both socially and economically' (Björkroth, 1995: 33). *Bygd* is a long-established term, whereas the concept of *hembygd* emerged in the 1870s. The *hembygd* movement was a response to industrialization and urbanization, and in the decades around the turn of the century a great deal of effort went into preserving or revitalising regional traditions. The legacy of the movement is still apparent: for example, the Swedish Local Heritage Federation was founded in 1916 and currently maintains 1400 centres through local societies (Sveriges Hembygdsförbund, 2018). The most famous product of the *hembygd* movement is Skansen, the world's first outdoor museum, which was conceived by Artur Hazelius (1833–1901) and opened in 1891. Agricultural buildings from across the country were reconstructed on an island in Stockholm, where some sense of the lived landscape has been preserved ever since by guides in regional dress and through demonstrations of traditional handicrafts. Hazelius had previously been responsible for founding the Nordic Museum in 1873, a project which initially relied on public donations. The museum's motto, 'Know yourself', is still the tagline for the webpage and shows just how deeply entwined regional heritage was (and is still perceived to be) with the formation of personal identity.

Since Rangström grew up in a culture preoccupied with regional belonging, it is not surprising that he defined his own identity in regional terms. In an interview towards the end of his life, he described himself as 'a native Stockholmer, and, moreover, a half-breed between Skåne and Sörmland' (Rangström, 1946). In this statement, Rangström placed a great deal of weight not only on his own birthplace, but also on that of his parents: his mother came from Södermanland, about 60 miles south of Stockholm, and his father from Skåne, Sweden's most southerly province. This reference to Skåne is anomalous in Rangström's reception and rather difficult to reconcile with the usual emphasis on *ursvensk* character and rugged landscape. (The region was part of Denmark until the Treaty of Roskilde in 1658, and so its cultural landscape is very different from that of the rest of Sweden, as is its physical landscape.) However, Rangström's sense of belonging to the region appears to have been negligible; in fact, even his father had lived in Stockholm since the age of ten. Perhaps the most pertinent observation to be made about Rangström's statement, then, is that it assumes a remarkably persistent regional identity, albeit a nominal one, based on birth and ancestry.

Crucially, though, Rangström's account did not end there. 'I am certainly a native Stockholmer, and, moreover, a half-breed between Skåne and Sörmland,' he said; 'but that does not stop me from adding to myself one more landscape for soul and heart. It is Gryt's beautiful archipelago in south Östergötland...' (Rangström, 1946). The landscape with which Rangström and his music were so strongly associated is thus an adopted one. An Östergötland newspaper expressed the situation in the following terms:

> He has lived in the district on weekdays and weekends, become like one of the natives. 'Ture Rangström is on shore. His boat is at the dock.' How many times have we heard that? He is as famous and well-known a person in Gryt as in Stockholm. And that is saying a lot for a non-native-born.
>
> (Jonsson, 1934)

Within just a few sentences, the author used two terms of regional belonging that are based on descent: *urinvånarna* and *infödding*. The first is translated above as 'natives', but literally means 'the *ur*-inhabitants', that is, the original residents; the second means 'native-born'. By these objective measurements of belonging, Rangström was necessarily excluded: he was not native-born and so he could only be *like* one of the natives. However, the concept of *hembygd* allows for more subjective means of belonging and can be applied to any place where one feels at home or to which one has a strong connection. It is through the concept of *hembygd* that the Östergötland author acknowledged Rangström's integration into the region's way of life: 'Perhaps it is the sounds from Gryt's skerries and the enchanting meadows or marshes inside the storm-lashed cliffs that are his true native tones [*hembygdstoner*]' (Jonsson, 1934).

The composite word *hembygdstoner* could be translated in various ways, because *ton* can mean tone, note, or even tune. In any case, it is clear that Rangström's sense of belonging in Östergötland was intimately bound up with his music. Jonsson suggests that Rangström's most natural musical sound-world was that of Gryt's seascape — that it was in Östergötland that Rangström found his true compositional voice. Rangström's description of the area as a landscape 'for soul and heart' confirms this impression of a deeply personal and emotional connection to the region and ties back into the ideas of sincerity and individuality that appear in his reception alongside landscape metaphors. In his wholehearted adoption of Gryt's landscape and its way of life, Rangström epitomized the ideals of the *hembygd* movement: he was a cultured city man who nevertheless treasured rural life; he 'knew himself' in terms of his heritage, identifying as 'an old sea-rover' (Rangström, 1946); and he preserved his experience of the landscape in his music, much as Hazelius had sought to do in his museums.

Given the preoccupation of the cultural elite with preserving Sweden's rural heritage, it is hardly surprising that Rangström's interaction with the Gryt archipelago prompted effusive comments about the authenticity and national character of his music. One critic wrote that 'his musical language has much of undressed stone, of bedrock, that Swedish granite, rather harshly, in both melody and harmony' (Stuart, 1941: 399). The metaphor of bedrock clearly conveys the fundamentally national character of Rangström's music, but the idea of authenticity appears more obliquely in the image of undressed stone. This picture conveys the romantic notion that Rangström's music was simply hewn out of the landscape and presented without the aesthetic compromise of being chipped and polished into a more acceptable form. In the context of the *hembygd* movement, this sense of raw and unrefined nature was commendable — and thoroughly national.

8.4 *Ursvensk*: 'Heartwood Through and Through as Person and Musician'

The term *ursvensk* is the linguistic equivalent of the German *urdeutsch* and both words are connected to ideas about idealized masculinity, national landscapes, and — at least by implication — racial purity. *Ursvensk* appeared in Rangström's reception in the 1930s, at which time *urdeutsch* was undergoing a dramatic increase in usage in Germany. Google Ngram (which charts word frequency in printed and digitized texts) shows a series of peaks in usage between 1850 and 1950; the three largest peaks coincide with the unification of Germany and the First and Second World Wars.[2] This term clearly flourished in eras when German identity was promoted in pursuit of a political agenda. There is currently no equivalent data with which to track the use of *ursvensk*, but Rangström's reception certainly supports the hypothesis that there was a similar increase in usage in 1930s Sweden. The term remained prominent until after Rangström's death in 1947, and the connotations of idealized masculinity came across strongly in many of his obituaries, in which parallels were also explicitly drawn between Rangström's physical stature and character, and the character of his music:

> There was something of Viking inheritance in Ture Rangström's tall figure with the straight back, but he also showed a distinctive blend of this Norse, harsh manhood and the finest and noblest chivalry. His musical language owns a similar mixture of harsh, defiant minor and romantic sensibility...
>
> (Broman, 1947)

Despite the similarities between *ursvensk* and *urdeutsch* (and without being disingenuous about the prevalence of pro-Nazi sentiment in 1930s Sweden), the terms need not be understood as ideological equivalents. In Rangström's reception, at least, *ursvensk* seems to be weighted so heavily towards nostalgia for the past that it has little sense of being directed towards a present political agenda.

---

[2]Google Books Ngram Viewer. Available online at https://books.google.com/ngrams/graph?content=urdeutsch&case_insensitive=on&year_start=1800&year_end=2000&corpus=20&smoothing=3&share=&direct_url=t4%3B%2Curdeutsch%3B%2Cc0%3B%2Cs0%3B%3Burdeutsch%3B%2Cc0%3B%3BUrdeutsch%3B%2Cc0

Indeed, its apparent increase in usage occurred in a country that had maintained neutrality through the First World War and would do so again during the Second.

This emphasis on nostalgia rather than political action may simply reflect the attitude of a nation that had already established and lost an empire. The seventeenth century had been Sweden's 'Age of Greatness' (*Stormaktstiden*): at various times Sweden had held sway over territory in modern-day Norway, Denmark, Finland, Russia, Latvia, Estonia, Poland, and Germany. In Rangström's lifetime, the term *storsvensk* (great-Swedish) was used to describe a belief in Sweden's ability to re-emerge as a significant European power; by way of contrast, *ursvensk* was more often associated with the remote past, usually the Viking era. When Rangström himself was described as Viking-like, it was usually with reference to his music, his personality, and landscape. One example appears in the tongue-in-cheek song written for his sixtieth birthday dinner:

> Hail, you Viking, late-born in time,
> bolt upright and stiff-necked and ready for battle
> — but always gentle before the heart's roar,
> the forests' darkness and the valleys' lightness...
> Oh Ture, oh Ture!
> As strong as granite and as weak as a reed!
> (Berco, 1944)

Here, Rangström is portrayed as a throwback rather than as the twentieth-century norm or even as an ideal. The same pattern of thought is evident — though the emotional context could hardly be more different — in Rangström's instructions that his sailing boat be burnt after his death. Burning the boat was a dramatic and deliberate echo of burial practices strongly associated with Viking chieftains; it was also a highly romantic gesture that was hardly normal for twentieth-century Swedes. It is this sense of anachronism that prevents *ursvensk* from carrying much sense of a political agenda: it is primarily associated with an era that is so remote as to be beyond reach or re-creation except in an individual and idiosyncratic way.

For all the historical distance of the Viking era, though, it had a kind of geographical proximity for Rangström through the concept of the lived landscape. Viking culture is intimately associated with sea travel — even the name is related to human interaction with the environment, being thought to derive either from *vik* (meaning 'bay' and probably relating to the pirate practice of ambushing ships from the concealment of the coast) or from *vika* (an Old Norse term referring to oarsmen changing shift). Rangström's experience of sailing the archipelago landscape gave him a strong affinity with those who had done so for centuries before. He relished the sense that long-dead mariners had weathered the same storms and sailed the same currents as he did, and his vivid imagination drew inspiration from their imagined presence:

> In the autumns [my island] is haunted: the old dead Ålanders and other seas' spectres wander in closed ranks between the junipers up towards the cottage, so that your heart sits in your throat.[3] The seals howling outside Harstena are foreboding. It is a grand place for music!
>
> (Rangström, 1946)

Rangström's lived experience of his adopted landscape thus gave him a connection to what was seen as the essence of Swedishness. In Swedish terms, he cultivated *ursvensk* character in the *landskap* of his *hembygd*. These three intertwined concepts were seen by Rangström's contemporaries as crucial factors in shaping both Rangström's personality and his compositions, which were, in turn, so closely connected as to be interchangeable.

8.5 Conclusion: 'Rangström Is National'

These ideas about landscape, belonging, heritage, character, and compositions meshed together very readily in the early twentieth-century Swedish mindset, as a fuller passage from Jonsson's 1934 article demonstrates:

> Anyone who met Ture Rangström in Gryt's archipelago perhaps understands more about his artistic creation than one who only

[3]'Ålanders' refers to inhabitants of the Åland islands, which lie midway between Sweden and Finland and which were ceded by Sweden to Russia in 1809. After the declaration of Finnish independence in 1917, the islanders petitioned unsuccessfully for reintegration with Sweden; a League of Nations ruling in 1921 affirmed Finnish sovereignty over the islands but protected the practice of traditional culture and Swedish as the official language. The status of Åland was therefore an ongoing political issue for much of Rangström's life, and his use of the term 'Ålanders' shows the importance that was placed on historic regional identity at the time.

> met him wearing the notorious, unchanging cravat customary in Stockholm. His music has strong Viking and bard features, but with such captivating lyrical traits as in 'Melodi' [one of Rangström's best-known songs]. Perhaps it is the sounds from Gryt's skerries and the enchanting meadows or marshes inside the storm-lashed cliffs which are his true native tones.
>
> (Jonsson, 1934)

Jonsson makes several significant claims in quick succession: he asserts that knowing how Rangström lives in Gryt helps one interpret his music; he implies that Rangström is more 'himself' in the archipelago than in the capital city (where he was indeed accused of wearing the same necktie for forty years); he refers to the Viking era in general and its musical tradition in particular; and he affirms that Rangström and his music belong in Gryt. To underline this belonging, Jonsson continues with an extended description of Rangström's physical interactions with the landscape:

> Ture Rangström has often wrestled with the wind and waves in his little open boat, of the pilot-boat type; he has sailed in the storm, rowed in the calms with strong arms, which scorned the motor's help; he has, during week-long trips with his boat, harboured in friendly coves and seen many nights fall silently and clearly around the islands and the water. But he has also anchored for tough weather and, in the storm's noise, heard the eternal shadows' voices become the noise of mighty wings in the night.
>
> (Jonsson, 1934)

The last phrase is lifted from one of Rangström's most popular songs, a setting of Bo Bergman's 'Vingar i Natten' ('Wings in the Night'), and thus emphasizes the idea that Rangström's compositions were direct re-expressions of an experienced landscape.

These ideas are not neatly compartmentalized: they feed into and reinforce one another in such a way that tracing cause and effect is not always possible or helpful. One noteworthy observation, however, is that each of these concepts contributes in some way to the expression of national identity. There was thus a kind of feedback loop at work in Rangström's reception that continually reemphasized the public's perception of Rangström as a truly national composer. The national character of Rangström's music has long been left unexamined, but this chapter has shown that the composer's relationship with the Gryt archipelago is indisputably central to it. This relationship embraced, firstly, the Nordic understanding of landscape as a lived experience and reflected the particularly Swedish preoccupation of the time with diverse regional ways of living. Secondly, it embodied the ideals of the *hembygd* movement, which valued local belonging as a tenet of national identity. Thirdly, it gave Rangström a sense of personal connection with Swedes of the past, which was increasingly seen, during the 1930s and 40s, as a crucial part of Swedish nationalism.

This nexus of interrelated ideas gives Rangström's musical evocations of landscape an exceptional depth of significance. They are not merely audible postcards of national scenes, but expressions of an experience of landscape that is in itself distinctively Swedish. The contemporary perception of Rangström's music as Swedish depended heavily on his activity within the Östergötland landscape and thus reflects the centrality of regional diversity, belonging, and heritage in conceptions of national identity in early twentieth-century Sweden.

References

Björkroth, M. 1995. '*Hembygd* — A Concept and its Ambiguities', *Nordisk Museologi* 2: 33–40.

Bladh, G. 2008. 'Selma Lagerlöf's Värmland: A Swedish *Landskap* in Thought and Practice'. In *Nordic Landscapes: Region and Belonging on the Northern Edge of Europe*, edited by K. Olwig, and M. Jones, 220–250. Minneapolis: University of Minnesota Press.

Germundsson, T. 2008 'The South of the North: Images of an (Un)Swedish Landscape'. In *Nordic Landscapes: Region and Belonging on the Northern Edge of Europe*, edited by K. Olwig, and M. Jones, 157–191. Minneapolis: University of Minnesota Press.

Helmer, A. 1998. *Ture Rangström: Liv och Verk i Samspel* [Ture Rangström: Life and Work in Interaction]. Stockholm: Albert Bonniers.

Olwig, K., and M. Jones. 2008. 'Introduction: Thinking Landscape and Regional Belonging on the Northern Edge of Europe'. In *Nordic Landscapes: Region and Belonging on the Northern Edge of Europe*, edited by K. Olwig, and M. Jones, 3–11. Minneapolis: University of Minnesota Press.

Other Media

Berco. 1944. 'Bordssång' [Table-Song] (song text, reproduced in magazine article, *Musikern*).

Broman, S. 13 May 1947. 'Ture Rangström Avliden' [Ture Rangström Deceased] (newspaper article, *Svenska Dagbladet*).

Jonsson, S. December 1934. 'Klanger från Östgötaskären' [Sounds from the Ostergotland Skerries] (newspaper article, *Östgötabygd Jultidning*).

N. 7 October 1946. 'Ture Rangström i Stan, har Snart Ny Opera Färdig' [Ture Rangström in Town, New Opera Ready Soon] (newspaper article, *Söderköpingsposten*).

Rangström, T. 17 December 1911. 'Svenskt Lynne — Svensk Musik' [Swedish Temperament — Swedish Music] (newspaper article, *Stockholms Dagblad*).

_____. 1942. 'Strindbergssymfonien' [The Strindberg Symphony] (magazine article, *Röster i radio* 15).

_____. 12 June 1943. 'Dolce far niente' [Doing Sweet Nothing] (newspaper article, *Svenska Dagbladet*).

_____. 1946. 'Fredagsbesök hos Ture Rangström' [Friday Visit to Ture Rangström's] (magazine article, *Röster i radio* 31).

Stuart, E. M. November 1941. 'Kortfattade Biografier över Samtida Svenska Tonsättare. II Ture Rangström' [Short Biographies of Contemporary Swedish Composers] (magazine article, *Musikern*).

Sundström, E. 12 May 1947. 'Ture Rangström — en Ursvensk som Människa och Musiker' [Ture Rangström — an Ur-Swede as Man and Musician] (newspaper article, *Stockholms-Tidningen*).

Sveriges Hembygdsförbund. 2018. Sveriges Hembygdsförbund (website). Available online at https://www.hembygd.se/shf/page/3420

Westberg, E. 8 January 1920. 'The War's Effect on Modern Swedish Music' (magazine article, *Musical Courier*).

Chapter 9

War, Folklore, and Circumstance: Dimitri Shostakovich's *Greek Songs* in Transnational Historical Context

Artemis Ignatidou
Institute of Modern Languages Research,
University of London, Senate House, Malet Street,
WC1E 7HU, London, UK
artemis.ignatidou@sas.ac.uk

9.1 Introduction

In August 1953, the Soviet folklorist Lev Kulakovsky took to the press to express his admiration for the heroic spirit of the (modern) Greeks.[1] Rather more precisely, he took to the Soviet musicological journal *Sovetskaya Muzyka* to describe in fiery words, and with the help of musical examples, the courage Greek fighters had demonstrated all the way from the Greek Revolution of 1821,

[1]The author would like to express her gratitude to Dr. Konstantinos Pantelidis for translating significant primary sources from Russian, and for sharing his insight on Shostakovich and his work, over tea and cookies.

Musical Spaces: Place, Performance, and Power
Edited by James Williams and Samuel Horlor

ISBN 978-981-4877-85-5 (Hardcover), 978-1-003-18041-8 (eBook)
www.jennystanford.com

which won independence from the Ottoman Empire, to their recent resistance against Nazi occupation and the civil war that followed. The basic premise of his article was that this continuous struggle against various forms of oppression was clearly depicted in the traditional, resistance, and communist songs of the Greek people (Kulakovsky, 1953: 92–95).

Fragmentary evidence and the broader historical circumstance suggest that Kulakovsky was the force behind Dmitri Shostakovich's little-known *Greek Songs* (*sans*. Op.), a set of four songs transcribed by Shostakovich between 1953 and 1954, first published as a set in 1982, and recorded as recently as 2001 (Hulme, 2010: 360–361; Shostakovich, 2010: 123). Although until recently most scholars and editors placed the composition of the songs between 1952 and 1953, this is disputed in the latest edition of the score (published in 2010), which regards the date of the manuscript as an approximation added later (Shostakovich, 2010: 123). Moreover, even though Hulme (2010) dates the premiere of the work to 1991 in France, there exists a recording of at least one earlier live performance by Heiner Hopfner (tenor) and Hartmut Höll (piano) at the Berlin Festival of 1986 (BBC Radio 3, 1986). Whatever the details, the transcription of this set of songs was — as will become evident in the course of this story — the outcome of three factors: Kulakovsky's passion for the theory of folk music, a purge, and a civil war; a set of processes that render these songs a testimony to the social life of music rather than an influential set among so many others in Shostakovich's oeuvre.

9.2 Folklore to Smooth the Edges

From 1930 onwards, Lev Kulakovsky (1897–1989) was employed at the State Academy of Art Studies (GAIS) in Moscow, where he focused on Russian folk song and folk polyphony (Zacharov, 2016: 47). During the first period of his work, before he joined the academy, he was an adherent of Boleslav Yavorsky's theory of modal rhythm and he wrote on form and rhythmic structure in folk song (*ibid*.: 42). After he was appointed to the academy, however, Kulakovsky joined the ranks of the 'anti-formalist' musicological critics, a development that according to Zacharov resulted in a shift in his approach from previously 'scientific' work to almost propagandistic analyses of 'expressiveness' in music (*ibid.*: 47). With the political climate turning just as much against musicologists and music historians as it was turning against composers, albeit less obviously, after 1948 the pressure upon musicologists to produce ideologically correct works intensified; music historians and critics were purged, and in 1950 a new set of objectives was announced for the advancement of Soviet musicology (Schwarz, 1972: 250–258). Consistently with this 'anti-formalist' ethnomusicological turn, a notion that had haunted Soviet music in all its appearances since the early 1930s, Kulakovsky's analysis of Greek songs in *Sovetskaya Muzyka* of August 1953 verged on outright propaganda.

In a sensationalist approach to modern Greek history, Kulakovsky conflated past and present struggles in an attempt to illustrate how the Greek revolutionary 'spirit' could be traced as an uninterrupted stream within and between traditional songs, songs of the resistance against the Nazis, and communist songs from the Greek Civil War of 1946–1949. The three historical events that came together in Kulakovsky's ethnomusicological observations about continuity through song were the 1821 Greek Revolution against the Ottoman Empire, which resulted in the creation of the modern Greek nation-state in 1830; the Greek resistance against Italian fascist and German Nazi occupation during World War II; and the civil war between the communist factions of the Greek resistance and the Greek government following the end of the war. The reason behind this conflation — and a key to the multiple stories that contributed to the transcription of the Greek songs by Shostakovich — is that the text accompanying the musical examples was an ideologically charged attempt to legitimize the idea of a broader historical continuity through a very specific genre of Greek song: songs of the communist wing of the resistance.

The ideological subtext here is obvious: as the songs of the Greek communists make clear, the indefatigable 'spirit' of the 1821 revolutionaries continued to serve to unify Greek fighters of the past and the present through the liberal messages of traditional and revolutionary songs. Accordingly, Kulakovsky swiftly connected his example of the Greek dance 'Zalongo' — a dance associated with courage and resistance — with the Greek communists singing this song of defiance in the concentration camps in which they were imprisoned following their defeat in the civil war (Kulakovsky,

1953: 92–93). Similarly, he had already linked the revolutionary songs of the Greek Enlightenment thinker Rhigas Ferreos (1757–1798) with the Greek communists by claiming that the communists were one group in a long line of Greek dissident fighters, a claim that puts the armed bandits of 1821 (*klephts*) in the same group as the communist guerrilla fighters of the civil war (*ibid.*). After establishing these basic connections between past glories and contemporary woes, Kulakovsky then presented another six songs of the Greek communist resistance — later the Democratic Army of Greece — before concluding that the songs of the fallen communists would forever carry the spirit of liberated humanity forward (*ibid.*: 95).

Where ideological narrative thrived, though, musicological integrity meant that these assertions had to come with a discreet disclaimer. Always keeping in mind that he was addressing a specialized musical audience through the Composers Union's musical journal, Kulakovsky made sure to insert a small clarification on page two of his analysis. Connected as these songs were in virtue of their revolutionary and patriotic national spirit, they were nevertheless 'diversified through their unique form, language, and emotional charge' (Kulakovsky, 1953: 93). In musicological terms, it appears as if the folklorist was half admitting that although traditional Greek song and the songs of the communist fighters were part of the same national heritage, they were otherwise quite diverse in form and content. Be that as it may, the ideological connection was enough to go on and, as will be seen shortly, it was probably the main reason these transcriptions were created in the first place.

Less than a year later, in May 1954, Dmitri Shostakovich's transcription of one of the communist songs presented by Kulakovsky was included in supplementary edition no. 5 of *Sovetskaya Muzyka*. The transcription, titled 'Forward!' ('Embros!'/ 'Εμπρός!' in Greek) and subtitled 'Song of the Greek Resistance', was accompanied by a footnote explaining that the original lyrics were by the revered Greek poet Kostis Palamas, an artist endorsed by the Secretary General of the Greek Communist Party (KKE) as 'the greatest poet of modern Greece' (Shostakovich, 1954: 17–19). The score also included some basic information on the Greek Enlightenment thinker Rhigas Ferreos, who is mentioned in the lyrics of the song, and it acknowledged Shostakovich as the transcriber and Sergei Bolotin and Tatyana Sikorskaya as the translators of the lyrics into Russian (*ibid.*). 'Forward!' was just one of the four Greek songs Shostakovich transcribed and, even though information on them is scarce, the broader social environment of Shostakovich's musical life during the years of their composition sheds some light on the mystery of their creation.

9.3 Circumstances That Shaped a Life's Work

As has by now been well established, between 1948 and 1954 Shostakovich went through a second rough patch in his relationship with the Soviet cultural authorities, and this had considerable consequences for his financial situation and his psychological condition. Transcriptions such as his *Greek Songs* are evidence of the coping mechanisms the composer had developed to navigate through the periods he was out of favour with the authorities. To better understand the function of such pieces, it is imperative to move away, at least briefly, from the 'totalitarian' model of political analysis of the Soviet Union and to follow in the footsteps of Sheila Fitzpatrick (1986) by investigating the social environment of the composer and by extending the social life of his music.[2]

While the Soviet Union was a society of increasing equality of opportunity — mainly through universal access to education — equality of income remained an as yet unachieved goal (Chapman, 1963: 179–180). The state was the only provider of employment, a fact that protected workers of all strata from unemployment and secured for them an array of quantifiable benefits, such as insurance, pension, and maternity cover, but that at the same time left them vulnerable to the ideological and political whims of the sole employer in the land: the Communist Party.[3] One of the effects of this inequality of income was that in a society that aspired to classlessness, it created a covert class system. In the case of the arts, Mervyn Matthews (1978:

---

[2]The shift to a more organic social history of the 'totalitarian' structure of the Soviet Union has been outlined concisely by Fitzpatrick (1986).

[3]Chapman (1963: 138, 178) has calculated the relationship of added income in the form of benefits to real gross wages to be 30.8 percent in 1948, 22.8 percent in 1952, and 22 percent in 1954 (the period of interest here).

97) has found the rates of pay for writers in 1944 to be between 3.3 and 6.6 times the average wage, a figure he suggests would have been similar, if not identical, for all of the arts. Moreover, artists were eligible for several sources of extra income and could compete for the prestigious and lucrative, if often artistically sterile, State Stalin Prizes (Frolova-Walker, 2004; Matthews, 1978: 92–98). En route to complete equality then, during the years we are interested in here, the Soviet state operated a system of differentiated treatment, with stratified incentives and varied lifestyles, a fact that is clear from the group relevant to this study: the artistic intelligentsia (Chapman, 1963: 179–180). This systemic privileging of the intellectual class was partly rooted in the mutually dependent relationship between the party and the intelligentsia: the two were entangled in a dance for power. After the Cultural Revolution (in the early 1930s) the intellectual elite would, through their institutions and unions, secure access to networks similar to those that served the communist administrators (Fitzpatrick, 1992: 2–6, 14). On the other hand, after the war, the party appears to have tightened its grip on the arts with Andrei Zhdanov manipulating the intelligentsia so that they engaged in venomous infightings that forced them into rapid denunciations and admissions of guilt in fear of grave harm (Schwarz, 1972: 205–206). Within this context of a highly complex relationship between the state and its artistic daughters and sons, a class structure based on income and access to networks of privilege, and the manipulative methods of the party, Shostakovich's various predicaments and the resulting riddle of his (in)famous compositional 'double face' will be regarded here as a manifestation of the composer's own relationship with the state and power.

After Zhdanov's 1948 decree against 'formalism' in music, all sources testify that Shostakovich faced an array of practical and psychological challenges, and he had to appease the establishment in order to survive. In financial terms, the denunciation resulted in a significant loss of income, following his humiliating dismissal from the conservatories of Moscow and Leningrad and a ban on the performance and publication of eight of his works (Fay and Fanning, 2001). The road to rehabilitation began immediately, and in the following years Shostakovich was paraded to represent the Soviet Union in international events, and he composed anodyne patriotic works (*ibid.*). Yet, as grave as all these facets of a life read, for Shostakovich — and for the guild of composers as a whole — the state made some room for negotiation and redemption, albeit only after a period of applying intense pressure and uncertainty. As Mstislav Rostropovich observed, the decree had 'the function of a biological experiment', and it had a variety of outcomes for the different people it targeted. In some cases, its effects were physical, as is clear from the case of Vissarion Shebalin, who suffered a stroke as a consequence of the immense psychological pressure. In other cases, it resulted in a loss of social status, as was the case with Khachaturian, and sometimes in financial losses — including for Shostakovich and Prokofiev (Wilson, 1994: 217–218). On the other hand, as Levon Hakobian has argued, such was the unusual position of music among the arts that during the various purges of the Soviet period, 'no significant composer perished in the GULAG, very few left the country, and almost no one was expelled from the Union of Composers (such expulsions were extremely severe punishment)' (2005: 219–220). In other words, in this struggle between the intelligentsia and the party, it appears as though the latter asserted its power through these almost masochistic experiments, but, aware of the special status of the composers, their class, and their valued ideological function for the regime, the party was also careful not to deplete its cultural capital through their complete (literal or artistic) annihilation. Gradually, in a multi-stage process that emerged after the first five-year plan (1928), the state succeeded, through stylistic submission and institutionalized boredom, in implementing 'socialist realism' in music (Frolova-Walker, 2004: 121).

Thus, although Shostakovich was affected psychologically by Zhdanov's 1948 decree, and although it afforded him a renewed awareness of the limits of his relationship with Soviet cultural politics, its effects on him were mainly financial. A rare glimpse of what this financial blow meant in practical terms for the Shostakovich family and their position in the social ladder is provided by Galina Vishnevskaya, renowned soprano and Rostropovich's wife, who described Shostakovich's personal distress and the hardship he faced in her memoirs. What she inadvertently shared, in providing this description, was a partial explanation of why, during the times of systemic hostility towards famous composers, the price for limited intellectual freedom was worth paying. Describing how Shostakovich had been made destitute after losing his position at the conservatories, and how surrender of part of his artistic freedom had caused his morale to plummet, Vishnevskaya detailed that, in times of extreme poverty, Shostakovich turned to his friends for short-term loans:

> All his life Shostakovich feared he wouldn't be able to provide for his family; it was a large one, and he was the only breadwinner. Both his children — his daughter Galina with a husband and two children, and his son Maxim, still a student, with his wife and a son — were in fact dependent on him. Besides them, there was the old nanny, who had been with him all his life, the maid in the Moscow apartment, and another maid and furnace-man at the dacha, plus his chauffeur and secretary. They all counted on him for wages. If we add Dmitri Dmitriyevich and his wife, that makes a total of fifteen persons to feed. He used to say, 'Just think. Tomorrow morning for breakfast we'll need three dozen eggs, two pounds of butter, six pounds of cottage cheese, and several quarts of milk! That's my family. What will happen to them if I stop composing?'
>
> (Vishnevskaya, 1984: 231)

And so compose he did. In light of the fact that Shostakovich was a patriotic Soviet citizen, even though tested and harshly disciplined by the regime, there are two immediate conclusions to be drawn here. Firstly, poverty is a relative measure, in this case conditioned by the social class the Shostakovich family belonged to in the Soviet structure: he was supporting nine family members and a staff of six between two households on a single salary. Secondly, even though he was unperformed and unpublished — and was thus deprived of his more profitable sources of income — he was nevertheless surviving this cat-and-mouse game that the regime was playing with him. For this period of 'discipline', Shostakovich had to produce compositions of a lower quality than his recognized masterpieces, a concession that was part of the bargain he had effectively struck with the regime for the maintenance of his long-term position and, ultimately, his life — something a significant number of other artists and ordinary citizens were not able to do. It is interesting to observe how the regime, via his colleagues — who denounced him publicly in the Composers Union and the conservatories — micromanaged his artistic output through a system of imposed poverty and selective rewards.

During this period of disfavour with the regime, Shostakovich pursued personal projects at home or performed privately — notably his First Violin Concerto (1947–1948), the song-cycle *From Jewish Folk Poetry* (1948), and his Fourth Quartet (1949) — but he was also allowed to work for the cinema (Norris, 1983: 1682). Between 1947 and 1953, when the film industry was itself looking to employ the best professionals possible to help it escape the party's unfair cultural persecution, Shostakovich composed seven works for the cinema, Vissarion Shebalin five, and Aram Khachaturian four, contributing significantly through their misfortune to the production of quality incidental music and to the revival of piano accompaniment (Egorova, 1997: 121–122). At the same time, on his path to rehabilitation, he worked extensively with texts by the conformist poet Yevgeny Dolmatovsky and won the prestigious and lucrative Stalin Prizes — worth up to 100,000 rubles apiece — for his *Song of the Forests* and *The Fall of Berlin* (First Class, 1950) and his *Ten Poems* (Second Class, 1952) (Fay and Fanning, 2001; Frolova-Walker, 2016: 233). Last but not least, he received royalties from the circulation of his work overseas, though significantly reduced after deductions by the state (Vishnevskaya, 1984: 231–232). In terms of his financial situation, then, Shostakovich indeed lost the main source of his income — performance and publication of his works and teaching — yet the state allowed him other sources (albeit reduced) and other rewards in kind: a state dacha in Bolshevo for the 1949 Peace Conference in New York, for example (Fay and Fanning, 2001). As Marina Frolova-Walker has remarked, in this potentially lethal and often coercive transactional relationship between important musicians and the state, composers had known their position in the system since the 1930s and cannot be seen uniformly as victims of intellectual oppression (2004: 103).

It is in this setting of coercion and submission that Shostakovich's need for rehabilitation and redemption, expressed partly through a promise to compose folklore-infused melodies, met with Kulakovsky's aforementioned 'anti-formalist' musicology and work on folklore (Fay, 2000: 167). And, indeed, sitting among Shostakovich's other possessions in the Shostakovich Archive in Moscow is the manuscript of Kulakovsky's handwritten collection of Greek songs, a collection compiled in the early 1950s, containing forty samples of Greek music, which the composer must have consulted when transcribing his own version (Kulakovsky, n.d.). Shostakovich's choice to transcribe the *Greek Songs* struck a perfect balance between the multiple fronts he had to negotiate during his predicament. Politically correct but not immediately relevant to Russian cultural politics at the time, in tune with the folklorist spirit of the times and in line with his promise to accentuate the melodic voice of the people, but at the same time so innocuous as to be inconsequential, Shostakovich's pairing with Kulakovsky's interest in folk songs resulted in a set of transcriptions that were harmless enough to bear his name, and insignificant enough to be forgotten and to pass by without prompting discussion. This is a fine example of a sort of music that was political in its original cultural setting being transformed into something apolitical through the highly political conditions in which it was recreated.

Three of Shostakovich's musical offerings to the state for his rehabilitation can be found in Kulakovsky's article in *Sovetskaya Muzyka*, and another one was not included for reasons that will be examined below:

(1) 'Forward!' (Εμπρός), a communist song of the Greek resistance
(2) 'Penthozalis' (Πεντοζάλης), a traditional dance from the island of Crete — not included in Kulakovsky's article
(3) 'Zolongo' (Ζάλογγο), a pseudo-folk dance with mixed origins
(4) 'Hymn of ELAS' ('Υμνος του ΕΛΑΣ'), an anthem of the communist wing of the Greek Resistance

In their original settings, these four songs tell a diverse set of Greek stories. Grouping the two political ones together, and the two folk songs similarly, we shall now examine their content and function in their original setting. The two folk songs — 'Penthozalis' and 'Zolongo' — represent two traditional dances, the first from the island of Crete, in the southern part of the Aegean Sea, and the other from the region of Epirus, in north-west Greece.

'Zolongo' refers to Mount Zalongo (Ζάλογγο) in the region of Epirus. It is said that, in 1803, between twenty and 100 women of Greek and Hellenized Albanian descent from the village of Souli danced to their deaths, falling off a local precipice with their children to avoid capture by the Ottoman army (Vranousis and Sphyroeras, 1997: 248–250). The political and artistic afterlife of this event, up to Shostakovich's transcription, is a fascinating story of a historical occurrence being transformed into a national myth through its association with song and dance. The story of the collective suicide and the dance was first recorded in 1815; it attracted pan-European interest and was included in various Western histories of the Greek Revolution thereafter (Politis, 2005: 37–39). In the various histories and narrations of the historical event throughout the nineteenth century, the dance was mentioned occasionally — depending on how much the historian in question trusted the original testimony — and it was never universally accepted as solid fact. At the height of philhellenism, the Souliot women and their courageous self-sacrifice — *sans* the dance — became the theme of Ary Scheffer's painting *Les Femmes Souliotes*, exhibited at the Salon of 1827 alongside other similar Romantic works, and the story became a symbol of Greek suffering under Ottoman oppression (Athanassoglou-Kallmyer, 1989: 102–107).

The first inclusion of a piece of music titled the 'Dance of Zalongo' in a collection of traditional songs came only in 1908, and thereafter this narration of a musical suicide was associated with the pseudo-traditional 'Farewell Bitter World' ('Εχε γεια καημένε κόσμε') (Loutzaki, 2006: 18; Politis, 2005: 43). A few years later, in 1913, there was the first reference to the creation of the song/dance of 'Zalongo', in the *Syrtos* style of traditional dancing, set to the aforementioned song in septuple metre (7/8) (Loutzaki, 2006: 19). From then on, the song/dance and the historical event came to be viewed as a single unit and to be featured in forms of popular entertainment — such as fiction and shadow puppet theatre — and it was widely re-enacted in school drama productions and, especially after 1950, in film (Loutzaki, 2011: 207–211). Consequently, a multilayered construct based on the events of 1803 became an integral part of the Greek national tradition and collective imagination. In the early 1950s, it travelled to the Soviet Union, and it was collected and placed among other Greek songs by Kulakovsky as a symbol of the enduring spirit of the Greeks. Then, between 1953 and 1954, it was transcribed by Shostakovich, and it was thus given a new (international) life, although in a significantly sorrowful and introspective style, without the element of dance. The lyrics in Shostakovich's transcription are an adaptation of the original by Tatyana Sirkoskaya, with the same references to the precipice as in the original. They lament the loss of life and celebrate the heroism of the Souliot women (Shostakovich, 2010: 53–55).

The 'Penthozalis' dance relates to a much more straightforward story. The lyrics accompanying this version are a love song, and it is probably for this reason that it was not included among Kulakovsky's revolutionary and patriotic presentation of Greek tradition in *Sovetskaya Muzyka*. As a dance, it originates on the island of Crete. It is a fast line dance, with local variations, in duple metre (2/4), and its name in the local dialect translates as 'Five Steps' (Holden and Vouras, 1965: 67; Hunt, 1996: 83–84). In western Crete, the dance is introduced with a slow section during which the dancers sing while holding hands, before the singing ceases as the tempo quickens and the dancers adopt a shoulder hold (Petrides, 1975: 89). Shostakovich's transcription is a straightforward harmonization of the melody at a slower tempo, and the lyrics are again a loose adaptation of the originals by Sergei Bolotin. Taking advantage of 'Pentozalis's' duple metre and reducing the tempo, Shostakovich removed the dance character of the folk dance/song and transformed it into his distinctive musical language, where it became something akin to a revolutionary march, with lyrics about the unfulfilled love of a man for a woman who will not return his affections (Shostakovich, 2010: 51–52). In the 1950s, at the composer's request, the verses of both 'Zalongo' and 'Penthozalis' were translated equirhythmically, that is, prioritizing the inherent metre and the melodic aspects of the language in relation to the music, rather than providing a literal translation of the meaning (Apter and Herman, 2016: 1; Shostakovich, 1982: vi). It is particularly interesting to note at this point that during the work's 1986 Berlin performance, the performers chose to sing both 'Penthozalis' and 'Zalongo' in Greek, using the original, untranslated lyrics (BBC Radio 3, 1986).

The second set of songs in Shostakovich's musical vignette of Greek music is intimately connected to the political identity of the person who transferred and gave them to Kulakovsky. 'Forward!' ('Εμπρός!') and the 'Hymn of ELAS' ('Ύμνος του ΕΛΑΣ') are both songs of the Greek communist faction of the resistance and later on of the Greek civil war. 'Forward!' is a song by the communist composer Alekos Xenos, and the lyrics come from the 1912 poem 'Forward' by the Greek poet Kostis Palamas, as the supplementary edition of *Sovetskaya Muzyka* in 1953 mentions (Shostakovich, 1954: 17–19). The original poem was created to verse a choral composition by Greek composer Manolis Kalomiris — father of the Greek National School of Music — and it pays homage to the motherland and to those who revolted against the Ottoman Empire in 1821 (Palamas, 1964: 371). In Bolotin and Sikorskaya's translation, the content and meaning remain broadly the same, including most of the same geographical references to important battles and landmarks, with some altered for rhyming purposes (Shostakovich, 2010: 49–50). In terms of tempo and style, Shostakovich kept the original common time signature (4/4), and he preserved the march-like character of the communist song (Shostakovich, 1954: 17–19). Lastly, the 'Hymn of ELAS', a 1940s song of the Greek resistance that uses the lyrics of Sofia Mavroidi-Papadaki and the music of Nikos Tsakonas, exalts the heroism of the members of the National Popular Liberation Army (ELAS), who, fighting for the liberation of Greece, embodied in their struggle the spirit of all those fighters of the past who had fought for the country's independence (Gazis, n.d: 11). Bolotin's translation appears to be a loose interpretation or adaptation of the original lyrics, in three verses instead of the original seven, which alters the exact content but maintains its original function as a war song for ELAS (Shostakovich, 2010: 56–58).

As mentioned briefly above, of the four transcriptions by Shostakovich, three are found in Kulakovsky's 1953 article, and they are all songs of the communist wing of the Greek resistance during World War II, with the exception of 'Zalongo', which is a traditional song that Kulakovsky nevertheless interpreted as a tribute to dissident fighters. The rest of the songs sampled in the same article (seven in total) are all communist or revolutionary songs, and thus their lyrics contributed to Kulakovsky's narrative of an omni-courageous Greek people. The last piece in this transnational puzzle of musical creation via politics and artistic compromise is a geographical displacement, since Kulakovsky handed the pieces to Shostakovich after collecting them from a Greek woman called Maria Beikou, who at the time resided in Moscow.

9.4 ...And War

Maria Beikou joined the Greek resistance and the Greek Communist Party in 1943 at the age of eighteen, and later the same year she took up arms with the ELAS, waging guerrilla war in the mountains against Nazi occupation (Beikou, 2011: sections 'At the EPON' and 'In the XIII Division of ELAS'). The day of liberation, in October 1944, found ELAS under the nominal command of General Ronald Scobie, a British general unaware of the fragile relations between the Greek government and the predominantly communist Greek Resistance Movement (EAM), while Greece overall was placed under British influence after the Stalin-Churchill 'percentages agreement' of 1944 (Close, 1995: 130–131). In a climate of mutual suspicion between the Greek government and the Greek communists, and amidst British military presence, the attempted demobilization of the 60,000 strong ELAS resulted in violence throughout the country and the so-called 'Battle for Athens' of 3 December 1944, when a peaceful demonstration by members of the resistance and civilians was fired upon by British soldiers (Clogg, 2013: 134). After the failure of both sides to reach an agreement for a political resolution to the tensions between the communists and the government, in 1947 Maria Beikou found herself again in the mountains, this time fighting with the communists in the full-blown Greek Civil War of 1946–1949 (Beikou, 2011: section 'With the Democratic Army of Roumeli'). The end of the Civil War in 1949 saw the surrender of the communists and the exile of, at a moderate estimate, 55,881 people — some sources suggest up to 130,000 refugees — who were accepted as political refugees in the countries of the Soviet bloc (Giannakakis, 2005: 10; Tsekou, 2013: 12). Maria Beikou fled to Albania in 1949, and from there she travelled through the Dardanelles to Poti in the eastern Black Sea through the Caspian Sea and Batoumi to Tashkent in Uzbekistan, one of the main cities receiving exiled communist fighters, where she settled and worked in a factory for three years (Beikou, 2011: section 'From Albania to Tashkent'). While in Tashkent she was informed that the Greek radio programme in Moscow was recruiting presenters, and in 1952 she managed to relocate to Moscow where she stayed until 1976 (Beikou, 2011: section 'In Moscow').

How exactly she came to pass the songs on she did not appear to remember. In her memoir, she recalls singing the songs to

Shostakovich himself, but it is improbable that such an event took place. As she herself admits, it was not until 2007 that she recalled this event at all, when a friend saw her name on the first recording of the *Greek Songs*, and her recollection does not fit with the rest of the evidence surrounding their creation. According to Beikou,

> One day they asked me to go sing songs of the Greek Resistance, as well as traditional Greek songs for a certain musician called Shostakovich! Of course I had no idea who that musician was. I don't remember if I sang at a house or at a [recording] studio. The only thing I know is that I sang 'Forward!', the hymn of ELAS, 'Penthozalis', and 'Zalongo'.
>
> (Beikou, 2011: section 'In Moscow')

She admits that it was only when she later met with the friend who made her aware of the recording, and she saw the CD (Shostakovich, 2002), that she remembered meeting Shostakovich and singing for him. She nevertheless suggested that the meeting would have occurred in 1953 or 1954, which is at least consistent with the broader chronology of the songs' collection and composition.

She never mentioned Kulakovsky at all, even though all other sources, as well as the folklorist's article for *Sovetskaya Muzyka*, suggest that he was the person who collected the songs and gave them to Shostakovich. Moreover, the fact that her only solid memory of the event was singing the same songs that later appeared in the recording suggests that she reconstructed the memory of meeting the composer after encountering the CD. Perhaps in reality she sang to Kulakovsky all the communist/resistance songs included in his 1953 article, out of which he then constructed the narrative in support of the Greek communists. This understanding of the story would also provide the missing link between the songs and the narrative by identifying the source of the songs, an exiled communist fighter who sang songs of the Greek resistance, alongside folk songs, and in this way embodied the very argument that Kulakovsky was attempting to make: that the Greek communist fighters were part of a long line of Greek dissident fighters who, since the creation of the state, had struggled for the motherland's 'liberation'.

Unfortunately, Maria Beikou passed away in 2011, and so a follow-up interview to establish whether she would have been able to recall Kulakovsky or the incident is now impossible. Nevertheless, there is no suggestion in the scholarship about Shostakovich that he received the pieces from Beikou rather than from Kulakovsky. The precise way they were transmitted will, for now, remain a mystery, and it is perhaps irrelevant whether Beikou in fact saw Shostakovich's face. At the same time that Beikou was relocating to Moscow, Shostakovich turned to the transcription of the songs, partly out of personal interest and partly to appease the establishment. His *Spanish Songs* (Op. 100) are seen, alongside the *Greek Songs*, as part of his second period of song production (1948–1966), a period during which the composer used easily digestible and widely acceptable material (Maes, 2008: 234, 244). Kulakovsky, taking part in the intellectual fights in his own guild, supplied Shostakovich with ideologically correct material to transcribe, and out of these three parallel processes a Greek song cycle was composed.

Ironically, while in the sphere of the arts, such folklore-inspired gestures were accepted as ways in which musicians could repent, while Kulakovsky was finding in song a continuous history of Greek courage, and while Beikou was becoming the new voice of the Greek radio programme in Moscow and (perhaps) Shostakovich's inadvertent song supplier, the Soviet state had been actively discriminating against and displacing ethnic Greek populations from Crimea, Georgia, Armenia, Azerbaijan, Kazakhstan, and other places since 1937. After 1949, 41,618 ethnic Greeks were relocated from the Black Sea to central Asia with devastating consequences for their livelihoods (Fotiadis, 2003: 128–130, 140). Nevertheless, as far as the cultural apparatus was concerned, part of the road to Shostakovich's musical redemption was paved with reimagined Greek melodies.

References

Apter, R., and M. Herman. 2016. *Translating for Singing: The Theory, Art and Craft of Translating Lyrics*. London: Bloomsbury.

Athanassoglou-Kallmyer, N. 1989. *French Images from the Greek War of Independence 1821–1830*. New Haven and London: Yale University Press.

Beikou, M. 2011. *Αφού με ρωτάτε, να θυμηθώ...* [Since You Are Asking, I Shall Remember]. Athens: Kastaniotis Publishers.

Chapman, J. G. 1963. *Real Wages in Soviet Russia Since 1928*. Cambridge, MA: Harvard University Press.

Clogg, R. 2013. *A Concise History of Modern Greece*. Third Edition. Cambridge: Cambridge University Press.

Close, D. H. 1995. *The Origins of the Greek Civil War*. London and New York: Longman.

Egorova, T. 1997. *Soviet Film Music: An Historical Survey*, translated by T. Ganf, and N. Egunova. Amsterdam: Harwood Academic Publishers.

Fay, L. 2000. *Shostakovich: A Life*. Oxford: Oxford University Press.

Fay, L., and D. Fanning. 2001. 'Shostakovich, Dmitry (Dmitriyevich)'. *Grove Music Online*. Available online at https://doi.org/10.1093/gmo/9781561592630.article.52560

Fitzpatrick, S. 1986. 'New Perspectives on Stalinism', *The Russian Review* 45(4): 357–373.

_____. 1992. *The Cultural Front: Power and Culture in Revolutionary Russia*. Ithaca: Cornell University Press.

Fotiadis, K. E. 2003. *Ο Ελληνισμός της Ρωσίας και της Σοβιετικής Ενωσης* [The Greek Populations of Russia and the Soviet Union]. Second Edition. Thessaloniki: Antonis Stamoulis Publishers.

Frolova-Walker, M. 2004. 'Stalin and the Art of Boredom'. *Twentieth-Century Music* 1(1): 101–124.

_____. 2016. *Stalin's Music Prize: Soviet Culture and Politics*. New Haven and London: Yale University Press.

Gazis, K. (n.d.) *Αντάρτικα Τραγούδια (Συλλογή)* [Partisan Songs (Collection)]. Athens: DAM Publishers.

Giannakakis, I. 2005. «Τό όπλο παρά πόδα»: η εγκατάσταση των προσφύγων στις σοσιαλιστικές χώρες [With the Guns at the Ready: The Settlement of Refugees in Socialist Countries]. In *«Το όπλο παρά πόδα»: Οι πολιτικοί πρόσφυγες του ελληνικού εμφυλίου πολέμου στην Ανατολική Ευρώπη* ['With the Guns at the Ready': Political Refugees of the Greek Civil War in Eastern Europe], edited by E. Voutyra, V. Dalkavoukis, N. Marantzidis, and M. Bontila, 3–17. Thessaloniki: University of Macedonia Press.

Hakobian, L. 2005. 'A Perspective on Soviet Musical Culture during the Lifetime of Shostakovich (1998)'. In *A Shostakovich Casebook*, edited by M. Hamrick Brown, 216–229. Bloomington: Indiana University Press.

Holden, R, and M. Vouras. 1965. *Greek Folk Dances*. Newark, NJ: Folkraft Press.

Hulme, D. 2010. *Dmitri Shostakovich Catalogue: The First Hundred Years and Beyond*. Fourth Edition. Plymouth, UK: The Scarecrow Press.

Hunt, Y. 1996. *Traditional Dance in Greek Culture*. Athens: Centre for Asia Minor Studies.

Kulakovsky, L. 19 August 1953. Греческие Песни Борьбы [Songs of the Greek Struggle], *Sovetskaya Muzyka* 8: 92–95.

Loutzaki, I. 2011. 'The Dance of Zalongos: An Invented Tradition on Canvas'. In *Imaging Dance: Visual Representations of Dancers and Dancing*, edited by B. Sparti, J. Van Zile, E. Ivancich Dunin, N. Heller, and A. L. Kaeppler. Hildesheim, Zürich: Georg Olms Verlag.

Maes, F. 2008. 'Between Reality and Transcendence: Shostakovich's Songs'. In *The Cambridge Companion to Shostakovich*, edited by P. Fairclough, and D. Fanning, 231–258. Cambridge: Cambridge University Press.

Matthews, M. 1978. *Privilege in the Soviet Union: A Study of Elite Life-Styles under Communism*. London: George Allen & Unwin.

Norris, G. 1983. 'Shostakovich, Dmitry'. In *The New Oxford Companion to Music*. Vol. 2, edited by D. Arnold, 1681–1684. Oxford: Oxford University Press.

Palamas, K. 1964. Απαντα [Collected Works]. Vol. 5. Athens: Biris-Govostis Publications.

Petrides, T. 1975. *Greek Dances*. Athens: Lycabettus Press.

Politis, A. 2005. «"Ο Χορός του Ζαλόγγου". Πληροφοριακοί πομποί, πομποί, πομποί αναμετάδοσης, δέκτες πρόσληψης» ['The Zalongo Dance': Transmitters of Information, Transmitters of Reproduction, Receptors of Understanding], *The Citizen* 139: 35–43.

Schwarz, B. 1972. *Music and Musical Life in Soviet Russia 1917–1970*. London: Barrie & Jenkins.

Tsekou, K. 2013. *Ελληνες πολιτικοί πρόσφυγες στην Ανατολική Ευρώπη 1945–1989* [Greek Political Refugees in Eastern Europe 1945–1989]. Athens: Alexandria Publishers.

Vishnevskaya, G. 1984. *Galina: A Russian Story*. London: Hodder & Stoughton.

Vranousis, L., and B. Sphyroeras. 1997. 'Revolutionary Movements and Uprisings'. In *Epirus: 4000 Years of Greek History and Civilization*, edited by M. B. Sakellariou, 244–251. Athens: Ekdotike Athenon.

Wilson, E. 1994. *Shostakovich: A Life Remembered*. London: Faber & Faber.

Zacharov, Y. 2016. Захаров Ю.К. У истоков анализа мелодии в СССР: Лев Владимирович Кулаковский [The Origins of Melody Analysis in the USSR: Lev Kulakovsky], *Man and Culture* 2: 41–51.

Other Media

BBC Radio 3. 9 September 1986. Shostakovich *Greek Songs* (radio broadcast of live recording, Berlin). British Library Sound Collection, B2719/2/2.

Kulakovski, L. (n.d.). Архив Д.Д.Шостаковича [Dmitri Shostakovich Archive], ф. 1, р.2, ед. хр. 52 (10 л.).

Loutzaki, R. September 2006. «Ο Χορός του Ζαλόγγου» [The Dance of Zaloggon: A Historic Image of Farewell to Life], *Archaeology and Arts* 100: 17–25.

Shostakovich, D. 1954. 'Forward!' (score). Supplementary Edition of *Sovetskaya Muzyka* 5: 17–19.

_____. 1982. D. Shostakovich: *Collected Works in Forty-Two Volumes* (score). Vol. 32. Moscow: State Publishers 'Music'.

_____. 2002. *Shostakovich: Complete Songs.* (CD). Vol. 1. Delos, DE 3304.

_____. 2010. *Shostakovich: New Collected Works* (score). Vol. 92. Moscow: DSCH Publishers.

Chapter 10

'O Monstrous! O Strange!': Culture, Nature, and the Places of Music in the Mexican Sotavento

Diego Astorga de Ita
Department of Geography, Durham University,
Lower Mountjoy, South Road, Durham, DH1 3LE
diego.astorga-de-ita@durham.ac.uk

'O monstrous! O strange! We are haunted.
Pray, masters; fly masters. Help!'
— A Midsummer Night's Dream, Act III, scene 1

'Afligido un poco canto
E invoco a Santo Tomás
Para curarme de espanto...'
— verse from *El Buscapié*
(traditional Leeward air)

Musical Spaces: Place, Performance, and Power
Edited by James Williams and Samuel Horlor

ISBN 978-981-4877-85-5 (Hardcover), 978-1-003-18041-8 (eBook)
www.jennystanford.com

10.1 Sotaventine Sounds

Son Jarocho is the music of the Mexican Sotavento (Leeward) Region, a region south of the Atlantic port of Veracruz consisting of the southern part of the state of the same name and extending up to the neighbouring states of Oaxaca and Tabasco (Barahona-Londoño, 2013; García de León, 2009; Kohl, 2010).[1] The region gets its name because of its position in relation to Veracruz and the dominant northern winds; it is to the leeward (Domínguez Pérez, 2015; García de León, 2011; Llanos Arias and Cervantes Pérez, 1995). It encompasses the basins of the Papaloapan, Coatzacoalcos, and Tonalá rivers, which are a central part of the Sotaventine landscapes and were historically important means of transportation in the region (Thiébaut, 2013; Velasco Toro, 2003).

El Sotavento was heavily influenced by the port city of Veracruz, which is one of the northernmost points of the region, and was the only transatlantic port of colonial Mexico during the three hundred years of Spanish rule. Through Veracruz, merchandise from Asia, America,[2] and Europe moved in and out of New Spain. Consequently, Veracruz became a place of encounter between sailors, soldiers, merchants, workers, and slaves from across the 'first world system' (Chaunu, 1960; Wallerstein, 1999). Even Antonio's 'argosy... at Mexico' from Shakespeare's *The Merchant of Venice* (2006: 1.3 ll. 15–18) would have been anchored in this port (de Ita Rubio, 2012: 164–165). It is not surprising, then, that it was in Veracruz and its hinterland that *son Jarocho* originated from the mixture of numerous cultures, predominantly Spanish popular music, its Baroque chordophones, and Andalusian dances; African and Afro-Caribbean rhythms, dance, and song patterns; and indigenous languages, spaces, and themes (García de León, 2009). The resulting Sotaventine sounds are played in fandangos;[3] that is, parties that revolve around the region's music and dance.

---

[1]I would like to thank the Mexican National Council for Science and Technology (Consejo Nacional de Ciencia y Tecnología, CONACYT) for the grant that has made this research possible, as well as the Sotaventine musicians who have taught me and talked to me throughout my research, especially Arcadio Báxin and his family.

[2]In this text, I use the word 'America', as it is understood in the Latin languages, to refer to the whole of 'the Americas', not to the USA. Likewise, I use 'American' in a broad sense.

[3]Also known as *huapangos*.

10.2 Fandango: Festive Charm and Fescennine Dance

Fandangos have been contested spaces ever since their origin, seen by some as pleasing and by others as profane, as the eighteenth-century Spanish *Diccionario de Autoridades* shows. This dictionary defines fandango as a very joyful and festive dance with origins in the new world; at the same time, it gives it a Latin name: *Tripudium fescenninum*; that is, 'obscene dance'. The second definition given by this dictionary is that of a banquet or celebration, this time accompanied by the Latin terms *Festiva oblectatio* (festive charm) and *Jucunditas* (delight) (Real Academia Española, 1732). Because of their obscene connotations, fandangos and *son Jarocho* were attacked by the Inquisition and the state during the colonial period (the sixteenth to the nineteenth centuries); they were frowned upon and considered indecent and sinful, often associated with racialized Others (Camacho, 2007; García de León, 2009; Ortiz, 2005). With Mexican independence in the nineteenth century, *son Jarocho* was used to develop a sense of regional identity and to protest against unpopular governments; however, the negative connotations attached to fandangos continued, although they were rooted more strongly in classism and racism than in religious conservatism (Pérez Montfort, 1991).

After the Mexican Revolution, in the first half of the twentieth century, the perception of *son Jarocho* changed as it went through a process of folklorization resulting from an increased media exposure, a heightened migration from rural to urban areas, and the commoditization of the music (Gottfried Hesketh and Pérez Montfort, 2009). This folklorization led to a diminishing repertoire and to the abandonment of fandangos (Stigberg, 1978). And while the fandango lost — to a certain extent — the connotations of licentiousness and sin, it was still considered a dangerous space in which political squabbles and personal vendettas often flourished violently (Báxin, 2017; Kohl, 2010; Pérez Montfort, 2010). All this furthered the decline of fandangos throughout the mid-twentieth century.

In the late 1970s, some young musicians started going back to older styles of *son Jarocho* still played in some rural communities, prompting a revival that became known as the *jaranero* movement (Cardona, 2009).[4] This movement started promoting the practice of the fandango, stressing its pedagogical and communitarian aspects, and reinterpreting it as an indispensable practice for the survival of *son* (Pérez Montfort, 2002).[5] Thus, the space and the practice of fandango or rather the *spatial practice* of fandango has survived up to today, though not without transformations. Nowadays there are modern fandangos, not only in Sotaventine towns, but also in cities across the globe where the dance and the delight survive, and where the histories and stories woven through centuries of music can still be heard (Figure 10.1). This creates a performative palimpsest, a living memory that is still changing. But the palimpsest not only reaches across time — past, present, and future — but also stretches across worlds. The 'festive charm' is not just quaint delight, but also literal enchantment, as musicians' stories, memories, and experiences make evident.

Figure 10.1 Fandango in a bar in Mexico City, 24 November 2017. Photo by Diego Astorga de Ita.

[4]The movement is named after the *jarana*, the main chordophone instrument played across different variants of *son Jarocho*.

[5]'*Son*' is a term used to refer to several traditional American music genres. Throughout this chapter, my use of '*son*' refers to *son Jarocho* in particular.

10.3 Prey of Enchantment

Fandangos often happen in open, public places — streets, gardens, or plazas — sometimes under the cover of a marquee and always around the *tarima,* the wooden idiophone that also functions as dance floor (Figure 10.2). The openness of the space means that people are welcome to join the party if they can sing, dance, or play an instrument, or if they just want to watch, eat, and drink. The fandango starts off with the *son* of 'El Siquisirí',[6] and it goes on for as long as there are people playing and dancing. The intensity of the fandango changes according to the hour and the spirits of dancers and musicians.

Figure 10.2 A *tarima* in Tlacotalpan, 1 February 2018. Photo by Diego Astorga de Ita.

The openness of the fandango acquires a different meaning when considering Sotaventine lore. Fandangos are often visited not only by musicians from neighbouring communities, but also by non-human beings with exceptional characteristics: a smiling man with a glowing red stone in the headstock of his guitar who impairs a musician's ability to play and move; a foreigner dressed in black, wearing silver spurs, who plays an unusually loud *jarana*; a strange singer with a vibrant voice, unbearable breath, and twisted, backwards

[6]The term '*son*' is also used to refer to particular traditional music pieces, akin to 'tune' or 'air' in English.

feet. These musicians are, respectively, the devil, Yobaltabant,[7] and a *chaneque*.[8] Moreno-Nájera (2009) magnificently captures these and other stories of otherworldly encounters in his book *Presas del Encanto* (Prey of Enchantment); here he gathers testimonies of old Sotaventine musicians from the region of Los Tuxtlas, who have interacted with these otherworldly beings. The inhabitants of the Sotaventine Otherworld join in the music and dance, especially when musicians happen to be in particular spaces (near rivers or ceiba trees),[9] or when they stray from the straight and narrow (literally or figuratively). These Others terrorize or enchant musicians; they punish misbehaviour, trick musicians, lure them into the forest, and leave a lasting mark on instruments and bodies (Moreno-Nájera, 2009: 68, 104).

Throughout these stories, the Others respond to misbehaviour that goes against the grain of fandangos and Sotaventine music, repaying evil with evil. Musicians who keep re-tuning so that other players cannot join the music, dancers who do not let other party-goers dance, people who are obsessed with fandangos and *son* to an unhealthy extent, or musicians who disrespect the effigies of Catholic virgins and saints: they all fall prey to enchantment as a consequence of their actions. In contrast, *jaraneros* (*jarana* players) who take part in music and fandangos with moderation, respect the communitarian logic of the party, revere virgins and saints, and 'cleanse' themselves with flowers from their altars are able to avert a bitter end. In some cases enchantment is a punishment, and in others it is a trick played on an unaware musician, not necessarily worthy of a reprimand: whether punished or pranked, musicians are prey to enchantment time and again (Moreno-Nájera, 2009).

Given these accounts, Sotaventine fandangos seem to be places where Weber's rational 'disenchantment of the world' (see Bennett, 2001; Greisman, 1976) has not yet taken its toll, or where it is temporarily arrested as the inhabitants of the Otherworld interact with human society.

---

[7]Yobaltabant is a character from the Sotaventine Otherworld, described variously as the devil by another name (Báxin, 2017), or an old deity from the region of Los Tuxtlas that tricks adults (Moreno-Nájera, 2009: 122).

[8]Another being of the Sotaventine Otherworld, associated with the wilderness, particularly with forests and waterways.

[9]There are two species of ceiba trees in Mexico: *Ceiba pentandra* and *Ceiba aesculifolia*.

10.4 Enchantment and 'the Topsy-Turvydoms of Faery Glamour'

The Otherworld of Sotaventine enchantment and its inhabitants are not unlike the people of Faery, of the British Isles. In Irish lore, the Sidhe are often in contact with musicians and are themselves the makers of the most beautiful music. They play fiddles and pipes, and interactions with them and their music can have both benefic and catastrophic consequences. They too enchant humans through their music and are found at particular places 'associated with the other world' (Uí Ógáin, 1993).

Some study of faery music has been undertaken relatively recently by Ríonach Uí Ógáin (1993), but accounts of faeries and their musical ways are found mainly in older texts. In Chaucer's *Wife of Bath's Tale* (written between 1387 and 1400), it is noted that 'Al was this land fulfild of fayerye' (cited in Holland, 2008: 22). In Shakespeare's *A Midsummer Night's Dream* (published in 1600), part of the cast of characters belongs to the Otherworld, and while this tale is set in ancient Greece, the otherworldly beings are clearly the people of Faery. Likewise, Yeats' *The Celtic Twilight* (first published in 1893) brings together stories of the Sidhe in nineteenth-century Ireland, and it describes how 'men and women, dhouls and faeries, go their way unoffended or defended by any argument' (Yeats, 2012: 3).

The idea of enchantment described by Jane Bennett — 'a state of wonder' and 'a condition of exhilaration' (2001: 5) — is certainly well suited to understanding the country of Faery and its people, for the Sidhe are often wonderful and exhilarating. They know 'the cure to all the evils in the world' and make 'the most beautiful music that ever was heard' (Yeats, 2012: 36, 88), while also inflicting fascinating fear on mortals. Even in the Shakespearean drama, where characters flee from the faeries, in the end the Sidhe are benevolent (Holland, 2008), as in Titania's declaration that 'Hand in hand with fairy grace/We will sing and bless this place' (Shakespeare, 2008: 5.1 ll.390–391).

But while, in the British Isles, 'the topsy-turvydoms of faery glamour' are 'now beautiful, now quaintly grotesque' (Yeats, 2012: 9, 56), the enchantment of El Sotavento seems to be dominated by fear; one falls prey to it and there is little beauty to be found in the stories of enchantment. Although it is similar in resisting (or surviving) the rational disenchantment of modernity — and it certainly is an 'uncanny feeling experienced through the senses' (Bennett, 2001: 5) — here enchantment is more monstrous and strange than in Shakespeare's play. So, while El Sotavento is — at least in memory — a land not yet disenchanted, the enchantment of this world does not correspond to textbook definitions. The term '*encanto*' (enchantment) corresponds rather to the spell-like effect of interactions with the Otherworld's inhabitants, or to the thing with which they interact, the Other itself. Here, I must stress that *el encanto* emanates from nature or from the *chaneques* (who some consider to be synonymous with enchantment) and is the protector of natural environments. Often, as is the case with the stories gathered by Moreno-Nájera, these charming interactions happen through music (or) in the fandango.[10]

Perhaps the most fitting way to understand Leeward enchantment and its whereabouts is Octavio Paz' concept of *la fiesta* (the party), which he explores in *The Labyrinth of Solitude*. Paz writes that 'above all [la fiesta] is the advent of the unusual. It is governed by its own special rules, that set it apart from other days, and it has a logic, an ethic and even an economy that are often in conflict with everyday norms' (1967: 42–43). The fiesta 'occurs in an enchanted world [where] time is transformed' and 'space, the scene of the fiesta, is turned into a... world of its own'; in the fiesta, 'everything takes place as if it were not so, as if it were a dream' (*ibid.*). Paz' idea of fiesta is an oneiric utopia of lived poetry; in his poetics, poetry is a lived act (Wilson, 1979). As in the fiesta, in the fandango, poetics and music are lived and embodied, and the world is enchanted, even if just briefly.

Akin to Paz' fiesta is Mikhail Bakhtin's notion of carnival, which is characterized by 'a ludic... and critical relation to official discourses' (Folch-Serra, 1990: 265). The carnival also inverts everyday norms, takes place in a subversive time, and is an inherently sensuous reality. Furthermore, the Bakhtinian carnival highlights the functions and

---

[10]These interactions are not exclusive to music. Arcadio Báxin, a Sotaventine musician, told me that he was enchanted once, as a child, when cutting wood for building a fence. Likewise, enchantment seeks retribution against hunters when animals are not killed properly (Báxin, 2017). Enchantment then has to do with man's relationship with nature — as if it were the forest's immune system responding to a disturbance. But, as shown in these examples, it is not always mediated through music.

attributes of bodies, particularly grotesque ones. Participants are both actors and spectators, and mésalliances arise (Vice, 1997). This theory is useful in understanding the fandango and the strange happenings that take place in it. In this chapter, I subsume carnival under the idea of la fiesta, as the fandango is quite literally a fiesta, and Bakhtin's and Paz' ideas are in line with each other when it comes to the festive poetics and spaces as well as to the question of Otherness. Furthermore, using the concept of la fiesta in this geographical-musicological chapter is an attempt to add a modicum of diversity to a wider theoretical dialogue — to use Bakhtinian terms, adding an utterance to the polyphony (Holloway and Kneale, 2000). Bringing an American voice to the mix not only enriches what might be an otherwise Eurocentric theoretical construction applied to an American festive phenomenon (the fandango), but it also echoes the intercontinental characteristics of the Sotaventine celebration.[11]

The fiesta time and space, and the impermanence of fandangos, also bring to mind Bakhtin's idea of 'the chronotope': a position or a place in time wherein time and space are conjoined (Bakhtin, 2002; Folch-Serra, 1990). In fandangos, space is transformed through performance: music, dance, and food make the garden, street, or plaza into another place; an elsewhere embedded in the time of music. Thus, a fandango is both place and performance: it transforms the abstract, Cartesian space into a phenomenological, lived place that exists while it is performed and dies out when the last song ends. Fandangos are like Tschumi's event-cities: 'conceived, erected, and burned in vain'; like fireworks that show the 'gratuitous

[11]This point also comes with a caveat. Although it is important to draw upon Latin American authors — even one as recognized as Paz — to supplement a mostly Anglo- and Eurocentric academic context, the polemic aspects of Paz' life and works should not be overlooked. While some have praised his critique of machismo in the *Labyrinth of Solitude* (1967), his cavalier treatment of rape and his disdain of Chicanos in this book, as well as his derisive conduct and writings towards women and the LGBT+ community, have been strongly criticized by many others (Gaspar de Alba, 2014; Vera Tudela, 2018). By using his work on the nature of la fiesta from *The Labyrinth of Solitude*, I do not mean to endorse the problematic parts of this book or of any of his other texts, simply to draw upon his insights on this particular topic. Ethical questions follow regarding the use of works and authors that are polemic and of the (im)possibility of separating authors from their works. While the present chapter is not the appropriate forum for a full discussion, I believe there are some answers to be found in Said's idea of contrapuntal readings (Said, 1994).

consumption of pleasure' (2000: 19). Although not precisely 'gratuitous' or 'vain', the fandango is an ephemeral chronotope built for enjoyment; it only exists as long as the music lasts. Once it stops, there is no longer a fandango; the place has ceased to be.

Music has an important role in the Otherworld of Faery and of Leeward enchantment. Faeries are keen on music and can play and dance nonstop. Their musical madness often comes into contact with humanity, as in the case of a woman who was stolen by faeries: 'after seven years she was brought home again... but she had no toes left. She had danced them off' (Yeats, 2012: 106). In spite of the faeries' musicality, there is a fundamental difference between the two Otherworlds: the fandango. While the Sidhe do not seem to have a particular place or time for their musical endeavours, in El Sotavento both human and non-human musicians come together in the chronotope of the fandango. Hence, it is important to understand that what makes the fandango is the praxis of music, the praxis of la fiesta. The important thing is that the chronotope is performed, regardless of where and when. Perhaps because of this, all theoretical considerations on fandangos are bound to be much ado about nothing; after all, 'theories are poor things at best' (Yeats, 2012: 79) and Leeward enchantment and faery glamour are things whose 'meaning no man has discovered nor any angel revealed' (*ibid.*: 18).

10.5 'Para Curarme de Espanto...'

In the performance of enchantment in fandangos, there are two groups of actors: the enchanters and the enchanted. Of the first, we have already talked: they play, dance, punish, and prank. Faced with their monstrous mischief, there are a number of ways — both material and immaterial — in which would-be prey protects itself. Some luthiers and *jaraneros* put amulets in their instruments. It is not uncommon to see *boconas* (bass guitars)[12] with a mirror embedded in their headstocks (Figure 10.3), or for images of virgins and saints to be kept among the tuning pegs of instruments or inside

[12]*Bocona* literally translates as 'big mouthed'. This type of guitar is more commonly known as *leona* (literally 'lioness'), but according to Joel Cruz Castellanos, this is a misnomer and the correct term is *bombona*, *vozarrona*, or *bocona* (personal communication, 28 March 2018).

their sound boxes. The violin is in itself an amulet as it makes 'the sign of the cross' when played. A red string can also bring musicians protection, and — according to the dancer Rubí Oseguera — people used to put rattlesnakes' rattles in each corner of the *tarima* to protect dancers against perilous enchantments. Similarly, in one of Moreno-Nájera's stories, a plectrum carved like a rattle saves a guitar player from the devil (2009: 23–25).

Figure 10.3 Fandango in Tlacotalpan, 1 February 2018. The Sotaventine guitar (left) and the *bocona* (centre) have round mirrors in their headstocks for protection. Photo by Diego Astorga de Ita.

The singing of verses invoking the divine — Catholic saints, virgins, or God — is another method used by musicians to guard themselves against enchantment (Báxin, 2017; Moreno-Nájera, 2009). Words are the amulets of musicians: when someone realizes that the singer has backwards, twisted, furry feet (a *chaneque*!), or that there is something wrong with the foreign virtuoso dancer (the devil!), people start singing verses to the divine, which prompts the otherworldly visitors to vanish or flee. Often these disappearances come when lightning strikes, in whirlwinds, or in smokey puffs of fire-and-brimstone; or they lead to the transformation of the stranger into some sort of animal that runs promptly into the safety of the forest (Moreno-Nájera, 2009).

Some *sones* are particularly prone to draw attention from certain Others, and so they require the performance of amulet-verses. '*El*

Buscapié' (The Firecracker) is a *son* that attracts the devil to the *tarima* when played, and hence one must call upon God or saints in verse (Hidalgo, 1978). This has led to the formation of a large corpus of verses 'to the divine' that act as protection, for example:

Afligido un poco canto *e invoco a Santo Tomás* *para curarme de espanto* *hermano de Barrabás,* *por el Espíritu Santo* *¡Retírate Satanás!*	I sing my song with affliction to Thomas the saint I pray to cure myself from this fright Barabbas' brother I say, by the power of the Spirit Satan, be gone! Go away!
Con todas las oraciones *de los santos milagrosos* *vencí las tribulaciones,* *los malos, los venenosos,* *la envidia, las tentaciones* *para salir victorioso.*	With the help of all the prayers of the miraculous saints I beat all the tribulations, evil, and venom, and pains, the envies, and the temptations victorious I overcame.

(Ochoa Villegas, 2015: 26)[13]

On the other hand, there are verses that show certain sympathy for the devil and that mock traditional religiousness; these I have heard sung in fandangos several times:

¡Ave María, Dios te salve! *¡Dios te salve, ave María!* *Así gritaban las viejas* *cuando el diablo aparecía.*	Oh hail Mary! May God save you! Hail Mary! God save you dear! That's what old ladies would shout when the devil did appear.
Salió a bailar Lucifer, *no canten a lo divino,* *mejor toquemos pa' ver* *¡que demuestre a lo que vino!*	Lucifer came out to dance, don't call on the divine fleet, let's play and look at him prance let's see if he can move his feet!

---

[13]This collection of verses is produced for didactic rather than academic purposes. Verse anthologies such as this one are resources often used and compiled by *son* enthusiasts and musicians. While I have no record or recollection of having heard these particular verses in fandangos — at least not the first one — they are good examples of verses 'to the divine' and are included in the aforementioned text under the section for '*El Buscapiés*'. The rest of the verses cited in this chapter were collected from fandangos and records. All verse translations in this chapter are my own, and the primary intention is to maintain the rhythm and rhyming scheme of the original Spanish.

In *'El Buscapiés'*, amongst salutes, love letters, and self-references,[14] there is often a poetic stand-off between the divine and the profane. The mixture of texts, addressed to the divine and to the demonic, as well as the bodily characteristics of otherworldly beings — the backwards, twisted feet of *chaneques* (Moreno-Nájera, 2009); the devil's feet, one human and one chicken-like, or his disappearance in a brimstony fart (Hidalgo, 1978) — speak of the fiesta-ness of the fandango. Furthermore, the dialogic poetics of '*El Buscapiés*' are an echo of the dialogic performance of the fandango and a testimony to the presence of Others in the Sotaventine party. While the Shakespearean characters scream and flee when they come across the Faery Otherworld, Sotaventine musicians sing and face the music when the Others come out to play.

10.6 Of Dancers and *Nahuales*

Another way in which enchantment is present, and in which the lines between culture and nature are blurred, is through the performance of *sones* that speak of animals. In '*La Guacamaya*' (The Macaw), '*La Iguana*' (The Iguana), '*El Toro Zacamandú*' (The Zacamadú Bull), '*El Palomo*' (The Pigeon), and '*El Pájaro Carpintero*' (The Woodpecker), performers take the mantle of animals by enacting their behaviours in dance and verse. Women 'fly and fly away' like the macaw; men become iguanas or brave bulls; singers declare their love to doves or declare themselves to be woodpeckers; and dancers peck at the *tarima* with their feet, imitating the woodpecker's sounds. The relationship with nature is embodied in these *sones*: a dialogue is established through song, and animality is enacted through dance. While this might seem an overtly naive or literal analysis of the song's texts (too 'representational'), this is not the case,[15] especially considering that *Jarocho* poetics and performance echo Mesoamerican cosmovisions in which animals and humans talk and interact or transform from one to the other (Estrada Ochoa, 2008;

[14]As in many other *sones*, there are several verses in '*El Buscapiés*' that talk of the *son* itself. There is an even clearer example in '*La Bamba*', in which instructions for dancing to it are sung: '*Para bailar la bamba / se necesita...*' ('To dance *la bamba* / you are going to need...').

[15]For more on non-representational theories, see Anderson and Harrison (2010), Lorimer (2008), and Thrift (2008).

Martínez González, 2010). These transformations are also a form of enchantment: Sotaventine stories speak of fandangos in the thick of the jungle in which men literally become bulls and women macaws when they step into the *tarima*; or of a cheating, shape-shifting enchantress who, when confronted (with extreme violence) by her cuckolded husband, becomes an otherworldly cow and flees (Hidalgo Belli, 2016). Furthermore, many of Moreno-Nájera's stories end with the otherworldly intruder turning into an animal and running away. These stories are similar to those of *nahuales* — sorcerers who can turn into animals (Brinton, 1894: 13–14) — that are also present in Leeward verse:

No te asomes vida mía,	Don't come out my dear, my lover,
no te vayan a espantar,	lest you be frightened to tears,
dicen que por esta esquina	I have heard that near this corner,
se te aparece el nahual.	at night, the *nahual* appears.

(Heard in a fandango in Santiago Tuxtla, Veracruz, 29 December 2017)

Considering the dancers' performance of animality, they can be thought of as *nahuales* of sorts, who cross the boundaries of humanity through music and dance. These frontiers are further transgressed in Leeward poetics through interactions with non-human Others that go from small talk, to messenger services, to marriage proposals. Thus, the borders between culture and nature begin to blur.

'*El Pájaro Carpintero*', the song of the woodpecker, is a good example of the overlapping and de-bordering of culture and nature in *son Jarocho* lyricism. The bird is, in a sense, a totem for some Soteventine peoples, particularly woodworkers. This song refers to woodpeckers and woodworkers almost interchangeably, not entirely surprising as the woodpecker is the original woodworker according to Leeward poetics.[16] The woodpecker of the song helps carpenters and luthiers in their endeavours and has magical powers that allow those who know 'the woodpecker's prayer' to open locked doors at will. The bird is connected to the divine, as verses attest to its relationship with Noah's ark and Christ's cross, and the woodpecker's body is a source of magic that can be used to show

[16]The Spanish for woodpecker literally translates as 'carpenter bird'.

true love or 'to find a cure for the enchantment/of a love that was not true' (Mono Blanco, 2013).[17]

Like that of the woodpecker, many other *sones* — both in performance and in text — explore and establish relationships between humans and non-humans, undermining the modern borders between culture and nature.

10.7 Forest Voices and Fandangos

The performance of nature in the fandango speaks volumes of its significance in El Sotavento. Much as in Descola's critique of modernity's division between culture and nature (2013: 61–63), in the Sotaventine fiesta, nature is not ontologically separate from humans, as it takes part in what might be considered a cultural practice, music. Nature's enchantment plays the same instruments and the same tunes as humans, even if they are not part of the human world. Hence, fandangos are chronotopes in-between human society and the natural world, in spite of them happening in a place some theorists would consider distinctly human (Toledo, 2008). The fact that nature plays music throws doubt upon thinking about music as a purely human phenomenon.

The significance of the natural dimensions of music is emphasized further when looking at Sotaventine instruments — the *jaranas*, *tarimas*, and guitars — and their botanic origins. Usually red cedar (*Cedrela odorata*, known in English as 'Spanish' cedar) is the species of choice for Leeward chordophones. Traditionally, a single piece of this wood was painstakingly carved into shape with axes, gouges, and machetes (Bearns Esteva, 2011). The resulting instruments are clearly the product of human labour, but they are not merely cultural artefacts; their connection with nature has not been severed by human work or culture, as some Marxist theories would suggest (González de Molina and Toledo, 2014; Toledo, 2008). Instead, instruments are seen as living things; they are described as having 'a voice' and 'a cedar heart' in Moreno-Nájera's tales, in Sotaventine poetics, and by current luthiers (Báxin, 2017; Moreno-Nájera, 2009: 61; Segovia, 1981: 612). Furthermore, musicians still

[17]'Para curar el encanto/de algún amor traicionero...'.

sing of them as '... a cedar [that] cries' and claim that 'the forest finds rest' in the melodies they make (Gutiérrez Hernández and Son de Madera, 2014). Likewise, some Leeward musicians consider trees, like humans, to have a soul. They have agency; old trees can help heal diseases if asked, and before cutting down a tree, one must seek permission from God (Báxin, 2017).

All this is relevant when considering instrument building. The work of the luthier is confounded with the work of nature, of the trees, and of the woodpecker. These sentiments are seen (or rather heard and imagined) in Patricio Hidalgo Belli's verse:

... y los carpinteros	... and the carpenters
sueñan tercerolas	dream of long guitars
por el monte viejo	through forests of old
de cedro y caoba	mahogany and cedar
para que la vida	so life may be filled
se cuaje de aromas.	with pleasant aromas.

(Hidalgo Belli and Son de Madera, 2009)

In the original Spanish, it is ambiguous whether 'the carpenters' are woodpeckers or luthiers. This is not at all strange considering the totemic nature of the woodpecker in the Leeward tradition.

Ultimately, Sotaventine instruments are hybrid things, or perhaps hybrid beings, akin to Latour's (1993) idea of hybrids or to Haraway's (2006) cyborgs. Instruments, however, are not the children of modernity or of science — as is the case in the Latourian narrative and with Haraway's cyborgs — but the survivors of an enchanted world.

10.7.1 Leeward Acousmatics

It is not only in cedar instruments that nature's voices are heard. In a Leeward tale, a man walks home alone at night after going to a fandango; he starts hearing music coming from the heart of the jungle, decides to leave the road, and gets lost in a phantasmagoria of musical oddities in which people turn into animals when they dance (Hidalgo Belli, 2016). Not all the stories of Sotaventine musical enchantment happen in the space of the fandango; many occur on the road when musicians are making their way to or back from a fandango. Moreno-Nájera tells the story of a man who has to cross a river on his way home from a fandango; near the river, he runs into a child who asks him to play the *son* of '*Los Enanos*' (The Little People), and when he does, a thick mist rises from the river and he loses sight of the child. The man tries to find him but cannot see because of the fog; when the haze rises, the child is gone and the man realizes it was a *chaneque* in disguise (Moreno-Nájera, 2009: 35).

In further stories, Others present themselves immaterially and in disembodied form, only through sound. Some accounts speak of men obsessed with fandangos who start hearing music coming from neighbouring ranches; they make their way there and when about to arrive, the music stops and starts again from a different location. They never can reach the origin of the sounds: their origin is enchantment. These men begin to lose themselves in the forest, trying to reach the acousmatic music (Hidalgo Belli, 2016; Moreno-Nájera, 2009: 83–84, 89–90).

George Revill writes that 'acousmatism is associated with the difficulties intrinsic to locating the specific source or point of production for sound' (2016: 249), a phenomenon that brings into question the sound's origins and authority. In the Leeward world, acousmatics are less a function of politics or authority and more a function of enchantment. Sotaventine music is ubiquitous and permeates the charmed listener like the sounds of nature that Revill describes (*ibid.*). This brings again into question the idea of music as a product of culture, for the sounds of *son* are like the sounds of nature: their origins cannot be located; they change or stop when musicians get near or when they cross a river, and they make musicians lose track of time or draw them into the forests.

Leeward acousmatism is a tool of enchantment that raises questions about the spatial nature of fandangos and about their uniqueness as the quintessential chronotope of music in El Sotavento. In these acousmatic tales, paths and rivers are central. It is not in the place of human music that the story develops, but in the spaces of the Otherworld. And yet the sounds are those of a fandango: voices, *jaranas*, and shoes striking the *tarima* are heard. All this adds a phantasmatic dimension to the fandango, plunging participants further into the strange dimension of the *fiesta* where everything seems to be but a dream.

10.8 Discussions and Disenchantment

What place does enchantment have in modernity? Arcadio Báxin — a Sotaventine *jaranero* I have cited throughout this chapter — says that enchantment is driven away by people (Figure 10.4). When they build houses and dwell near rivers or where there were fields, the *chaneques* go away (Báxin, 2017); 'In the great cities we see so little of the world', says Yeats (2012: 19). On the other hand, enchantment plays its tricks even in high-tech situations: '*Those* chaneques *are messing with us*!', half-jokingly exclaims Tereso Vega — *jaranero* from the group Son de Madera — in a concert in Mexico City when they cannot get the sound equipment to work. There are no people with twisted feet in sight though (Figure 10.5).

Figure 10.4 Arcadio Báxin, El Nopal, 10 February 2018. Photo by Diego Astorga de Ita.

If Arcadio Báxin is to be believed, there is indeed a disenchantment of the world with the advent of modernity. In this case, the stories and experiences of enchantment are but remnants of a waning past. Furthermore, Moreno-Nájera draws a connection between disenchantment and general unbelief: to believe there is a devil you must believe there is a God (Moreno-Nájera, 2018). Nowadays, the divine and the demonic are apparently mere curios in the lyrical universe of *son*. The abundance of demonic verses in urban fandangos seems to point in this direction and, while divine verses might be sung to keep tradition alive, in urban *son* — like in rock 'n' roll — a sympathy for the devil prevails.

Figure 10.5 Son de Madera in concert, Mexico City, 4 November 2017. From left to right, Oscar Terán, Ramón Gutiérrez, and Tereso Vega. Photo by Diego Astorga de Ita.

Alternatively, belief in enchantment might be a form of resistance to some of the consequences of modernity. The modern's clean-cut divide between culture and nature described by Latour (1993: 10–11) has no place in the enchanted world of Leeward music. And although *son* is a human endeavour, it is coproduced by nature's Otherworld and its inhabitants. Sotaventine musical enchantment reveals a social nature that interacts with human society, bringing forth a reality in which the bodies of instruments, trees, and monstrous beings come together in the fiesta-chronotope of the fandango. The fiesta-ness of the space allows for these entanglements and mésalliances to happen. In that sense, the eighteenth-century definitions of fandango strike a chord: *Festiva oblectatio* (festive charm). In this chronotope of fiesta and charm, the boundaries between culture and nature, and between the earthly and otherworldly, vanish with the encounter of Others.

Thus, Sotaventine enchantment — its performance and beliefs, as well as its ontological results — acts as a political tool against a modernity that has little or no place for the otherworldly or for Otherness. This enchantment offers a different path to the absolute domestication of nature; it questions the privileged position that humankind holds in the world according to modern reason, and it offers alternative behaviours and attitudes in dealing with the environment and human and non-human Others. All of this is mediated through the performance of *son Jarocho* and the fandango. What is more, Leeward enchantment eludes the elitism that Greisman critiques in other responses to disenchantment. Sotaventine *encanto* is not the result of a 'training that can be provided only by privilege and affluence', nor it is 'the illusion of a Nature free from struggle' (Greisman, 1976: 504–505). It resides in a music played historically by subaltern working classes and is characteristic of a nature that is monstrous and strange and not yet declawed by romantic ideals.

Still, the wider question of enchantment versus disenchantment remains, and it should be pondered and discussed further to complement the excellent commentary that already exists (Bennett, 2001; Ingold, 2013; Yeats, 2012).

I end this chapter with one last episode, perhaps simply a strange coincidence, perhaps a trace of enchantment that made its way to the North East of England.

At a friend's house in County Durham, I was playing '*El Buscapiés*'. We sat with a few friends at a small table in candlelight after dinner. The candles were nearly burnt out and, while I played, the wooden candlestick on the table caught fire. My friend grabbed it and ran to the kitchen tap to put it out. The incident was over quickly, and there was no brimstone. But there was fire.

References

Anderson, B., and P. Harrison, 2010. 'The Promise of Non-Representational Theories'. In *Taking-place: Non-Representational Theories and Geography*, edited by B. Anderson, and P. Harrison, 1–36. Farnham: Ashgate.

Bachelard, G. 2014. *La Poética de la Ensoñación*. Seventh Edition. México D.F.: Fondo de Cultura Económica.

Bakhtin, M. 2002. 'Forms of Time and of the Chronotope in the Novel: Notes toward a Historical Poetics'. In *Narrative Dynamics: Essays on Time, Plot, Closure, and Frames*, edited by B. Richardson, 15–24. Columbus: The Ohio State University Press.

Barahona-Londoño, A. 2013. *Las Músicas Jarochas ¿De Dónde Son? Un Acercamiento Etnomusicológico a la Historia del Son Jarocho*. México: Testimonios Jarochos A.C. / Consejo Nacional Para la Cultura y las Artes.

Bearns Esteva, S. 2011. *A la Trova Más Bonita de Estos Nobles Cantadores: The Social and Spatiotemporal Changes of Son Jarocho Music and the Fandango Jarocho*. PhD thesis, University of California, Santa Cruz.

Bennett, J. 2001. *The Enchantment of Modern Life: Attachments, Crossings, and Ethics*. Princeton and Oxford: Princeton University Press.

Brinton, D. G. 1894. 'Nagualism. A Study in Native American Folk-Lore and History', *Proceedings of the American Philosophical Society* 33(144): 11–73.

Camacho, E. D. 2007. '"El Chuchumbé Te He de Soplar:" Sobre Obscenidad, Censura y Memoria Oral en el Primer "Son de la Tierra" Novohispano', *Mester* 36(1): 54–71.

Cardona, I. 2009. 'Los Fandangos, la Música de Escenario y los Festivales: La Reactivación del Son Jarocho en Veracruz'. In *Veracruz. Sociedad y Cultura Popular en la Región Golfo Caribe*, edited by Y. Juárez Hernández, and L. Bobadilla González, 47–68. México D.F.: Universidad Michoacana de San Nicolás de Hidalgo, Universidad Veracruzana, Instituto Veracruzano de Cultura, Universidad Nacional Autónoma de México.

Chaunu, P. 1960. 'Veracruz en la Segunda Mitad del Siglo xvi y Primera del xvii', *Historia Mexicana* 9(4): 521–557.

De Ita Rubio, L. 2012. 'Piratería, Costas y Puertos en América Colonial y la Organización del Espacio Novohispano'. In *Organización del Espacio en el México Colonial: Puertos, Ciudades y Caminos*, edited by L. de Ita Rubio, 163–205. Morelia, México: Universidad Michoacana de San Nicolás de Hidalgo.

Descola, P. 2013. *Beyond Nature and Culture*, translated by J. Lloyd. Chicago and London: University of Chicago Press.

Domínguez Pérez, O. 2015. 'El Correo del Sotavento: Una Mirada a la Cuenca del Papaloapan'. In *La Fuente Hemerográfica en la Diacronía: Variedad de Enfoques*, edited by M. F. García de los Arcos, T. Quiroz Ávila, E. Ramírez Leyva, J. Rivera Castro, M. A. M. Suárez Escobar, and Á. E.

Uribe, 124–146. México D.F.: Universidad Autónoma Metropolitana, Unidad Azcapotalco.

Estrada Ochoa, A. C. 2008. 'Naturaleza, Cultura e Identidad. Reflexiones desde la Tradición Oral Maya Contemporánea', *Estudios de Cultura Maya* 34: 181–201.

Folch-Serra, M. 1990. 'Place, Voice, Space: Mikhail Bakhtin's Dialogical Landscape', *Environment and Planning D: Society and Space* 8(3): 255–274.

García de León, A. 2009. *Fandango. El Ritual del Mundo Jarocho a Través de los Siglos.* Second Edition. México D.F.: Consejo Nacional Para la Cultura y las Artes / Programa de Desarrollo Cultural del Sotavento.

_____. 2011. *Tierra Adentro, Mar en Fuera. El Puerto de Veracruz y su Litoral a Sotavento, 1519–1821.* México D.F. and Xalapa: Fondo de Cultura Económica and Universidad Veracruzana.

Gaspar de Alba, A. 2014. *[Un]framing the 'Bad Woman': Sor Juana, Malinche, Coyolxauhqui, and Other Rebels with a Cause.* Austin: University of Texas Press.

Gottfried Hesketh, J., and R. Pérez Montfort. 2009. 'Fandango y Son Entre el Campo y la Ciudad, Veracruz-México, 1930–1990'. In *Veracruz: Sociedad y Cultura Popular en la Región Golfo Caribe*, edited by Y. Juárez Hernández, and L. Bobadilla González, 69–94. México D.F.: Universidad Michoacana de San Nicolás de Hidalgo, Universidad Veracruzana, Instituto Veracruzano de Cultura, Universidad Nacional Autónoma de México.

González de Molina, M., and V. M. Toledo. 2014. *The Social Metabolism. A Socio-Ecological Theory of Historical Change.* New York and London: Springer.

Greisman, H. C. 1976. '"Disenchantment of the World": Romanticism, Aesthetics and Sociological Theory', *The British Journal of Sociology* 27(4): 495–507.

Haraway, D. 2006. 'A Cyborg Manifesto: Science, Technology, and Socialist-Feminism in the Late 20th Century'. In *The International Handbook of Virtual Learning Environments*, edited by J. Weiss, J. Nolan, J. Hunsinger, and P. Trifonas, 117–158. Dordrecht, The Netherlands: Springer.

Holland, P. 2008. 'Introduction'. In *A Midsummer Night's Dream*, edited by P. Holland, 1–118. Fourth Edition. New York: Oxford University Press.

Holloway, J., and J. Kneale. 2000. 'Mikhail Bakhtin: Dialogics of Space'. In *Thinking Space*, edited by M. Crang, and N. Thrift, 71–88. London: Routledge.

Ingold, T. 2013. 'Dreaming of Dragons: On the Imagination of Real Life', *Journal of the Royal Anthropological Institute* 19(4): 734–752.

Kohl, R. 2010. *Escritos de un Náufrago Habitual: Ensayos Sobre el Son Jarocho y Otros Temas Etnomusicológicos.* Xalapa, Veracruz: Universidad Veracruzana.

Latour, B. 1993. *We Have Never Been Modern*, translated by C. Porter. Cambridge, MA: Harvard University Press.

Llanos Arias, J, and J. Cervantes Pérez. 1995. Vientos Máximos en el Estado de Veracruz', *La Ciencia y el Hombre* (21): 185–207.

Lorimer, H. 2008. 'Cultural Geography: Non-representational Conditions and Concerns', *Progress in Human Geography* 32(4): 551–59.

Martínez González, R. 2010. 'La Animalidad Compartida: El Nahualismo a la Luz del Animismo', *Revista Española de Antropología Americana* 40(2): 256–263.

Moreno-Nájera, A. B. 2009. *Presas del Encanto: Crónicos de Son y Fandango.* N.p.: Ediciones del Programa de Desarrollo Cultural del Sotavento.

Ortiz, M. A. 2005. 'Villancicos de Negrilla', *Calíope* 11(2): 125–137.

Ochoa Villegas, C. 2015. *Un Apoyo de Son Jarocho Versada Recopilada de Aquí y de Allá.* Recursos Fandangueros. Available online at https://recursosfandanguerosenglish.files.wordpress.com/2015/01/como-la-brisa-del-viento.pdf

Paz, O. 1967. *The Labyrinth of Solitude. Life and Thought in Mexico.* London: The Penguin Press.

Pérez Montfort, R. 1991. 'La Fruta Madura (el Fandango Sotaventino del XIX a la Revolución)', *Secuencia* 19: 43–60.

_____. 2002. 'Testimonios del Son Jarocho y del Fandango: Apuntes y Reflexiones Sobre el Resurgimiento de una Tradición Regional Hacia Finales del Siglo XX', *Antropología: Boletín Oficial del Instituto Nacional de Antropología e Historial* 66: 81–95.

_____. 2010. 'Desde Santiago a la Trocha: La Crónica Local Sotaventina, el Fandango y el Son Jarocho', *Revista de Literaturas Populares* 10(1–2): 211–237.

Real Academia Española. 1732. 'Fandango'. In *Diccionario de Autoridades.* Tomo III. Available online at http://web.frl.es/DA.html

Revill, G. 2016. 'How is Space Made in Sound? Spatial Mediation, Critical Phenomenology and the Political Agency of Sound', *Progress in Human Geography* 40(2): 240–256.

Said, E. W. 1994. *Culture and Imperialism*. New York: Vintage Books.

Segovia, F. 1981. 'La Versada de Arcadio Hidalgo', *Acta Poética* 26: 605–613.

Shakespeare, W. 2006. *The Merchant of Venice*, edited by J. R. Brown. Second Edition. London: Arden Shakespeare.

_____. 2008. *A Midsummer Night's Dream*, edited by P. Holland. Fourth Edition. New York: Oxford University Press.

Stigberg, D. K. 1978. 'Jarocho, Tropical, and "Pop": Aspects of Musical Life in Veracruz, 1971–72'. In *Eight Urban Musical Cultures: Tradition and Change*, edited by B. Nettl, 260–295. Urbana: University of Illinois Press.

Thiébaut, V. 2013. 'Paisaje e Identidad: El Río Papaloapan, Elemento Funcional y Simbólico de los Paisajes del Sotavento', *LiminaR: Estudios Sociales y Humanísticos* 11(2): 82–99.

Thrift, N. 2008. *Non-Representational Theory*. New York: Routledge.

Toledo, V. M. 2008. 'Metabolismos Rurales: Hacia una Teoría Económico-Ecológica de la Apropiación de la Naturaleza', *Revista Iberoamericana de Economía Ecológica* 7: 1–26.

Tschumi, B. 2000. *Event-Cities: Praxis*. Cambridge, MA: MIT Press.

Uí Ógáin, R. 1993. 'Music Learned from the Fairies', *Béaloideas* 60/61: 197–214.

Velasco Toro, J. 2003. 'Cosmovisión y Deidades Prehispánicas de la Tierra y el Agua en los Pueblos del Papaloapan Veracruzano', *Boletín del Archivo Histórico del Agua* 25: 5–17.

Vera Tudela, E. S. 2018. 'The Tricks of the Weak: Sor Juana Inés de la Cruz and the Feminist Temporality of Latina Literature'. In *The Cambridge History of Latina/o American Literature*, edited by J. Morán González, and L. Lomas, 74–92. Cambridge: Cambridge University Press.

Vice, S. 1997. *Introducing Bakhtin*. Manchester: Manchester University Press.

Wallerstein, I. 1999. *El Moderno Sistema Mundial: La Agricultura Capitalista y el Origen de la Economía-Mundo Europea en el Siglo XVI*. Vol. 1. México D.F.: Siglo Veintiuno.

Wilson, J. 1979. 'Mentalist Poetics, the Quest, "Fiesta" and Other Motifs'. In *Octavio Paz: A Study of His Poetics*, edited by H. Bloom, 44–80. Cambridge: Cambridge University Press.

Yeats, W. B. 2012. *The Celtic Twilight*. Kindle Edition. N.p.: Amazon.

Other Media

Báxin, A. 30 December 2017. *Interview with Arcadio Báxin*. El Nopal: Interviewed by Diego Astorga de Ita.

Gutiérrez Hernández, R., and Son de Madera. 2014. 'El Coco' (from the CD album *Caribe Mar Sincopado*, Fonarte Latino).

Hidalgo, A. 1978. 'El Buscapiés' (radio broadcast, Radio Educación). Available online at http://www.fonotecanacional.gob.mx/index.php/escucha/secciones-especiales/semblanzas/arcadio-hidalgo

Hidalgo Belli, P. 12 March 2016. *Interview with Patricio Hidalgo*. Mexico City: Interviewed by Diego Astorga de Ita.

Hidalgo Belli, P., and Son de Madera 2009. 'Jarabe Loco' (from the CD album *Son de Mi Tierra*, Smithsonian Folkways Recordings).

Mono Blanco. 2013. 'El Pájaro Carpintero' (from the CD album *Orquesta Jarocha*, Casete Upload).

Moreno-Nájera, A. B. 2 February 2018. *Interview with Andrés Moreno Nájera*. Tlacotalpan: Interviewed by Diego Astorga de Ita.

Chapter 11

Journeys to *Plastic Beach*: Navigations Across the Virtual Ocean to Gorillaz' Fictional Island

Alex Jeffery*

dralexjeffery@gmail.com

11.1 Introduction

For at least two decades, a literature has formed that looks at popular music in geographical terms, connecting popular music studies with approaches from cultural geography in particular. These directions have been largely restricted to real locations, however, at the expense of considering how fictional geographies in popular music might be mapped and traced culturally. Analyses of the encoding of geography within popular music have thus far tended to focus on how music is produced and performed, searching for connections 'between sites and sounds, for inspirations in nature and the built environment' (Connell and Gibson, 2003: 91). Much of this is focused on what

*Alex Jeffrey is an independent scholar.

Musical Spaces: Place, Performance, and Power
Edited by James Williams and Samuel Horlor

ISBN 978-981-4877-85-5 (Hardcover), 978-1-003-18041-8 (eBook)
www.jennystanford.com

happens at a local level (Bennett and Whiteley, 2004), as well as on flows from the local to the diasporic (Daynes, 2004) or towards the transnational. Cultural geographers like Lily Kong (1995) have also interrogated how popular music participates both in the hegemonic construction of national identity and in the resistance against such constructs. Much of this attention is directed towards real places, yet the creation of fictional spaces and locations within popular music culture is a widespread, if largely unexplored, phenomenon that can be observed in many songs and concept albums. In other examples, more overarching geographical conceits can span entire careers, such as that of the fictional/virtual group that is the focus of this chapter — Gorillaz.

It is not uncommon for popular songs themselves to carry references to local places in many ways; references to American towns, cities, and states, for example, have been exhaustively mapped, if only on a superficial level, by Hayes (2009). When a named place forms the subject of the song, this can help to further reify mythologies of locations that are already well known and well visited. London, New York, Paris, and Rio de Janeiro have all inspired musical odes to the cities as a whole, and urban locations within them, in turn contributing to each one's mythology. Someone who has never been to New York is nonetheless likely to have a highly mediated preconception of the city as large-scale metropolis with a grungy street culture that is tough, but that can make dreams come true if conquered. These images both condition our understanding of the songs and are conditioned *by* them, as anyone who has grown up listening to 'New York, New York' can confirm. This is further enshrined on the level of genre, where punk and disco stand in for the sketchy, dangerous street life and a shiny world of glamour and sexual permissiveness, respectively. In the case of João Gilberto's 'Garota de Ipanema' (Girl from Ipanema), it is not simply lyrical references to Rio's Ipanema beach and the beautiful, gently swaying female who frequents it that conjure the setting, but the indexical links to the city through *bossa nova* — the musical style that sprung up there. In Kid Creole and the Coconuts' 1981 album *Fresh Fruit in Foreign Places*, this indexical quality of genre goes even further by providing a structuring device that helps the listener to navigate the loose narrative of the album — an odyssey around the Caribbean in search of a lost lover. The style of the music in each song (calypso, salsa, ska, etc.) indicates on which island the song-story takes place (Palmer, 1981). Both of these are examples of Tagg's genre synecdoche where a part (musical style) can stand in for a whole ('the rest of the culture to which that "foreign" style belongs' (1992: 374)).

When pop songs are set on fictional remote islands, they generally retread well-worn clichés borrowed from other popular media. 'Bali H'ai' from *South Pacific* (1949), Blondie's 'Island of Lost Souls' (1982), Madonna's 'La Isla Bonita' (1987), and The Beach Boys' 'Kokomo' (1988) are all pop songs about fictional tropical islands that are painted in broad strokes as colourful places of escapism in which to find love or a break from the mundane. These clichés are present for a reason — without the connection to a real location, it is far more difficult to convey a sense of place to a listener. If they are to achieve any foothold in listener imaginations beyond the archetypal or clichéd, the fictional location in music must often be represented visually somehow as well as heard. It must be said here that location is merely one element in immersion in some kind of *world*, which may or may not be narrative or perceived as such. Generally of greater interest to the audience within this *worldness* (Klastrup, 2003) are the star's performance persona(e), which may be variations upon their recognized public persona (Björk in *Utopia*), characters more akin to the roles of an actor (David Bowie in *1.Outside*) or, in the case of Gorillaz, purely virtual constructions that have become untethered from the musicians themselves. However, in addition to providing a narrative setting, location may offer a conceptual focus or merely ground the listener by suggesting that the music happens *somewhere*.

This chapter is concerned with one fictional setting — Gorillaz' *Plastic Beach* — and the attempts made to aid immersion in a particular chapter of Gorillaz' storyworld through this location. As the location exists largely in virtual space, I explore the consequences of virtuality for users of the *Plastic Beach* project and assert that due to its limitations, fans are encouraged to transcend this virtuality through their own creative fan practices. By remediating the fictional environments in their own ways, users manage to enact desires to insert themselves into the world, achieving unusual methods of metalepsis, or crossing ontological borders, in the process.

11.2 Gorillaz and *Plastic Beach*

The cartoon band Gorillaz first arrived in the public consciousness in early 2001 with the video for the international hit single 'Clint Eastwood'. Founded by Blur musician Damon Albarn and comic book artist Jamie Hewlett, the motivations behind the Gorillaz were manifold. Conceptually, the project would allow their creators a means to provide a sharp, parodic critique of the vapidity of the music industry, while simultaneously being able to hide behind their fictional creations. Having been the subject of often intense media scrutiny as a member of the rock group Blur, Gorillaz allowed Albarn in particular a musical output that could deflect attention away from himself as a performer. Gorillaz' embrace of ongoing new technological developments has also led to them developing a reputation as agents of audiovisual innovation. Their singles and hugely popular animated music videos helped launch them to global recognition and significant record sales for their albums *Gorillaz* (2001) and *Demon Days* (2005). With the creators of Gorillaz choosing to remain mostly hidden behind the characters they had created, the fictional construct therefore operated largely in virtual space, where the records were promoted, and a fanbase built.

The believability of Gorillaz, and by extension *Plastic Beach*, involves successfully achieving the transportation of consumers across the boundary between the real and the virtual. A conceit used in early Gorillaz shows was a screen separating the audience from the real musicians, who were only visible in silhouette. Graphics and video depicting the virtual band were then projected on the audience's side of the screen, enacting a physical boundary or 'virtual curtain'. Although this device was soon abandoned when it became clear how it limited audience interaction in live shows, the idea of the permeation of this curtain remains as a compelling metaphor; not only is it the audience that must make the cognitive leap through the curtain, but the band themselves are frequently agents of transition and transgression. The mechanics of how Gorillaz perform live particularly fascinated writers in their earlier career (from 2001 to 2005). Holographic representations of the band stood in for flesh and blood performers in several award shows, although technological limitations (the tendencies of the holographic image to be disrupted by high amplitude sound) meant that Gorillaz' tours were more conventional. Expanding on ideas by Ryan (2004), Roberta Hofer views this in terms of metalepsis, where ontological boundaries are crossed. Citing Ryan's insistence that real metalepsis never really affects the 'real world level', she notes that with Gorillaz, 'cartoon characters will never REALLY step out of their comics and into our real world' (Hofer, 2011: 239). However, rather than simply consume Gorillaz' media passively by listening to and watching their media, it is also possible for us to step into their world and engage with it through our own *active* consumption practices. As I will argue later, how we do this can also have significant consequences for how we interface with location.

Viewed by the media as Gorillaz' most ambitious project yet, their third album *Plastic Beach* (2010) was a loose concept album with the overarching theme of human waste and the man-made artificiality of what we produce.[1] Beyond just the songs on the album, these themes were extended in a large-scale transmedia narrative followed by its users beyond the album into a series of interconnected videos, games, and other media. These put a virtual location at the heart of the project — the titular island. Transmedia storytelling has been defined by media scholar Henry Jenkins as that which 'unfolds across multiple media platforms, with each new text making a distinctive and valuable contribution to the whole' (2006: 95–96). It is rare that transmedia narratives are rooted within popular music culture, but when they are, as has occurred with Gorillaz, this implies — indeed necessitates — the presence of visual texts. Other transmedia projects in popular music have a similar reliance on a sense of location in suturing the user to the storyworld. David Bowie's *1.Outside* sets much of its action in Oxford Town, a fictional city located in New Jersey. Thomas Dolby's transmedia project *The Floating City* (2011), meanwhile, maps its future earth setting with the continents of Urbanoia, Oceania, and Amerikana.

For *Plastic Beach*, to help furnish the narrative with a setting that encapsulated the album's ecological theme, an island was envisioned in the remote Pacific. The inspiration for this island was the real-life

[1]This theme arose from observation during a seaside holiday of Damon Albarn that the small pieces of sea-smoothed plastic proliferating among the beach pebbles were as much part of 'nature' as the rest of the beach.

phenomenon of the Great Pacific Garbage Patch, an area of marine debris within the North Pacific gyre thought to cover millions of square kilometres of aggregated marine waste, including plastics and chemical sludge. The fictional *Plastic Beach* island was consequently visualized as an aggregation of various human detritus, including obsolete appliances and, in keeping with the musical metanarratives of the piece, outdated music technology. To enhance the reality of the location, a real-life eight-foot model was constructed from small plastic toys, glued together and sprayed pink with a studio complex built on top. This model was then filmed in a large water tank at Shepperton Film Studios, with footage debuting online via YouTube at the start of promotion in January 2010 as the 'Orchestral Trailer'. This highly cinematic piece allowed the island exterior to be viewed from multiple angles, which could subsequently be explored further via arcade-style games available through Gorillaz' website. The interior of the studio complex built on the island could also be explored in a separate point-and-click adventure game, with players gaining access to new rooms by solving puzzles. The island could also be found remediated in further drawings used as sleeve art or magazine covers, several animated videos, and other digital art offered as screensavers.

Why did so much effort and expense go into visualizing the location, when, ultimately, the main commercial product of the *Plastic Beach* project was a collection of audio recordings? It is largely Gorillaz' virtuality that might provide the answer. The fact that Gorillaz' virtual cartoon characters are unable to appear in person to promote or tour their products, as would happen with a non-virtual band, has led to a number of ingenious promotional strategies that rely on technological novelty. Examples of such technology that Gorillaz have embraced over the years include puppetry, holographic live performance, and, increasingly, augmented reality. A constant feature of their first three albums was the creation of an inhabitable, gamified band website where a strong sense of location was key to maintaining engagement from users. From 2000 to 2007, this location was Kong Studios, an important part of Gorillaz' mythology that has featured in many of their videos. On the website, Kong Studios became an explorable three-dimensional space with rooms and corridors that housed Gorillaz' audiovisual media as well as offering various puzzles and games to promote repeat visits. The intensity of engagement by users, who would often visit the site every day, is well documented in web forums, and a recent YouTube thread demonstrates how former devotees of the website still express nostalgic loss and mourn its passing (TWG, 2017). The emotional attachment evident here has been noted as important not only for contributing to a 'sense of place' but also for experiencing full spatial immersion in a location (Ryan, Foote, and Azaryahu, 2016: 54). It was clearly hoped that the same attachment would occur for the new location, a 'beachsite' created for *Plastic Beach*, which arrived in 2011 to replace the old Kong Studios interface. In the new site, the island and the studio complex could be explored through gameplay, helping to set the scene for a narrative that would unfold through the music videos, games, and other media. Like Kong Studios, this interface was later removed after Gorillaz became inactive in the long gap between albums, and it has been replaced by a more conventional, non-navigable feed-based website. As well as deepening emotional engagement, the provision of a hub website based around the island's topography helped situate *Plastic Beach's* users (the term I will employ to describe Gorillaz' fans and consumers during this specific period) within a highly complex, and potentially confusing, constellation of media objects. Around these objects, it was evidently hoped that story immersion would also lead to successful commercial outcomes for the project. In addition to sales of CDs and vinyl LPs, the 'innovative 18-month multimedia campaign' was hoped to attract paying subscribers to the website, which would provide them with access to online content, subscription-only live events, ticket priority, and a toy (Leonard, in Barrett, 2010: 15). When the campaign was curtailed by the record company EMI roughly halfway through the projected period, and further animated music videos cancelled, it can be assumed that both streams of revenue (music sales and subscription fees) proved to be insufficient.

However, beyond providing a focus for these commercial aspirations, the choice of a remote Pacific island as the main location for *Plastic Beach* has deeper consequences that situate *Plastic Beach* within a particular tradition of storytelling. What is the extraordinary mythical power that the island holds as a fictional location, and why does it often result in outlandish fantasy worlds like *Plastic Beach*?

11.3 Islands in Popular Music

In *Narrating Space/Spatializing Narrative* (2016), Ryan, Foote, and Azaryahu point out how it is islands in particular that are favoured when the imagination engages in world-creation. They list several reasons why this might be, including how islands are able to represent the real world in geographical microcosm, as well as how their limited size makes them 'knowable and mappable' (*ibid.*: 57). The limited space of *Plastic Beach* gives it an intimacy that is appealing to those who want to engage with it as a location, a quality it shares with other fictional examples from film and literature, including *Swiss Family Robinson*, *Treasure Island*, *Jurassic Park*, and *The Island of Dr. Moreau*. Its location in the middle of the world's largest ocean — the Pacific — goes further in linking it intertextually with other fictional islands such as Skull Island from *King Kong* or Bali Ha'i from *South Pacific*. Two cult TV shows set on mysterious Pacific Islands are in fact clearly referenced in *Plastic Beach's* visual media — *Fantasy Island* (1978–1984) and particularly Gerry Anderson's 'Supermarionation' show *Thunderbirds* (1964–1966), whose Tracy Island has an obvious (and confirmed) influence on the design of the *Plastic Beach* island complex (Lamacq, 2010). More recently, the mysterious island has also emerged as a popular trope within gaming, such as in the CD-ROM game *Myst* (1993), the beach levels of *Mario Kart* (1992 onwards), the *Tropico* series (2001 onwards), and *The Sims 2: Castaway* (2007). The geographical fantasy of these game-world islands bears a striking resemblance in places to *Plastic Beach*, and this encourages further intertextuality in fan videos, which have remediated *The Sims 2* locations into *Plastic Beach* machinima.[2]

As I have previously stated, when songs are set on tropical islands, the treatment of the setting can be rather clichéd. Occasionally, though, popular musicians have also conjured more deeply thoughtout mythical islands in the service of ambitious

[2]In machinima, gameplay from software engines is screen-captured and later edited into videos that may relate a narrative or be set to music to create fan-made music videos. This will be explored later in examples.

and expansive storyworlds.[3] Björk's recent album *Utopia* (2017) is inspired by a thread running through folk tales found across several continents, where women break out with their children from a society that is oppressing them, steal flutes, and escape to a new place (Sawyer, 2017). Unsurprisingly, these flutes join in the musical narrative in the album, where the location is loosely conceived as a wild, blissful, and matriarchal island, perhaps the result of an eco-disaster. Although this location is not evinced for the listener particularly strongly through the lyrics, the digitally created landscapes of *Utopia's* music videos, particularly that for 'The Gate' and the title track, bring the world of Björk's imagination vividly to life. Motifs used in the videos include grass meadows, fluttering mythical winged creatures, and outlandishly pink-hued beaches, themselves not dissimilar from those to be found on *Plastic Beach*.

Decades prior to this, British electronic band the KLF placed a fictional island at the heart of their philosophy and mythology from 1988 to 1994, naming it Mu. Namechecked in a number of songs, Mu was a partial retreading of the Atlantis myth. The island was particularly inspired by Bill Drummond's prior involvement in a stage adaptation of Robert Shea and Robert Anton Wilson's *Illuminatus* trilogy of novels, which itself took inspiration from a number of fanciful fictional treatments of the Atlantis myth (Fitzgerald and Hayward, 2016: 51–52). In 1991, in order to stage a press event on Mu, the KLF borrowed the Isle of Jura in the Inner Hebrides to stand in for it, with journalists and media industry figures invited to the island during the summer solstice. The invitees then participated in a series of ritualistic events modelled on those taking place in another fictional location Summerisle (*ibid.*: 59), the mysterious western Scottish island upon which the British cult horror film *The Wicker Man* (Robin Hardy, 1973) was set. The KLF's use of the mythology surrounding the film, which subsequently developed a strong cult following, is further evidence that fictional islands are highly intertextual by nature; both their appearance and mythologies exert strong influence on other works and worlds.

[3]World-building in storyworlds is differentiated from the construction of *narrative* by Jenkins (2006: 114), who highlights the expansive contents and complex interrelations between characters and stories that storyworlds contain, as well as their inability to be bound within a single medium.

The design behind all of these islands is clearly at least partly utopian. Mu's temporary layering over Jura provides an interesting twist on Foucault's reading of utopias as having 'a general relation of direct or inverted analogy with the real space of Society' (1986: 24). However, such locations are perhaps really closer to the concept of the heterotopia, an 'effectively enacted utopia in which all the real sites, all the other real sites that can be found within the culture, are simultaneously represented, contested, and inverted' (*ibid.*). Heterotopias are often unbound from fixed locations, as in the 'heterotopia par excellence' of the ship or boat, conceived as 'a floating piece of space, a place without a place, that exists by itself, that is closed in on itself and at the same time is given over to the infinity of the sea' (*ibid.*: 27).

Despite having a fixed rather than a transient location, *Plastic Beach* might nonetheless qualify as heterotopian in a number of ways. Its own media contain an encyclopaedic array of nautical references from a number of other media types, including the previously mentioned *Fantasy Island* and *Thunderbirds*, the 1998 film version of *Godzilla*, and literary classics such as *Moby Dick* and *The Old Man and the Sea.* In keeping with Gorillaz' transnational style, which borrows from Japanese culture in particular, a cover for Q Magazine also references the highly recognizable Japanese woodblock print 'The Great Wave off Kanagawa'. Among countless other intertextual references, one of the rooms of the beachsite features a sub-game that challenges users to name all the faces displayed on a 'captain's wall'. These include musicians (Captains Beefheart and Sensible) and literary characters (Captains Hook, Flint, Nemo, and Haddock). Decades, even centuries, of cultural signs are therefore collapsed into an atemporal fictional space that recalls Alan Moore's *The League of Extraordinary Gentlemen* series.[4] *Plastic Beach* has also been framed on several occasions as a musical/multicultural 'microcosm of the globe', as the album featured collaborations across several continents, including contributions from New York, Syria, and Lebanon, as well as the British and European artists. Many of these musical collaborators, such as Lou Reed and Snoop Dogg, were cartoonized and depicted journeying to the island in a flotilla of submarines in a segment of the video for 'On Melancholy

[4]Moore in fact had already collaborated with Albarn and Hewlett on another project, which eventually became the *Dr. Dee* opera in 2011.

Hill', positing *Plastic Beach* as an imagined space for borderless, cosmopolitan musical harmony. Within this space, international musical relationships that never actually took place during the recording of the album can be perceived, as parts from sessions conducted separately in different countries and cities were spliced together by Albarn at a later date.[5] Heterotopia here carries some political weight. It is implied that the collapsing of borders (whether real or imagined) is viewed as positive and beneficial for music in general, reflecting statements Albarn has made both in interviews and at political rallies, such as the 2003 'Stop the War' march.

Plastic Beach relies on a high degree of intertextuality to situate its users culturally and geographically on the island and help them identify personally with the setting. This bears similarities to Mu, where to intensify identification for journalist participants, the KLF combined an existing physical location with a fictional/mythic one. By doing this, they managed to create 'a temporary space in which their mythos and logic systems briefly prevailed' (Fitzgerald and Hayward, 2016: 59). This performative resignification of place is recognized by Fitzgerald and Hayward as part of a wider folkloric tradition within a number of cultures (*ibid.*: 63). Fantastic Islands are more often visited than they are inhabited, though, and this necessitates often lengthy journeys that, as we are dealing with the virtual rather than the real, require some approaches to aid transportation to the destination. Returning to the subject of the island in *Narrating Space / Spatializing Narrative*, Ryan, Foote, and Azaryahu argue: 'the ocean that separates islands from the continents provides an allegory of the ontological difference that separates fictional worlds from the real world' (2016: 57). The vast and empty geographical space that must be traversed in order to reach fictional islands correlates to a change of state that the user must undergo. It therefore follows that in order to jump across the ontological gap and change state, some kind of *mental* journey must be undertaken. Journeys to the islands are an important part of the enjoyment of most fiction set on remote islands, and identification and the plight and wonder of characters arriving in strange new worlds are an important element in stories as diverse as *Treasure Island* and *Swiss Family Robinson*.

---

[5]A subsequent tour did, however, finally unite many of these musicians within the same physical space.

Perhaps it is no surprise, then, to find that in *Plastic Beach*, journeys through geographical space to reach the destination are prominent within its media; various agents can be seen to undergo them, including the band members, within character 'ident' videos and three music videos, and the real musical guest collaborators on the album. From ocean liner and lifeboat to hang-glider and several submarines, the means of transport here are diverse and often eccentric.[6] It was only nine months into the campaign that this set of journeys finally ended with the planned music video for 'Rhinestone Eyes', and the 'action' could truly be said to begin *on* the island. Unfortunately, when funding for music videos dried up after poor sales for the album, fans were then left frustrated by the lack of satisfactory progression and conclusion of the story. The ontological transition is particularly important with *Plastic Beach* because the location is not just outlandish, but it also exists largely in cyberspace. In the various film versions of *King Kong*, there is a veil of mist that shrouds Skull Island through which the American protagonists of the story must pass. An analogy could well be made here to the 'virtual curtain' that separates Gorillaz' real world and virtual loci. To aid the mental transportation further, in several levels of *Plastic Beach's* computer games, Gorillaz themselves become playable avatars. Here, identification with the characters, as well as a haptic engagement with the journey by using physical controls, contributes to a sense of reality for the player.

There is another method by which the ontological gap could be bridged; if the fictional world visited requires an act of mental transportation, then time spent in that destination might consequently be viewed as an act of virtual tourism, where the reader, audience member, listener, or player can vicariously experience a location that does not in fact exist.

11.4 Touristic Frames

'Virtual tourism' involves an application of Urry's influential concept of the 'tourist gaze' (1990) to fictional virtual reality environments,

[6]To untangle the order and meaning of these multiple journeys, which often seemed to contradict each other in terms of story and chronology, a long video was eventually created for YouTube entitled 'Journey to Plastic Beach'. This spliced the idents and music videos together with other artwork and pieces of video, and a voice-over narratively sutured all this heterogeneous content.

where the user may be seduced by the landscapes (and townscapes) of the setting. Christian Krug uses the examples of mid-1990s CD-ROM game *Myst* and its sequel *Riven* to demonstrate how 'the semiotics of visual consumption have already spilled over into products that are not normally associated with tourism' (2006: 269), such as games and films. In such games, the act of revelling in the beautiful scenery of the islands is viewed as 'pure unadulterated tourist gaze' (*ibid.*) and more of a draw to the world than gameplay itself. As mentioned earlier, *Plastic Beach* bears much similarity to the digitally rendered landscapes of games like *Myst*, albeit with an inverted aesthetic sense that seeks to make the ugly and artificial visually appealing, rather than the conventionally pretty or natural. *Plastic Beach* explicitly invites a tourist gaze since the location, and by extension the act of mental transportation, is framed *in terms of* tourism in a number of ways, in particular two elements of the media campaign. A set of four *Plastic Beach* 'postcards' was created and used as a promotional item, each with the caption 'Greetings from Plastic Beach' (as well as a digital image of a 'passport' with Plastic Beach stamps on it). The circulation of promotional ephemera such as posters and postcards, sometimes given away as a bonus with record releases, has played a constant role in situating popular music within material culture; their 'thingishness' acts as a counterbalance to issues surrounding the intangible nature of music and music cultures in general (Leonard, 2007; Straw, 1999). The postcards can also be seen partly as arising from a satirical twenty-first century culture of celebrating mundane and toxic images of mid-twentieth century civilization, such as Martin Parr's *Boring Postcards* book for Phaidon (2004). This is well in keeping with Hewlett's style, which is both highly intertextual and revels in sarcasm and irony. Within popular music culture, there is also something of a tradition of using postcards as images for albums that somehow wish to evoke a real place: Bruce Springsteen's *Greetings from Asbury Park, N.J.* (1973) is one of a number of rock albums that simply use a vintage picture postcard of the city (in this case circa 1940) as their cover.[7] This tradition is effectively pastiched by Gorillaz' postcards, highlighting

[7]More sardonically, Tim Buckley's *Greetings from L.A.* (1971) employed a deliberately ugly postcard image of a smog-covered tangle of freeways for its cover. This reflected the themes of sexual depravity, hopelessness, and masochism in its songs, which were set in seedy Los Angeles bars and bathhouses.

the surreality and artificiality of the island, which appears toxic and unnatural, yet curiously attractive in its garish colours.

In an attempt to penetrate the media further with touristic frames, in the week of *Plastic Beach's* launch, British newspaper The *Guardian* staged a 'Gorillaz Takeover Week' containing several travel-related media objects.[8] As well as a pictorial 'virtual travel guide', two travelogues were commissioned from real authors. One written by the literary journalist and travel writer Harry Ritchie (2010) was relayed as a straight travel piece; the other, far more psychedelic piece, came from Howard Marks, the Welsh drug smuggler turned best-selling novelist (2010), whose purple prose was more in line with Gorillaz' house style. While drawings and videos of the island showed new users what it looked like, the significance of the *Guardian* articles is that they may have encouraged them to feel that the fictional location was *real*. This has much to do with the interface through which the articles were accessed: the website of a well-regarded broadsheet newspaper. Instead of posting a token article or interview, the newspaper dedicated a significant amount of space that reflected the regular sections of the newspaper's coverage (travel, food, news, reviews etc.). Despite the ironic distance still clearly present in the articles, the coverage integrated *Plastic Beach* completely into a mediascape that, through its indexical connection to real life, makes it all the more believable, helping further collapse ontological difference. The collusion of the media has always been crucial in maintaining Gorillaz' virtual address to the audience. Journalists and radio or television broadcasters have generally tended to present articles and interviews as if the band were real, without flagging up the knowledge that they are not. This media complicity, shared with the audience, was extended by The *Guardian*'s travel pieces well beyond the characters and their backstory to the virtual location of the island. Belief in the island as somehow real was also encouraged by real-world co-ordinates in the middle of the Pacific Ocean that were provided in Ritchie's fake travel article. The co-ordinates — 48% 52' 36" S, 123% 23' 36" W — mythologize the island as 'Point Nemo', the remotest spot on the planet, being the 'furthest point from any other bit of land' (Ritchie, 2010). In addition to these touristic frames, and the emphasis on

[8]Still available online at
https://www.theguardian.com/music/series/gorillaz-takeover

journeys, another way in which the storyworld can become more real for users is by exchanging ideas socially about the island with other users within fan forums. This happened particularly within gorillaz-unofficial.com, the (currently defunct) main site where fans congregated to share information about their fandom and activities.

However, the fact that all these virtual activities occur via online portals suggests a lack of user mobility, an issue discussed by Sarah Gibson (2006). However, while Gibson is concerned with the immobility of the cinematic experience, computing has a more complex relationship with mobility. As mobile technology develops, games like *Myst,* experienced on desktops in the home or office, are giving way to new types of environmental gaming. These exploit the capabilities of mobile devices to access and utilize real-world location data. When this data is transformed into a gamified experience for the player, they are encouraged outside to explore the streets and landmarks of the city. In addition to the well-known example of the augmented reality (AR) game *Pokemon Go*, a rash of immersive, location-based games like *Geoglyph* and *Ingress* have quickly sprung up to become the new popular frontier of gaming. Although mobility is not necessarily a prerequisite for augmented reality (Bimber and Raskar, 2005: 5), this is nonetheless forcing a fairly rapid *volte-face* in the perception of gaming as essentially sedentary, closeted, and immobile. Gorillaz have adapted to this technological shift with customary canniness, adopting AR technologies in their music videos and interviews. One mobile app for their 2017 album *Humanz* superimposed elements from their music videos, including Kong Studios, over real life. A second, called 'The Lenz', would reveal 'exclusive Gorillaz content' when pointed at anything magenta (a sponsored tie-in with the T-Mobile brand, whose logo is the same colour). Although harnessing such technology helps Gorillaz to appear current, this strategy can also be viewed as a way of keeping down promotion costs, which have traditionally been high for Gorillaz. A pivot can be felt here away from the creation of immaculately crafted virtual environments like *Plastic Beach*, with their high development costs, and towards the user's own environment, aided by the recycling of Gorillaz' previous media. Indeed, both music videos and interviews for *Humanz* have focused heavily on other AR techniques of overlaying real human actors with digital characters — a fast and cheap route to character animation for the band. Fans need not rely on Gorillaz to perpetuate, recreate, or extend their favourite environments, though, as they have frequently taken the ball into their own court by recreating them themselves.

11.5 Remediating *Plastic Beach*

Although we might speculate about what the virtual geographical space of *Plastic Beach* may mean to users and fans, and trawl forum discussions for evidence of reception, perhaps the most convincing evidence of this is not what they say in response but what they *make*. Surprisingly, the drawing and redrawing of maps is almost entirely absent within fan-made media. This activity would be an obvious signifier of deep engagement with a fictional topography, and with many storyworlds, map-drawing can even be an aid to understanding narrative features like plot (Ryan, Foote, and Azaryahu, 2016: 62). However, the presence of *Plastic Beach* as a three-dimensional environment in cyberspace does not lend itself as well to graphical map-drawing as it does to three-dimensional mapping with software. This occurs most readily in the form of game mods. Various digital remediations of the island posted on YouTube have been created by fans using the sandbox game *Minecraft* and the previously mentioned *Sims 2*. Most impressive are two fully playable gamified versions of the island created within the god game *Spore* by YouTube user THEdragon.[9] These allow the player to control a small dragon (the player's own creation, with no connection to the official canon of Gorillaz characters), which can fly or walk around the island, interacting with the band members and facing off against antagonists from *Plastic Beach* music videos.

It is useful at this point to again consider practices that are so embedded within digital cultures from the perspective of materiality, from which these practices seem to have become untethered. This is particularly key with *Plastic Beach* as, thematically speaking, it is very much *about* materiality, from the startling design of the island

[9]God games are artificial life games in which the player is cast in the position of controlling the game on a large scale, rather like an entity with divine and supernatural powers.

to lyrical references to plastic waste and artificial materials in songs such as 'Some Kind of Nature', 'Rhinestone Eyes', and the title track. The aesthetic and exploration of waste and junk in *Plastic Beach* have strong resonances with what Will Straw has documented as 'museums of failure' — accumulations of unwanted CDs and vinyl in second-hand stores as 'cultural waste' (1999). The title track of *Plastic Beach* not only references the obsolete technology of the Casiotone keyboard in its lyrics ('it's a Casio on a plastic beach'), but also in the musical arpeggio patterns that denote the cheap and cheesy early 1980s sounds of the Casiotone's preset rhythms. A dialectic emerges here between so-called digital and analogue cultures, where material cultures (the crafting of the *Plastic Beach* model, the manufacture of CDs, vinyl, and other promotional artefacts) collide with digital cultures (the remediation of the island in Gorillaz' videos and computer games, as well as those created by their fans). I have argued more extensively elsewhere (Jeffery, 2017) that *Plastic Beach*, arriving within a dematerialized music culture, provides a teasing 'look but don't touch' to its audience, who are provided with visual material promise, but with little in terms of actual material or haptic objects to handle. This dialectic extends to the locations of *Plastic Beach*, and how fans of the project choose to explore and inhabit them within their own practices. Fannish responses to *Plastic Beach* that embrace materiality span the crafting of replicas, customized toys, and stop motion videos that remediate the island. However, the most compelling way fans insert themselves materially into these spaces is through the act of cosplay, which is simultaneously a craft and performance practice.

Returning to the idea of metalepsis, Astrid Ensslin coins the term 'participatory metalepsis' (2011: 12) to describe one form of interactional metalepsis, where fans of media narratives bring aspects of the fictional world into the extratextual world. As well as *vidding* (the making and posting online of fan videos, which is another core Gorillaz fan practice), she discusses cosplay in this context. Both of these forms of UGC (user generated content) are often brought to life by the affordances of digital media, and the sharing of such content online is a key motivator for its creation. Dressing up as Gorillaz' characters is a fairly widespread activity and can be witnessed on media-sharing platforms such as YouTube, DeviantArt, or cosplay.com. A particularly popular choice for this is the female Japanese guitar player Noodle, fetishized largely by female fans for her playful and mildly sexualized embodiment of the Japanese principle of *kawaii* (loosely translated as 'cuteness'). However, in many images of Noodle cosplay, it is not just the styling details of costume, make-up, and hair that aid the cosplayers in achieving their fantasy, but also the locations chosen for the photoshoots. The role of location has surprisingly attracted little writing from the scholars of cosplay, who, when they do engage with location, are mostly interested in cosplay conventions, where cosplayers gather to engage in carnivalesque public parades of their inclinations and creativity.[10] However, the creative possibilities of how cosplay shoots can transform the locations themselves prompt the question of whether it is the location that makes the *cosplay* come alive or the cosplay that brings the *location* to life.

Langsford notes that locations in this practice are 'hybrid spaces produced through the endless interpretations of globalized imagery' (2016: 20), which can often give rise to an uncanny quality. She gives as an example her own practice of photographing *Game of Thrones* cosplay on the steps of Parliament House in Adelaide, Australia. The 'strange assemblage of European symbolism and local materials' of the building (*ibid.*: 15) has to stand in for the 'remixed European history' of *Game of Thrones* (*ibid.*: 20), bringing to mind again the KLF's fictional/real composite island of Mu. Gorillaz' cosplay locations, however, are curiously placeless; re-creations of *Plastic Beach's* beach are most popular, although the beaches used could be anywhere. The beach, often viewed as a liminal, heterotopian space itself (Andriotis, 2010), is given a digital makeover in these images, where either the sand or rocks on the beach (or both) are given bright pink filters in Photoshop that approximate the look of the island model. Other cosplayers have gone further in seeking out more unusual locations, where scenes from the videos or computer games can be played out. These include a public aquarium, where the circular viewing windows act as a stand-in for the portholes of Gorillaz' submarine, a location used across several music videos and

[10]The cosplay convention has itself been traced back to American science fiction conventions in the 1960s and 1970s (Lamerichs, 2013: 158).

computer games. It is, of course, less the specifics of the location that are active in transporting the agent to the location than the embodied acts of impersonation themselves. The true fantasy can only be fulfilled when the images are viewed at a later date, after they have been touched up with digital colour saturation in Photoshop. The process of image creation therefore charts a progress from the virtual (Gorillaz' images and videos of *Plastic Beach*) to the real (human agents cosplaying on a beach) and back into the virtual (digitalized remediations of these images posted online).

11.6 Conclusion

In complex transmedia based in popular music, it is not enough for location to be simply mentioned in the lyrics if an ongoing engagement with a (narrative) setting is required. Elaboration in the visual realm, via drawings, animation, and detailed three-dimensional models, activates an inhabitable, memorable spatial environment in ways that music, despite its own spatial organization, cannot easily provide by itself. Within the set of texts that represent *Plastic Beach*, multiple devices are used to help situate users on the island, including the creation of an engaging, colourful location that is recognizable through intertextual links to other fictional islands, and touristic frames such as postcards. Beyond the merely touristic, mediated and sometimes haptically assisted journeys to the location help to bridge the ontological gap between the real and the fictional/virtual. As entertainment media are increasingly erasing the borders between media types (cinema, animation, video games etc.), dichotomies are collapsed between the analogue and the digital, the real and the virtual, creators and fans. Within complex transmedia, how such borders are traversed, and the metaleptic journeys often involved in these processes, demand attention that must focus not only on frames provided by the original creators, but also on the fan-made media that are created in response. These may give a more accurate picture of reception, particularly how users craft their individual journeys through the transmedia world.

References

Andriotis, K. 2010. 'Heterotopic Erotic Oases: The Public Nude Beach Experience', *Annals of Tourism Research* 37(4): 1076–1096.

Bennett, A., and S. Whiteley, eds. 2004. *Music, Space and Place: Popular Music and Cultural Identity*. Aldershot: Ashgate.

Bimber, O., and R. Raskar, 2005. *Spatial Augmented Reality: Merging Real and Virtual Worlds*. Boca Raton: CRC Press.

Connell, J., and C. Gibson, 2003. *Sound Tracks: Popular Music Identity and Place*. London: Routledge.

Daynes, S. 2004. 'The Musical Construction of the Diaspora: The Case of Reggae and Rastafari 1'. In *Music, Space and Place: Popular Music and Cultural Identity*, edited by S. Whiteley, A. Bennett, and S. Hawkins, 25–41. London: Routledge.

Ensslin, A. 2011. 'Diegetic Exposure and Cybernetic Performance: Towards Interactional Metalepsis'. Unpublished plenary, presented at *Staging Illusion: Digital and Cultural Fantasy*, University of Sussex, 8–9 December 2011.

Fitzgerald, J., and P. Hayward. 2016. 'Chart Mythos: The JAMs' and The KLF's Invocation of Mu', *Shima* 10(2): 50–67.

Foucault, M. 1986. 'Of Other Spaces', translated by J. Miskowiec, *Diacritics* 16(1): 22–27.

Gibson, S. 2006. 'A Seat with a View: Tourism, (Im)mobility and the Cinematic-Travel Glance', *Tourist Studies* 6(2): 157–178.

Hayes, D. 2009. '"From New York to LA": US Geography in Popular Music', *Popular Music and Society* 32(1): 87–106.

Hofer, R. 2011. 'Metalepsis in Live Performance: Holographic Projections of the Cartoon Band "Gorillaz" as a Means of Metalepsis'. In *Metalepsis in Popular Culture*, edited by K. Kukkonen, and S. Klimek, 232–251. Berlin: De Gruyter.

Jeffery, A. 2017. 'Marketing and Materiality in the Popular Music Transmedia of Gorillaz' *Plastic Beach*', *Mediterranean Journal of Communication* 8(2): 67–80.

Jenkins, H. 2006. *Convergence Culture: Where Old and New Media Collide*. New York: New York University Press.

Klastrup, L. 2003. *Towards a Poetics of Virtual Worlds: Multi-user Textuality and the Emergence of Story*. PhD thesis, Southern Cross University.

Kong, L. 1995. 'Popular Music in Geographical Analyses', *Progress in Human Geography* 19(2): 183–198.

Krug, C. 2006. 'Virtual Tourism: The Consumption of Natural and Digital Environments'. In *Nature in Literary and Cultural Studies: Transatlantic Conversations on Ecocriticism*, edited by C. Gersdorf, and S. Mayer, 249–274. Amsterdam: Rodopi.

Lamerichs, N. 2013. 'The Cultural Dynamic of Doujinshi and Cosplay: Local Anime Fandom in Japan, USA and Europe', *Participations* 10(1): 154–176.

Langsford, C. 2016. 'Floating Worlds: Cosplay Photoshoots and Creation of Imaginary Cosmopolitan Places', *Sites: A Journal of Social Anthropology and Cultural Studies* 13(1): 13–36.

Leonard, M. 2007. 'Constructing Histories through Material Culture: Popular Music, Museums and Collecting', *Popular Music History* 2(2): 147–167.

Parr, M. 2004. *Boring Postcards*. London: Phaidon.

Ryan, M. 2004. 'Metaleptic Machines', *Semiotica* 150(1–4): 439–470.

_____, K. Foote, and M. Azaryahu. 2016. *Narrating Space / Spatializing Narrative: Where Narrative Theory and Geography Meet*. Columbus: The Ohio State University Press.

Straw, W. 1999. 'The Thingishness of Things', *Invisible Culture: An Electronic Journal for Visual Studies* 2. Available online at https://urresearch.rochester.edu/handle/1802/976

Tagg, P. 1992. 'Towards a Sign Typology of Music'. *Secondo Convegno Europeo di Analisis Musicale*, edited by R. Dalmonte, and M. Baroni, 369–378. Trento: Università degli Studi di Trento.

Urry, J. 1990. *The Tourist Gaze. Leisure and Travel in Contemporary Societies*. London: Sage.

Other Media

Barrett, C. 3 June 2010. 'Releases: The Beach Boys' (magazine article, *Music Week*: 15).

Lamacq, S. 2 March 2010. 'Talking Shop: Gorillaz' (interview, BBC News). Available online at http://news.bbc.co.uk/1/hi/entertainment/8545490.stm

Marks, H. 2 March 2010. 'Gorillaz Took My Mind to the Second Dimension' (newspaper article, The *Guardian*). Available online at https://www.theguardian.com/music/2010/mar/02/gorillaz-mind-second-dimension

Palmer, R. 1981. 'The Pop Life; Kid Creole: He Mixes a Heady Brew of Styles' (newspaper article, The *New York Times*). Available online at http://www.nytimes.com/1981/06/10/arts/the-pop-life-kid-creole-he-mixes-a-heady-brew-of-styles.html

Ritchie, H. 4 March 2010. 'Heading off the Beaten Track to Plastic Beach' (newspaper article, The *Guardian*). Available online at https://www.theguardian.com/music/2010/mar/04/plastic-beach

Sawyer, M. 12 November 2017. 'Björk: "People Miss the Jokes. A Lot of It Is Me Taking the Piss out of Myself"' (newspaper article, The *Guardian*). Available online at https://www.theguardian.com/music/2017/nov/12/bjork-utopia-interview-people-miss-the-jokes

TWG [The World of Gorillaz]. 2 May 2017. 'Gorillaz — Kong Studios Ruins Website Tour' (online video). Available online at https://www.youtube.com/watch?v=w81r9S8VvZo (last accessed 6 May 2020)

Part II

Music-Making Environments

Chapter 12

Person–Environment Relationships: Influences Beyond Acoustics in Musical Performance

James Edward Armstrong
School of Film, Media and Performing Arts,
University for the Creative Arts, UCA Farnham,
Falkner Road, Farnham, Surrey, GU9 7DS UK
james.armstrong@uca.ac.uk

12.1 Introduction

The environment in which a musical performance takes place is highly influential over a musician's playing and performance and over their experience while doing so. Existing research within music performance studies tends to focus on the effects of the acoustic characteristics within an environment on a musician's playing. From such studies, it has been established that tempo, dynamic range, articulation, and musical expressiveness are shaped, and sometimes dictated, by the acoustical characteristics within a space (Kawai *et al.*, 2013: 1; Schärer Kalkandjiev, 2015). But when focusing

Musical Spaces: Place, Performance, and Power
Edited by James Williams and Samuel Horlor

ISBN 978-981-4877-85-5 (Hardcover), 978-1-003-18041-8 (eBook)
www.jennystanford.com

entirely on acoustical characteristics, the experiential influence of an environment often goes unnoticed.

A small number of studies have indicated that a musician's perception of their surroundings on a psychological and emotional level is important when investigating the environment within a music performance context. The purpose and everyday uses of an environment, or its place in culture, do in fact impact upon a musician's playing and performance experience. These affective environmental qualities are carried into musical performance experiences from the everyday existence of the environment, rather than emerging as the result of musical activity, and are thus referred to as 'non-musical'. The surrounding environment's attributes contribute greatly towards the manifestation of a 'person–environment relationship'; this term describes the interactions between a person and their surroundings in relation to both their immediate physical setting and the wider socio-cultural setting. The aim of this chapter is to reveal how person–environment relationships can be influential in the context of a musical performance, focusing on themes of behaviour-settings, socio-cultural significance, and personal meaning.

12.2 Existing Studies

A majority of existing studies on the subject of environment and space in musical performance tend to focus only on the influence of acoustical characteristics over a musician's playing. In the interest of providing a controlled environment to ensure scientific accuracy and to minimize potential variables (Woszczyk and Martens, 2008: 1043), the experiments conducted by scholars such as Sato, Kamekawa, and Marui (2011), Ueno, Kato, and Kawai (2010), and Ueno, Tachibana, and Kanamori (2004) take place within an anechoic chamber. With the absence of an acoustical response within anechoic chambers, acoustical characteristics are instead simulated via convolution processing. This provides the participating musician with artificial reverberation that responds in real time to their playing, in a similar way to that in a real-world space. Researchers can observe changes in a musician's playing in response to a range of simulated acoustic environments, encompassing attributes such as reverberation, early reflections, resonance, and echoes. One advantage this approach offers is the ability to test a large number of acoustic environments without having to physically relocate the experiment (Galiana, Llinares, and Page, 2016: 110). It also provides the researchers, and all of those involved, with the convenience of conducting their experiments in a fixed setting that does not introduce additional environmental variables.

Although studies of this nature are conducted to a high level of accuracy, they examine a very limited set of variables; rarely, if at all, are real-world environments outside of the laboratory setting considered for investigation (Schärer Kalkandjiev, 2015: 117). A church, a cathedral, a concert hall, a rock venue, a rehearsal room all are defined by much more than their acoustical properties and are likely to impart more than their acoustical identities into a musician's playing and performance. The personal meaning and individual associations a musician may have in connection to each environment transcend the limitations of what an artificial reverberation effect can be expected to create. Indeed, this is acknowledged in fields concerned with simulated audio environments and auditory virtual reality. Anechoic chambers are unrealistic performance environments, and this fact is a likely cause of discomfort for musicians participating in experiments (Noson *et al.*, 2002: 474). A small number of existing studies have also implied that there are emotional and psychological influences attached to real-world environments that cannot be replicated in a simulation (Lenox and Myatt, 2007: 209), and these studies acknowledge the need for research that factors in these influences (Sato, Kamekawa, and Marui, 2011: 4).

A final point to note is that existing studies typically analyse the performances of classically trained musicians unless investigating performers from a specific genre. None of the authors of the studies mentioned above offer reasoning for favouring this musical background. There is, however, criticism of recent music performance research retaining a focus on classical music (Gabrielsson, 2003: 222). Perhaps the preference is based on a generally accepted idea of there being a higher level of instrumental proficiency and professionalism amongst classically trained musicians, and it is suggested that future studies include musicians of alternative genres as well. Similarly, many of the studies regarding the influence of environment over a musician's playing and performance suffer from a lack of repetition and replication. These publications regularly conclude by calling for further investigation.

12.3 Introducing Environmental Psychology into Music Performance Research

This research is conducted using an interdisciplinary approach combining music performance studies and environmental psychology. Environmental psychology is a discipline concerned with the study of the interactions between humans and their surroundings; how behaviour-settings, social instruction, and place attachment affect the ways we perceive and experience our built environment. The aim of an environmental psychologist is to understand how and why we interact with our surroundings in the ways that we do, with the overall goal to improve the environment and our attitudes towards it. In the context of a musical performance, the knowledge and methodologies afforded by the field of environmental psychology offer an understanding of a musician's relationship with their environment that is under-researched, and that remains unavailable in current music performance research. A relatable study from within psychology by Aarts and Dijksterhuis investigated the situational norms and behaviours attached to a library setting, suggesting that the built environment has the potential to prime the behaviour of those within it, even if there are no direct social stimuli providing encouragement (2003: 20). Similarly, Cassidy indicates that the behavioural demands of a church environment are also likely to remain in effect even when users are engaged in activities beyond the space's main function of hosting a service or other religious activity (1997: 40). Unlike with the behavioural influences exerted by social stimuli, a built environment remains fixed, creating boundaries of influence to a specific location that extend to include wider social factors. Furthermore, environmental psychology is yet to be applied in a music performance context in any detail.

'Behaviour-settings' is a term frequently mentioned in this chapter, particularly with reference to the behaviour-settings theory first outlined by ecological psychologist Roger Barker (1968). This approach is often found in environmental and ecological research fields, and therefore it requires an explanation before application in a music performance study. Behaviour-settings theory approaches the relationship between a person and their surrounding environment as a series of behavioural interactions: 'in many cases the behavior outcomes of individuals can be predicted more accurately on the basis of the situation in which they are located' (Popov and Chompalov, 2012: 19). To contextualize, using the example of a church: from a young age, people are taught not to run and shout in this kind of building; that disruptive actions are disrespectful; and that there is a clearly defined etiquette to follow here. This is an example of socio-normative expectation, and it instructs the learned behaviour associated with a specific environment; a behaviour-setting is established.

12.4 Methodology

The methodological approach, developed for a wider research project from which this chapter's findings are extracted, builds upon those used in existing music performance studies. Participating musicians in the wider project are tasked with playing a selection of musical excerpts of their own choosing in a real-world performance environment, a simulated acoustic environment, and an acoustically dry setting. For example, a participating musician performs in a church, followed by in an acoustic simulation of the same church, and finally in an acoustically deadened studio. This process is repeated with eight musicians performing on variations of the acoustic guitar; they each come from different musical backgrounds and have different skill and experience levels (see Table 12.1). Specifically focusing on the performance experiences for players of the same instrument, the guitar, is a decision made for the purpose of consistency within the study and also to benefit from access to the network of musicians around the International Guitar Research Centre (IGRC) at the University of Surrey.

Table 12.1 Participants within the study

Participant	Age	Gender	Instrument type (acoustic guitar)	Playing style
Guitarist A	30	Male	Classical (nylon-string)	Classical
Guitarist B	37	Male	Classical (nylon-string)	Classical
Guitarist C	44	Male	Steel-string	Folk
Guitarist D	38	Male	Classical (nylon-string)	Latin/Flamenco
Guitarist E	32	Female	Steel-string resonator	Folk and Blues
Guitarist F	27	Male	Steel-string	Pop
Guitarist G	29	Male	Steel-string	Pop
Guitarist H	30	Male	Steel-string	Pop

The study uses the 'three-stage method', the name referring to the three different environmental settings (real-world, simulated, unmediated) encountered by participating musicians (see Figure 12.1). This method provides similar results to existing studies as to how a musician's playing is altered in response to the acoustic feedback of their surroundings. The addition of the real-world setting provides another level of comparison to explore changes in a musician's performance, with the potential to highlight that differences in performances do not necessarily result from changes in acoustical conditions. Indeed, the present chapter focuses specifically on the musicians' experience in this category of environmental setting, real-world performance environments. As in previous studies, no audience attends any of the performances in these experiments, and all performance environments are closed to the public to reduce the risk of interruption and the creation of additional variables. It must be acknowledged that the absence of an audience while performing in environments where one would usually be found may lead to some unusual experiences for the musicians involved (Kartomi, 2014: 190).

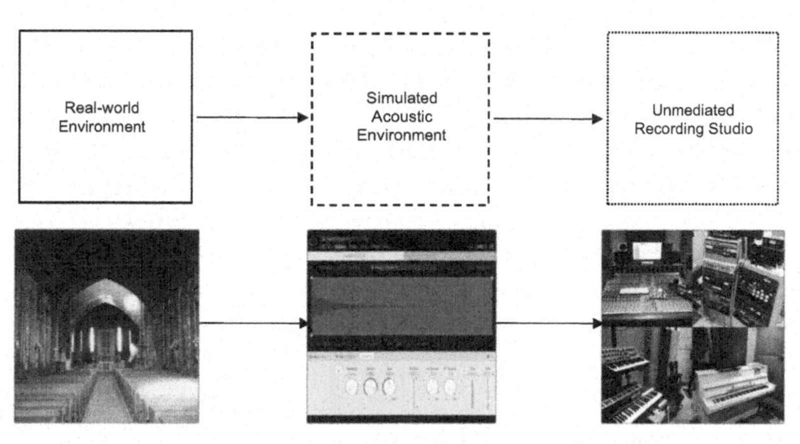

Figure 12.1 Real-world, simulated, and unmediated performance environments within the three-stage method.

Following from the practical experiments, participants are interviewed about their experiences while performing during the study. They are initially questioned on their level of awareness of acoustical characteristics within different environments, and how these can influence their instrumental playing. The main focus of the post-experiment interviews, however, is on how participants experience their performances as a result of environmental qualities, including behaviour-settings, socio-cultural significance, and personal meaning. It is at this point that qualitative data can be obtained about the emotional, psychological, and experiential qualities of responding to a music performance's surrounding environment; these data are unavailable through analysing differences in playing exclusively (Holmes and Holmes, 2013: 72–73).

The responses of each participant were analysed separately, and common themes were extracted manually for presentation in the research findings. The experiment was conducted eight times with the addition of a pilot study for testing the methodology; the findings of the pilot study do not contribute to the overall findings. Locations in the South East of England included Guildford Cathedral, St. Martha's Church, and the Studio One hall at the University of Surrey, all located in Guildford, Surrey. Locations in the North East of England included St. Mary's Church in Horden, County Durham, and the Sage Gateshead and Cluny 2 live venues in Tyne and Wear.

This method contributes to understanding of how musicians experience and interact with their environment beyond its acoustical characteristics by offering insights into behaviour-settings, socio-cultural significance, and personal meaning.

All of the practical performance experiments and post-experiment interviews took place between February 2016 and July 2017. The following research findings form a case study into the influence of behaviour-settings, socio-cultural meaning, and personal meaning experienced by a selection of the participating musicians within a real-world church environment.

12.5 Research Findings

This research indicates that a number of environmental qualities have the potential to significantly influence a musician's performance and their experience in performing. Excerpts of the post-experiment interviews conducted with the participating musicians are included in this section in support of the identified outcomes. All of the research findings in the rest of this chapter focus on the real-world church setting.[1] The church was chosen as a music performance environment to investigate due to the widely accepted understanding of how to act and behave within this place of worship, established through social expectations and learned during previous experiences (Cassidy, 1997: 35). The behavioural expectations attached to a church setting are also applicable outside of music research fields. The following section of the chapter discusses themes of behaviour-settings, social and cultural significance, and personal meaning.

12.5.1 Behaviour-Settings

Participating musicians adhere to the behavioural demands of their surroundings, and this becomes a factor in their performances within a church setting. All instances of the experiment showed that the associated expectation for a person's behaviour in a

[1]Three different church environments were used during the practical performance and recording experiments in this study. Guitarists A and B performed in Guildford Cathedral's Lady Chapel and the Church of St. Martha-on-the-Hill, both in Guildford, Surrey. Guitarists C to H performed in St. Mary's Church in Horden, County Durham.

church remains intact in a performance situation. For example, the musicians remained quiet and noticeably reserved once they had entered the building. Upon reflection, some of the participants recalled a sudden awareness of a change in environment that was, in part, signalled by the abrupt contrast in acoustics; suddenly, the smallest movements and sounds were exaggerated by the highly reverberant space, and this caused a hyperawareness of one's own actions and the subconscious imposition upon oneself of the requisite restraint. Some of the participants were able to elaborate further, having acknowledged their awareness of the behavioural demands attached to their surroundings, during their reflection on their involvement in the study. However, the acoustical conditions are unlikely to be the sole factor in directing a behavioural change; dense reverberation is an environmental characteristic widely associated with churches that may further enhance the perceived need for a change in behaviour.

All but one of the participating musicians behaved as might be expected in a place of worship; they were calm and peaceful, acting with respect to the beliefs presented in the surroundings, speaking quietly, and minimizing any playing between recording takes. The one anomaly, Guitarist B, appeared to be highly excitable, was energetic in performing, and acted in ways that might be deemed disruptive. An explanation for Guitarist B's actions is found in personal meaning associated with the specific church environment he was performing in at the time and is discussed in Section 12.5.3.

12.5.2 Social and Cultural Significance

During the post-experiment interviews, a number of interesting details about the participants' individual experiences are revealed. All of the guitarists involved in this research indicated that they felt performing in a church produced conflicts with the belief systems associated with the environment. Guitarist A remarked: 'You wouldn't play Flamenco in a church', and Guitarist D agreed: 'It feels slightly wrong to be playing Flamenco in a church'. These two guitarists are from different countries, play different styles of music, are at different skill levels, and have different religious beliefs, yet they shared the same concerns over what is appropriate musical behaviour for this environment. Guitarist D commented further on his concerns about the cultural expectations and restrictions perceived to be in play in the church setting:

> I think having been in a church a lot, I know what churches are. I have an idea of how you're meant to behave in a church. I spent many hours of my life in a church. I just thought it was appropriate that I make music in church that was right for the environment, and that's more of a spiritual type of thing. One of the songs is quite percussive and it didn't even feel quite right playing it there. It's about an attractive Latin lady, I was very conscious of what the environment wants of you.
>
> (Guitarist D)

Despite the churches used in this experiment having impressive acoustical qualities with expansive reverberation and musical resonances, the musicians involved approached their performances with a sense of caution due to their concerns around behaving disruptively, despite there being no audience in attendance to view any perceived disruption. In the case of Guitarist D, this came from a feeling that his chosen musical excerpts contrasted with the belief systems attached to the building. The same sense of unease was not experienced in the recording studio when the simulated acoustic environment of the church was introduced; the recording studio setting clearly did not carry the same social expectations. This further indicates that an environment is much more than its acoustical characteristics within the context of a musical performance.

12.5.3 Personal Meaning

A sense of personal meaning in relation to the surrounding environment is often unique to the individual. For example, the one participant whose behaviour proved an anomaly was not purposefully trying to act disruptively or show disrespect to his environment. During the interview, he revealed a particular interest in horror films. The 1976 movie *The Omen* was filmed in Guildford Cathedral, and this explained the musician's excitement; his behaviour was an attribute of personal meaning and its significance could not have been identified through performance analysis alone. A sense of personal meaning is more likely to affect a person's emotional state as opposed to their behavioural actions. For example, Guitarist G spoke of the church setting with a great sense of sadness since his previous experiences in churches often revolved around funeral services:

> In the church I probably... to be honest, think of times of mourning. That was probably the first that sprung to mind, and that kind of, like, affects you, because it's a place where you're meant to be quite sad, and therefore that's definitely the way I felt in that environment.
>
> (Guitarist G)

This association altered Guitarist G's emotional state during his time in the church, including throughout his performances, and this contributes to his overall performance experience.

12.6 'Classical Divide' and Localness Amongst Participating Musicians

In addition to insights into the relationships between individual musicians and their surroundings during a performance, two further areas of interest emerged from the post-experiment interviews. The first is a distinct difference in how a musical performance is approached and experienced when comparing musicians with classical training to those with other backgrounds. In this study, two self-identified classical guitarists (Guitarists A and B) provided responses regarding their experiences that were significantly different from those given by Guitarists C to H, who are not classically trained and do not play in a classical style. The second additional area of interest is the difference in the individual participants' experiences during a performance depending on whether they were local to a specific performance environment or were visiting. This comparison reveals interesting generalizations in actions and behaviour stemming from association and expectation. The following section provides an overview of these two areas of additional findings and discusses how these outcomes contribute to a more holistic understanding of the person–environment relationship in music performance contexts.

12.6.1 Classical Versus Pop

This research indicates a significant divide between participants with formal training and a classical music background when compared to participants from pop, folk, and experimental genres. For those with classical training, there was an emphasis on delivering a precise performance and 'playing to the score' throughout their participation in the study. Guitarists A and B both revealed a sharp awareness of their acoustic environment, noting a focus on playing to accommodate the space whilst also trying to remain faithful to the notated music. In highly reverberant spaces, one of these classical guitarists expressed frustration at not being able to tune his instrument or play at the tempo that the score stipulated. Although they indicated recognizing an obligation to adhere to the expectations of a church setting, their main anxiety was with feeling unable to perform as desired. For the non-classical musicians, the acoustical characteristics of a space seemed to offer grounds for experimentation, and the necessity to alter their playing was not considered a negative. One possible explanation for the divide is that the musicians without classical training are more likely to experiment, compose, and improvise in their music-making more generally and are therefore more likely to take advantage of the opportunity to be creative with varying environmental conditions (Sovansky *et al.*, 2016: 34). Although this study did not aim specifically to demonstrate differences between participating musicians based on their musical backgrounds, the divide here suggests that previous studies of this nature using solely classically trained musicians may have resulted in an unforeseen bias.

12.6.2 Local Versus Visiting Musicians

The results of these experiments also reveal a divide between visiting and local musicians in how they respond to the performance environment. A local musician is likely to be familiar with the cultural significance of an environment, whereas a visiting musician may approach it with an outsider's perspective. Where a church environment provides a behaviour-setting that is typically universal across all places of worship (Cassidy, 1997: 35), the personal significance of performing in a church in the North East of England, such as St. Mary's in Horden, is much less likely to affect musicians that are not local to the region. In this experiment, Guitarists C and D, both from the North East of England, expressed a sense of sadness when performing in St. Mary's Church. Guitarist C stated:

> I felt quite upset there, in that town, Horden, and the beach, because this is how my father has lived, and forefathers lived, and I felt, you know? How much hardship and suffering... It still goes on today. The conditions were appalling when there was work, and now it's appalling because there is no work. It wasn't just a church in a village, but it was in a town that was... A town in distress, you know? It's ailing, suffering.
>
> (Guitarist C)

The building is situated in an ex-mining village, apparently neglected since the closure of the mining activity on which the vast majority of its residents once relied for their livelihoods. When a participant who lives in a similar ex-mining village, Guitarist E, was asked about the influence of local history and heritage, she replied, 'Well, that's just the North East'. This points to an indifference that may have emerged as a result of a gradual normalization of this apparently depressed situation.

12.7 Conclusion

This research project reveals a number of environmental attributes that influence a musician's playing and performance experience beyond acoustical characteristics. The adaptation of methods typically found in the field of environmental psychology has allowed for a wider range of aspects of environment and space to be introduced to existing research in music performance studies. Rather than disregarding research currently available, this project provides an extension by including real-world environments and investigating participants' individual experiences throughout the various stages of performance. Behaviour-settings and social expectations change a musician's approach to a performance in ways not replicated within a simulated environment.

Although the acoustical characteristics of church settings are often impressive and conducive to (certain) music, the musician is likely to be influenced in ways beyond simply adjusting their playing to acoustical responses. The differences between musicians with classical and other kinds of musical training or background suggest that some musicians may be more susceptible to the effects of the person–environment relationship on an experiential and emotional level than others. Owing to the qualitative nature of this research project, there is not enough data to confidently make conclusions on how a musician's training impacts upon their interaction with a performance environment beyond acoustics. This does, however, highlight a possible area for further research, and one that would require dedicated case work. While previous studies suggest that musicians play in certain ways in response to their acoustic environment, this research reveals that there are more environmental attributes to take into consideration.

References

Aarts, H., and A. Dijksterhuis. 2003. 'The Silence of the Library: Environment, Situational Norm, and Social Behavior', *Journal of Personality and Social Psychology* 84(1): 18–28.

Barker, R. 1968. *Ecological Psychology: Concepts and Methods for Studying the Environment of Human Behavior.* Palo Alto, CA: Stanford University Press.

Cassidy, T. 1997. *Environmental Psychology: Behaviour and Experience in Context.* Hove, East Sussex: Psychology Press.

Gabrielsson, A. 2003. 'Music Performance Research at the Millennium', *Psychology of Music* 31(3): 221–272.

Galiana, M., C. Llinares, and Á. Page. 2016. 'Impact of Architectural Variables on Acoustic Perception in Concert Halls', *Journal of Environmental Psychology* 48: 108–119.

Holmes, P., and C. Holmes. 2013. 'The Performer's Experience: A Case for Using Qualitative (Phenomenological) Methodologies in Music Performance Research', *Musicae Scientiae* 17(1): 72–85.

Kartomi, M. 2014. 'Concepts, Terminology and Methodology in Music Performativity Research', *Musicology Australia* 36(2): 189–208.

Kawai, K., K. Kato, K. Ueno, and T. Sakuma. 2013. 'Experiment on Adjustment of Piano Performance to Room Acoustics: Analysis of Performance Coded into MIDI Data'. In *Proceedings of International Symposium on Room Acoustics, Toronto* (available online).

Lenox, P., and T. Myatt. 2007. 'Concepts of Perceptual Significance for Composition and Reproduction of Explorable Surround Sound Fields'. In *Proceeding of the 13th International Conference on Auditory Display, Montreal.*

Noson, D., S. Sato, H. Sakai, and Y. Ando. 2002. 'Melisma Singing and Preferred Stage Acoustics for Singers', *Journal of Sound and Vibration* 258(3): 473–485.

Popov, L., and I. Chompalov. 2012. 'Crossing Over: The Interdisciplinary Meaning of Behavior Setting Theory', *International Journal of Humanities and Social Science* 2(19): 18–27.

Sato, E., T. Kamekawa, and A. Marui. 2011. 'The Effect of Reverberation on Music Performance'. Presented at the 131st Audio Engineering Society Convention, New York. Available online at http://www.aes.org/e-lib/browse.cfm?elib=16582

Schärer Kalkandjiev, Z. 2015. *The Influence of Room Acoustics on Solo Music Performances: An Empirical Investigation.* PhD thesis, Technical University of Berlin.

Sovansky, E., M. Wieth, A. Francis, and S. McIlhagga. 2016. 'Not All Musicians Are Creative: Creativity Requires More than Simply Playing Music', *Psychology of Music* 44(1): 25–36.

Ueno, K., K. Kato, and K. Kawai. 2010. 'Effect of Room Acoustics on Musicians' Performance. Part 1: Experimental Investigation with a Conceptual Model', *ACTA Acustica United with Acustica* 96(3): 505–515.

Ueno, K., H. Tachibana, and T. Kanamori. 2004. 'Experimental Study on Stage Acoustics for Ensemble Performance in Orchestra'. In *Proceedings of the 18th ICA*. Available online at https://www.icacommission.org/Proceedings/ICA2004Kyoto/pdf/We2.B2.4.pdf

Woszczyk, W., and W. Martens. 2008. 'Evaluation of Virtual Acoustic Stage Support for Musical Performance', *The Journal of the Acoustical Society of America* 123(5): 3089.

Chapter 13

The Social and Spatial Basis of Musical Joy: Folk Orc as Special Refuge and Everyday Ritual

Thomas Graves
Department of Music, Durham University,
Palace Green, Durham, DH1 3RL, UK
thomas.a.graves@durham.ac.uk

13.1 Introduction

Halfway through the regular session, there was a tea break. I paid £6 for the session and stayed in the rehearsal room to chat for a while before making my way to the cafe. I asked for a coffee and dropped 50p into the honesty box. We chatted for a while until the musicians began to head back to the rehearsal room. As the guitarists checked their tuning, Chris began to make some announcements — concerts, the new website, folk day, and my research. We began to play the more familiar songs, and I found myself tapping my foot along with the others as I strummed without much thought. One of the songs we played in

Musical Spaces: Place, Performance, and Power
Edited by James Williams and Samuel Horlor

ISBN 978-981-4877-85-5 (Hardcover), 978-1-003-18041-8 (eBook)
www.jennystanford.com

> the second half was Billy Bragg's version of 'The Hard Times of Old England (Retold)', about which I had a short conversation with the man next to me, as well as 'Blackleg Miner', which someone mentioned was sung during the miners' strike. At the end, there was a brief chat about day-to-day feelings of being out of place as politically left-of-centre in a local context where conservative values prevail.
>
> (Field notes, 30 March 2017)

For nine months in 2015 and 2016, and again for three weeks in 2017, I attended regular sessions of Folk Orc, a casual group of musicians who meet weekly at two locations in New Milton and Swanage, Dorset, UK. I originally joined as part of my everyday life, and it was the joy and sense of community I felt at that time which led me to an interest in the processes by which these strong positive feelings become manifested, their relevance to longer term well-being, their connection with a sense of belonging or not belonging to a place, and the phenomenal experience of other Folk Orc participants.

Much has been written about emotional responses to music; however, the merits of group music-making in the experience of joy has often focused on religious trance, such as in Becker's (2004) and Rouget's (1985) overviews of the subject, rather than on music-making in day-to-day life, and psychological studies rarely focus solely on a single emotion. The word 'joy' is used here to separate the short-lived nature of emotion, as 'brief but intense affective reaction' (Juslin and Sloboda, 2010: 10), from the longer term affective state of happiness. This conceptualization of happiness is drawn from Aristotle's *eudaimonia*, which denotes a more permanent state of well-being, as '"[w]ell-being' and "blessedness" are good translations of Aristotle's *eudaimonia*' (Newby, 2011: 104) (although it must be noted that Aristotle also stressed the importance of material comfort for *eudaimonia*). For the purposes of this chapter, 'joy' refers to the short-term, positively valenced, high-arousal emotion characteristic of an effervescent or enthusiastic experience, whereas 'happiness' refers to longer term feelings of well-being. Therefore, the emotion of joy that can present itself as the result of a particular group of people making particular musical sounds together in a particular place can be conceptualized as one of many factors contributing to a prolonged state of happiness, similarly to how Collins, after

Durkheim, conceptualizes 'collective effervescence' as one element of an interaction ritual that leads to 'emotional energy' (Collins, 2004).

It has been suggested that geography, in particular the socio-economic environment (such as unemployment levels or social cohesion) of an area, plays a vital role in subjective happiness (Ballas and Tranmer, 2012: 71–72), and while this may be true of happiness (conceptualized as long-term well-being), exact location may not be integral in the immediate experience of *joy* as induced by music and social context as reported by my research participants: 'Swanage has exactly the same feel to it as here [CODA music centre, New Milton]' (Michelle, interview, 6 April 2017).

This chapter draws on existing literature in the fields of music psychology, sociology, and ethnomusicology to suggest possible (if speculative) ways in which this joy and longer term happiness may be a result of social, musical, and spatial factors. It will examine experiences of joy in the folk 'orchestra' Folk Orc in their rehearsals at CODA music centre in New Milton, Dorset. Folk Orc is a group of non-professional musicians led by a professional one playing simplified versions of traditional folk songs of the British Isles. It started in November 2009 and the only criterion for membership is the £7 fee for each session. Sessions are weekly at CODA, and fortnightly at another location, in Swanage, which has a partially overlapping membership. During my time of attendance, each regular session at CODA typically had between ten and twenty participants, consisting of a core membership of five to ten musicians and others who attended with varying degrees of frequency (although the Folk Orc website, as of 2020, indicates a regular membership of around twenty). They mostly play guitar, tenor guitar, mandolin, or fiddle, but the ensemble also often includes bodhrán, bass guitar, or flute. Songs are sung in unison or improvised harmony by the musicians. The group plays publicly at seasonal local festivals and community gatherings, but these were, during my period of membership, somewhat infrequent.

This chapter will discuss how Folk Orc fits into everyday lives, with either a continuous link or as a space cognitively separated from the everyday. As such, enjoyment (here conceptualized as the process of becoming full of joy) is examined not as 'embodied meaning' generated only by musical content, as suggested by Meyer

(Hesmondhalgh, 2013: 92), but as closer to the somatic emphasis of Keil's 'engendered feeling' (*ibid.*), contingent on social relations and the inclusive intention of the event, suggesting an aesthetic close to Turino's participatory performance (2008: 26). Beyond joy, other positive outcomes of Folk Orc for longer term happiness are manifested in formation and maintenance of friendships and utilization of social context in musical 'enskilment', where one is 'caught in an ever-flowing stream of practical acts' (Pálsson, 1993: 34). Such positive effects of this form of musical practice exist within a web of value judgments and expectations relating to musical taste, perceptions of musicality, and the 'amateur'/'professional' distinction (Finnegan, [1989] 2007: 15).

13.2 Methodology and Positionality

The personal associations this context has for my own life indicates a 'fieldwork at home' approach (Stock and Chiener, 2008: 108), conceptualized as 'a homecoming that [brings] with it familial and social obligations quite distinct from those confronting a visiting researcher' (*ibid.*: 109). As such, parts of this research may be framed in Conquergood's terms as 'ethnography of the ears and heart that reimagines participant-observation as coperformative witnessing' (2002: 149), that is, as shared experience.

Experiences of Folk Orc members are somewhat heterogeneous. This heterogeneity is reflected in the varied lengths of interviews and of answers within them and in different questions attracting the interest of different interviewees. The focus on lived experience also reflects the phenomenological turn in ethnomusicology, in which perception, as directed 'intentionality of consciousness' (Berger, 2008: 69), is 'the key place, where meaning in music is made' (*ibid.*), a perspective relevant to the experience of joy in musical practice.

This 'field' is particularly close to 'home', as the music centre at which the rehearsals take place is a ten-minute drive from where I spent my childhood. I was introduced to the group by a former schoolteacher and have played folk music since the age of sixteen. In this sense, I am an 'insider', giving me a clearer understanding of the 'emic' viewpoint (Stock and Chiener, 2008: 113). However, it must be emphasized that '"home" is as constructed as the "field." It may be multiple, as we add new "homes" to older ones as our lives progress' (*ibid.*: 113). As such, moving away from Dorset has separated me from longer term inclusion in Folk Orc.

During my own membership of Folk Orc, sessions gave me a release from my day-to-day stress, and the joy I felt in collaborating on relatively uncomplicated music with a group reignited my passion for playing the guitar. Other members expressed similar sentiments, and it is for this reason that I became curious about what, in the opinions of its members, was behind the joyful nature and long-term effects of membership of this ensemble.

As well as reflection on my subjective experience of Folk Orc and participation and observation of rehearsals, this research involved face-to-face semi-structured interviews with five Folk Orc musicians of ages 45 to 65, playing a variety of instruments and from a variety of occupations. Interviewees were selected based on conversations during two regular sessions at CODA on 31 March and 6 April 2017. Therefore, there may be bias towards people I personally know or who were more confident to speak about their musical practices. Interviews varied in length, depending on how much informants had to say, and whether answers to questions from the question list led to further enquiries, with the shortest interview (Michelle's) lasting twenty minutes and the longest (Roy's) lasting one hour and three minutes. Interviews were recorded and uploaded to my laptop. The selection of quotes to include was based on their relevance to the themes of joy, happiness, space, place, and socialization. These quotes were transcribed.

13.3 Everyday or Special? Folk as Accessible Music of Place

While some interviewees expressed that the *mood* of Folk Orc was the same at CODA as in Swanage — its other version, which occupies a pub — other aspects of place are vital in the existence and constitution of Folk Orc, experiences of joy, and its contribution to happiness. One noted that he 'got into Folk Orc because of CODA being there' (David, interview, 10 April 2017), recounting how after taking his daughter for guitar lessons there, he also began lessons and his guitar teacher encouraged him to join. Thus, while exact location may not immediately affect musical joy, it may be instrumental in creating *social conditions* that allow for the event in which musical joy is induced. Furthermore, lack of reported differences may be due to similarity in conceptualization of each rehearsal space, as both Swanage and CODA are examples of 'transforming the everyday into the special' (Sloboda, 2010: 497); both are 'public places characterized by the freedom to move through them at will' (*ibid.*), transformed by the presence of instruments, the ritualistic temporal structure of the event, and seating configuration.

A determinant of whether a musical emotion is everyday or special, other than a greater degree of attention paid to music in special situations, may be choice: 'Lack of choice tends to generate negative emotions as a response to a thwarting of goals or values' (Sloboda, 2010: 498), referring to the goal-directedness of emotions (Sloboda and Juslin, 2010: 77). While repertoire is decided by the bandleader, participants attend Folk Orc by choice and are often given collective choice of the final song. This can lead to the most positive responses occurring at the end of the session, thus influencing more positive longer term appraisal,[1] as events conceptualized as a series exhibit recency; that is, the effect of more recent events being remembered better than those in the middle of the series:

> ... long-term recall of a series of inputs to memory will exhibit an effect of recency only if those inputs, from the standpoint of the subject at the time of recall, constitute a well-ordered series.
>
> (Bjork and Whitten, 1974: 188)

Moreover, positive emotions are more likely to be remembered, as 'positive life events are remembered slightly better than negative life events' (Levine and Bluck, 2004: 559). This ongoing appraisal process, if it remains positive, may contribute to improvement of happiness, as 'appraisal emotions have the potential to last for years' (Huron, 2006: 16).

[1]In the ITPRA theory of musical emotion as triggered by expectations, appraisal is the final post-outcome response: 'feeling states are evoked that represent a less hasty appraisal of the outcome' (Huron, 2006: 17). The ITPRA theory splits the response into five stages: the two pre-outcome stages of imagination response and tension response, as well as the three post-outcome responses of preparation response (whether the expectation was correct), reaction (immediate) response, and appraisal response (*ibid.*: 16).

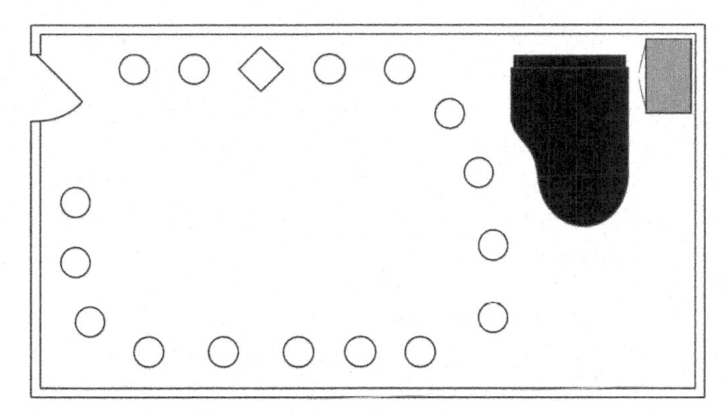

Figure 13.1 Seating positions of musicians at a rehearsal of Folk Orc at CODA music centre. Circle = musician, diamond = bandleader.

The role of seating position in joy and happiness is related to communication. As Folk Orc participants sit in a circle facing each other (Figure 13.1), they are able to see and communicate with all present in the room. This ease of communication facilitated by seating position likely improves circulation of emotion in 'affective economies where feelings do not reside in subjects or objects, but are produced as effects of circulation' (Ahmed, 2004: 8). Furthermore, the circular arrangement may carry representational Peircean semiotic meanings for its constituents, for example, as iconic legisign of equality (iconic of a perfect circle which, as it has no angles, accords no greater importance to any point on its circumference; legisign as circular arrangement is a general type exemplified by specific arrangements (sinsigns) at specific rehearsals) (Turino, 2014: 213–216). [2] This sign of equality may contribute to happiness, as 'there is an intimate connection between "well-being" at the personal level and the furtherance of communal well-being' (Newby, 2011: 154). Beyond its significance as a sign, the seating arrangement at rehearsals may also reflect physical effects

[2]Turino refers to Peirce's semiotics in which a sign's meaning is divided into three 'trichotomies', the first representing the object itself (sinsign, legisign, or qualisign), the second describing the direct relationship between the sign and its object (icon, index, or symbol), and the third describing the way in which this relationship is interpreted (dicent, semiosis, or argument). Here, the first is a legisign, a general type. The second is an icon, with a connection between sign and object found in their resemblance. The third is semiosis, the fact that a sign is being interpreted (Turino, 2014: 213–216).

regarding the way individuals interact. This may be interpreted with reference to Collins' theory of interaction ritual chains. The bodies of musicians are each close to both of their neighbours, sometimes close enough to cause guitars to collide. This is one element that may contribute to Durkheimian 'collective effervescence' (see below). Collins explains the importance of physical copresence and closeness in an interaction ritual:

> ... society is above all an embodied activity. When human bodies are together in the same place, there is a physical attunement: currents of feeling, a sense of wariness or interest, a palpable change in the atmosphere... Once the bodies are together, there may take place a process of intensification of shared experience, which Durkheim called collective effervescence, and the formation of collective conscience or collective consciousness.
>
> (2004: 34–35)

This physical closeness, along with the large range of possible interactions created by the ability to make eye contact across the room with almost any member of the orchestra, and the shared attention on a single object (i.e. the music), may contribute to 'heightened intersubjectivity' (Collins, 2004: 35); that is, the feeling of having a group focus, which Collins suggests enhances shared emotions and leads to collective effervescence (or joy). This, Collins theorizes, may lead to a longer lasting 'emotional energy', which continues beyond the immediate ritual situation, must be periodically 'topped up' as it fades over time, and 'is a feeling of confidence, courage to take action, boldness in taking initiative' (*ibid.*: 39). In short, emotional energy generated by shared attention on music performance, physical closeness, shared joy, and eye contact may be one of the driving forces of longer term 'happiness'.

These equal seating positions contrast the hierarchical arrangement of traditional Western classical orchestras, whose placement reflects both hierarchies of sections and of individual musicians within sections (Small, 1998: 68). In Folk Orc, however, musicians sit where they please, regardless of instrument played or perceived proficiency (although, in practice, musicians sit in certain places for proximity to friends or by habit). This egalitarianism extends to complexity of parts, as musicians are free to make parts as simple or complex as they wish, due to 'acceptance that there were people there at all levels and you played at your level' (David, interview, 10 April 2017). As such, the organization of seating at regular Folk Orc sessions is different from the hierarchically determined seating positions observed by O'Shea at an Irish Pub session, in which her informant told her: 'anybody who's new has to sit by the fire, and you earn your place away from it!' (O'Shea, 2006: 5). These observations are consistent with Stock's (at an English pub session), both that musicians 'employ their respective locations to generate musical clues' (2004: 62) and that English folk musicians prize egalitarianism more than their Irish counterparts (*ibid.*: 43).

While Michelle recognized no difference between the two venues, another interviewee, Roy, discussed Swanage as a location rooted in social history:

> The pub where it's organized is probably four or five hundred years old, so the history of the building in itself has an influence and is obviously different to the buildings here which are twentieth century, and obviously for a folk musician to be playing songs that are maybe three hundred years old in a building that is five hundred years old is very rooting and is very very poignant and you really do feel an intense historical connection that you are continuing a thread that somebody else maybe started three or four hundred years ago... you are part of that history.
>
> (interview, 6 April 2017)

Connection with history was important to other members too, with David noting 'a satisfying historical element' to songs (interview, 10 April 2017), suggesting that connection to wider history through lyrics plays a role in musical joy. However, this does not necessarily mean that the overall joyful effect is improved by lyrics; Ali and Peynircioğlu's study found that 'lyrics detracted from the emotions elicited by happy or calm emotions' (2006: 528–529). The satisfaction felt by David and Roy in historical aspects of both music and place, framed in the social terms above, as well as in terms of physical geography — 'Swanage is kind of on a promontory really and, like a lot of places in that geographical location, it's a bit of an enclave' (Roy, interview, 6 April 2017) — illustrates that 'social and natural history combine in often unpredictable ways to engender emotional attachments' (Smith, 2007: 221).

The accessibility of folk music was cited as an important reason for Folk Orc attendance, with its raison d'être being 'to make it accessible to as wide range of abilities as possible' (Chris H, email communication, 8 April 2017). Likewise, informants attributed reasons for attending to flexibility:

> It's what you make of it, for some people it's the bestest favourite thing they've ever done, for some people it's a chance to be in a band if you want that you might not have necessarily got, and for others it's a chance to get out of the house for two hours every week.
>
> (Craig, interview, 6 April 2017)

On the surface, these three reasons for attending Folk Orc seem unrelated; however, each suggests an affective state that is lifted or separated from the everyday into the 'special' (Sloboda, 2010: 497), whether in the form of especially intense joy, as for Chris A who stressed 'enjoyment' (interview, 5 April 2017); novel opportunity, as for Roy who called it a 'bridge between sitting on the end of your bed learning to play an instrument and the next step' (interview, 6 April 2017); or refuge from the day-to-day, as for David, who identified 'an opportunity in the week to do something completely unlike anything else I do' (interview, 10 April 2017).

Also lying behind the rhetoric of accessibility may be perceptions of 'professionalism' and 'amateurism' as expressed in terms of perceived musical ability rather than income source, recalling Finnegan's observation that 'when local musicians use the term "professional" they often refer to evaluative rather than economic aspects' ([1989] 2007: 15). A statement reminiscent of this was made by Craig, who said: 'I've never had a lesson, I'm completely self-taught, but there are people who have had years and years of musical training' (interview, 6 April 2017).

13.4 Listening Practices, Degrees of Integration, and 'Folk Performativity'

For some members, such as Michelle, Craig, and Roy, Folk Orc was greatly integrated into their lives and a key part of their identities, and for others, like David and me, it was a separate area, a refuge from everyday life. This was reflected in the congruence between listening practices and performance practices. Those who attended at both CODA and Swanage, formed bands with other Folk Orc members, and attended more extra Folk Orc-related events often reported listening practices broader than performing practices, but giving folk music a prominent position: 'It really is a mixed bag but I do listen to a lot of folk' (Michelle, interview, 6 April 2017). Those who separated Folk Orc from the everyday tended to give less emphasis to folk music in listening practices; David, for example, expressed 'a liking for both punk and prog rock' and admitted: 'I don't listen to much folk at all' (interview, 10 April 2017). However, these two modes of engaging with folk music and Folk Orc did not seem to affect emotional experiences at rehearsals, with all research participants reporting intense emotions of joy during rehearsals and often positive mood alteration as a result: 'I don't think I've been to one yet where I didn't feel uplifted or better in myself at the end of it' (Michelle, interview, 6 April 2017).

With different levels of life-Folk Orc integration may come different levels of what may be called 'folk performativity', drawing on Turner's version of performativity in which 'human life was necessarily performative, in the sense of being a set of active processes' (Loxley, 2007: 151). In informal conversation, a musician for whom Folk Orc was highly integrated with everyday life joked that finding clothes to wear for a pirate festival at which we had performed in summer 2016 was easy for Folk Orc members, as they wore those sorts of clothes anyway. Such items of clothing as waistcoats were thus seen as visual representations of what it was to be 'folk', as were drinking ale and playing folk music. As such, the co-occurrences of non-musical, performative aspects of Folk Orc with each other and with the music act as Peircean (metonymic) indices, creating an indexical cluster (Turino, 2014: 214–215), which demonstrates 'the reiterative power of discourse to produce the phenomena that it regulates and constrains' (Butler, 1993: 2). This is relevant to musical joy, as the maintenance of an in-group via 'folk performativity' likely promotes the 'feeling of complete solidarity' (David, interview, 10 April 2017) that was the most valued part of Folk Orc for most interviewees.

13.5 Music Performance, Social Relationships, and Joy

The sentiment, expressed by most interviewees, that social aspects of Folk Orc were the most important and the largest contribution to joy, with fairly little importance assigned to *place* in the production of joy, is congruent with the findings, published in the article 'Happy People or Happy Places? A Multilevel Modeling Approach to the Analysis of Happiness and Wellbeing', of Ballas and Tranmer:

> Area differences in happiness are not statistically significant, or in other words and in response to the question posed in the title of this article, it is 'people' and not 'place'.
>
> (2012: 95)

When asked 'Why do you come to Folk Orc?', almost all answers revolved around socializing, enjoyment, and fun. Laughter was often highlighted as vital, with Craig answering: 'Because it's a good laugh' (interview, 6 April 2017). Another word used was 'camaraderie' (Michelle, interview, 6 April 2017), combining elements of joy and socializing. The emphasis on laughter reflects that it 'is fundamentally a social phenomenon' (Panksepp, 2000: 183). Furthermore, as 'laughter is most certainly infectious and may transmit moods of positive social solidarity' (*ibid.*: 184), it is likely to compound positive affect talked about as a result of playing music together, and the solidarity mentioned by David as arising from the singing of the *a cappella* verse of the song 'Blackleg Miner'. While much laughter at Folk Orc is not musically induced, there are some cases in which it is; for example, in the third line of every other verse of 'Billy Boy', a B7 chord is held for an undefined period of time before a count of four ushers in the final line of the verse. During the B7, some guitarists pick out the notes of the chord to extend the undefined period, and at one rehearsal this went on for much longer than usual, eliciting laughter from the orchestra. This is an example of how 'musically induced laughter is one of the responses a listener can experience when surprised' (Huron, 2006: 27). Craig's observation that at rehearsals musicians 'mess about a bit because there's no audience' (interview, 6 April 2017) suggests that the absence of a presentational aspect to rehearsals is conducive to laughter, thus to joy and, in turn, happiness, vindicating Turino's advocacy of 'the value of participatory music' (2008: 231).

The structure of a Folk Orc rehearsal can be described in terms of a ritual. The group meets weekly at the same time, day, and place, with seasonally determined hiatuses. Each session is presided over by the bandleader, who determines most songs and pieces; time of start, break, and end; and arrangement of the pieces' structures. Each rehearsal follows a predetermined temporal structure: arrival, tuning up and chatting; welcome and introducing the piece or song; playing music with breaks for instruction, or chatting between pieces or songs; a tea break during which a 'money offering' is placed on the bandleader's violin case; playing more pieces or songs; clearing away chairs; and musicians leaving separately. The temporal architecture of Folk Orc rehearsals can be seen as structurally cyclic, as it always starts and ends with a song, often a new song at the beginning and a well-known one to finish. This suggests that different mechanisms of musical emotion are forefronted at different times of the 'ritual', with more familiar songs likely to rely more on episodic memory or evaluative conditioning (Juslin, 2013) and appraisal responses of musical expectancy (Huron, 2006: 14),[3] and less familiar songs more likely to rely on brain-stem reflex, emotional contagion,[4] visual imagery (Juslin, 2013), and reaction responses of musical expectancy (Huron, 2006: 13).

The social, ritual nature of Folk Orc rehearsals and the joy that accompanies them may be thought of in terms of Durkheim's effervescence, which explains how music is a necessary condition for expression of a 'collective sentiment':

---

[3]Episodic memory: a vivid memory of a specific incident in an individual's past. Evaluative conditioning: a behaviour or response that is learned from repeated exposure to two stimuli simultaneously, such that the response to one also becomes associated with the other.

[4]Brain-stem reflex: 'an emotion is induced by music because one or more fundamental characteristics of the music are taken by the brain stem to signal a potentially important and urgent event' (Juslin and Västfjäll, 2008: 564). Emotional contagion: a process whereby an individual mimics an emotion they perceive in another person or object. In the context of music, Juslin suggests this may be due to internal mimicry of musical expression.

> Since a collective sentiment cannot express itself collectively except on condition of observing a certain order permitting co-operation and movements in unison, these gestures and cries naturally tend to become rhythmic and regular; hence come songs and dances.
>
> (Durkheim, 1915: 216)

Beyond shared joy, other 'collective sentiments' expressed at Folk Orc rehearsals include that of national identity, being 'part of the heritage of the country as a whole' (David, interview, 10 April 2017), and folk music is thus an expression of 'imagined political community' (Anderson, 1991: 6). Another sentiment expressed in clothing and personality rather than music is that of difference from or opposition to the local hegemony of conservative values: 'Most people within Folk Orc are a little bit alternative... it's a means of personal expression... you see the clothes becoming a bit more flamboyant' (Roy, interview, 6 April 2017). Durkheim notes, though, that for effervescence 'the human voice is not sufficient for the task' (1915: 216), and this is reflected in the presence of Folk Orc's instruments.

13.6 Longer Term Effects: Flow and Enskilment

Beneficial effects for happiness have been attributed to the phenomenon of 'flow', characterized by Csikszentmihalyi as 'a state of heightened concentration, when one is so intent on the activity at hand that all other thoughts, concerns, and distractions disappear and the actor is fully in the present' (Turino, 2008: 4). In Folk Orc, just as lyric songs lead to feelings of solidarity and collective expression, instrumental tune sets can lead to states of flow, an experience described by one participant as:

> ... an enjoyable tension in playing it in that everybody's just kind of concentrating on playing without saying anything and so when it finishes you tend to get a big smile and a laugh of relief goes around the room, and enjoyment.
>
> (David, interview, 10 April 2017)

This response illustrates both the concentration required for flow and its immediate positive effects as 'both a temporary transcendence and a cumulative expansion of the self' (Turino, 2008: 233). Furthermore, it discusses 'tension', the second pre-outcome response of Huron's ITPRA theory of musical expectancy (2006: 9–10), raising the prospect of links between musical expectancy and flow. The conditions for flow, particularly the requirement for a balance between challenge and skill level (Turino, 2008: 4–5), are fulfilled by Folk Orc's commitment to accessibility that it can 'be as technical as you want' (Michelle, interview, 6 April 2017).

Members expressed how Folk Orc helped improve musical skills, with David reporting that 'you find yourself doing stuff you know that six months, a year ago you couldn't do' (interview, 10 April 2017), and Michelle advocating it as ideal method of learning at 'entry level with any instrument', and as a motivating force, saying: 'If I hadn't got involved with folk music it's actually questionable whether I'd still be playing' (interview, 6 April 2017). As musicians are caught in an 'ever-flowing stream of practical acts' (Pálsson, 1993: 34), musical capabilities are improved in processes of *enskilment*; that is, improving one's skills as a result of being immersed in the practical activities of life. However, as explored above, musicking together requires 'becoming attentive and responsive to our relations with others' (*ibid.*), and thus it likely also acts as social enskilment, giving practical experience of the lives of others. David, a teacher, also situated the educational value of musical activity in the wider political climate: 'I find it shameful that the state system doesn't provide what CODA provides' (interview, 10 April 2017).

13.7 Conclusion

From the case of Folk Orc, it is clear that experience of musical joy in group music-making takes myriad forms and mechanisms. Most sources of musical joy were identified as '"people" and not "place"' (Ballas and Tranmer, 2012: 95), with Craig calling Folk Orc a 'community resource if you like rather than a band' (interview, 6 April 2017). This sociality of musical joy can be approached as individually activated mechanisms of musical emotion and states of flow. These are produced, maintained, and modulated by social context in 'affective economies' (Ahmed, 2004: 8), facilitated by spatial configuration and ritual temporal structure, existing in shared systems of semiotic meaning, national history, and subcultural identity as expressed in 'folk performativity'. As such, the participatory nature of the rehearsal carries much of its affective power. Just as 'listeners are highly sensitive to the prosodic expression of anger, joy, and sadness' (Thompson and Balkwill, 2010: 777), it is possible that in Folk Orc this is compounded by performance, as musicians simultaneously create and listen, participating in the production of their own emotional responses.

Various elements in the process of musical joy contribute to longer term happiness. Enskilment, both musically and socially, also contributes to subjective happiness. However, an interpretative, ethnographic approach such as that presented here can only give an overview at a surface level. A more detailed appraisal of musical joy in community folk music requires more rigorous study in the forms of both empirical field experiments and the use of video footage to allow greater observational details on a microsociological level (Collins, 2004).

References

Ahmed, S. 2004. *The Cultural Politics of Emotion.* Edinburgh: Edinburgh University Press.

Ali, S. O., and Z. F. Peynircioğlu. 2006. 'Songs and Emotions: Are Lyrics and Melodies Equal Partners?', *Psychology of Music* 34(4): 511–534.

Anderson, B. 1991. *Imagined Communities.* London and New York: Verso.

Ballas, D., and M. Tranmer. 2012. 'Happy People or Happy Places? A Multilevel Modeling Approach to the Analysis of Happiness and Wellbeing', *International Regional Science Review* 35(1): 70–102.

Becker, J. 2004. *Deep Listeners: Music, Emotion, and Trancing.* Bloomington and Indianapolis: Indiana University Press.

Berger, H. M. 2008. 'Phenomenology and the Ethnography of Popular Music: Ethnomusicology at the Juncture of Cultural Studies and Folklore'. In *Shadows in the Field: New Perspectives for Fieldwork in Ethnomusicology*, edited by G. Barz, and T. Cooley, 62–75. Oxford: Oxford University Press.

Bjork, R. A., and W. B. Whitten. 1974. 'Recency-Sensitive Retrieval Processes in Long-Term Free Recall', *Cognitive Psychology* 6(2): 173–189.

Butler, J. 1993. *Bodies That Matter: On the Discursive Limits of 'Sex'.* New York and London: Routledge.

Collins, R. 2004. *Interaction Ritual Chains.* Princeton: Princeton University Press.

Conquergood, D. 2002. 'Performance Studies: Interventions and Radical Research', *TDR/The Drama Review* 46(2): 145–156.

Durkheim, E. 1915. *The Elementary Forms of Religious Life*, translated by J. W. Swain. London: George Allen & Unwin.

Finnegan, R. [1989] 2007. *The Hidden Musicians: Music Making in an English Town.* Middletown, CT: Wesleyan University Press.

Hesmondhalgh, D. 2013. *Why Music Matters.* Chichester: Wiley-Blackwell.

Huron, D. 2006. *Sweet Anticipation: Music and the Psychology of Expectation.* Cambridge, MA and London: The MIT Press.

Juslin, P. N. 2013. 'From Everyday Emotions to Aesthetic Emotions: Towards a Unified Theory of Musical Emotions', *Physics of Life Reviews* 10(3): 235–266.

Juslin, P. N., and J. A. Sloboda. 2010. 'Introduction'. In *Handbook of Music and Emotion: Theory, Research, Applications*, edited by P. N. Juslin, and J. A. Sloboda, 3–12. Oxford: Oxford University Press.

Juslin, P. N. and D. Västfjäll. 2008. 'Emotional Responses to Music: The Need to Consider Underlying Mechanisms', *Behavioural and Brain Sciences* 31(5): 559-621.

Levine, L. J., and S. Bluck. 2004. 'Painting with Broad Strokes: Happiness and the Malleability of Event Memory,' *Cognition and Emotion* 18(4): 559–574.

Loxley, J. 2007. *Performativity.* London and New York: Routledge.

Newby, M. J. 2011. *Eudaimonia: Happiness is Not Enough.* Leicester: Matador.

O'Shea, H. 2006. 'Getting to the Heart of the Music: Idealizing Musical Community and Irish Traditional Music Sessions', *Journal of the Society for Musicology in Ireland* 2: 1–18.

Pálsson, G. 1993. 'Introduction: Beyond Boundaries'. In *Beyond Boundaries: Understanding, Translation, and Anthropological Discourse*, edited by G. Pálsson, 1–40. Oxford and Providence, RI: Berg.

Panksepp, J. 2000. 'The Riddle of Laughter: Neural and Psychoevolutionary Underpinnings of Joy', *Current Directions in Psychological Science* 9(6): 183–186.

Rouget, G. 1985. *Music and Trance: A Theory of the Relations Between Music and Possession.* Chicago and London: University of Chicago Press.

Sloboda, J. A. 2010. 'Music in Everyday Life: The Role of Emotions.' In *Handbook of Music and Emotion: Theory, Research, Applications*, edited by P. N. Juslin, and J. A. Sloboda, 493–514. Oxford: Oxford University Press.

Sloboda, J. A., and P. N. Juslin. 2010. 'At the Interface Between the Inner and Outer World: Psychological Perspectives'. In *Handbook of Music and Emotion: Theory, Research, Applications*, edited by P. N. Juslin, and J. A. Sloboda, 73–97. Oxford: Oxford University Press.

Small, C. 1998. *Musicking: The Meanings of Performing and Listening.* Middletown, CT: Wesleyan University Press.

Smith, M. 2007. 'On "Being" Moved by Nature: Geography, Emotion, and Environmental Ethics'. In *Emotional Geographies*, edited by J. Davidson, L. Bondi, and M. Smith, 233–244. Aldershot and Burlington, VT: Ashgate.

Stock, J. P. J. 2004. 'Ordering Performance, Leading People: Structuring an English Folk Music Session', *The World of Music* 46(1): 41–70.

Stock, J. P. J., and C. Chiener. 2008. 'Fieldwork at Home: European and Asian Perspectives'. In *Shadows in the Field: New Perspectives for Fieldwork in Ethnomusicology*, edited by G. Barz, and T. J. Cooley, 108–124. Oxford: Oxford University Press.

Thompson, W. F., and L. Balkwill. 2010. 'Cross-Cultural Similarities and Differences'. In *Handbook of Music and Emotion: Theory, Research, Applications*, edited by P. N. Juslin, and J. A. Sloboda, 755-788. Oxford: Oxford University Press.

Turino, T. 2008. *Music as Social Life: The Politics of Participation.* Chicago and London: University of Chicago Press.

_____. 2014. 'Peircean Thought as Core Theory for a Phenomenological Ethnomusicology', *Ethnomusicology* 58(2): 185–221.

Chapter 14

Echoes of Mongolia's Sensory Landscape in Shurankhai's 'Harmonized' *Urtyn Duu*

Sunmin Yoon
School of Music, University of Delaware,
Amy E. Du Point Music Bldg., Newark, DE 19716, USA
syoon@udel.edu

14.1 Introduction: Mongolia's Contemporary *Urtyn Duu* Scene Since 2009 and the Formation of the Group Shurankhai

In December of 2009, during my field research, young people I met in Mongolia's capital Ulaanbaatar (hereafter UB) constantly asked me about Shurankhai, when they found out that I studied and had been researching the traditional folk genre called *urtyn duu* or long-song (hereafter I will use both terms interchangeably). The three female members of the *urtyn duu* vocal group called Shurankhai had come together formally in 2008, after their first concert together, and since then their first album had been released, followed by two

Musical Spaces: Place, Performance, and Power
Edited by James Williams and Samuel Horlor

ISBN 978-981-4877-85-5 (Hardcover), 978-1-003-18041-8 (eBook)
www.jennystanford.com

more. Knowing that recordings in Mongolia are released in limited numbers, usually through private funding, I had a feeling that I should quickly buy the band's new album, also called *Shurankhai* (2009). I purchased their CD and listened, sitting in my old Soviet-style apartment in the winter cold, and found myself amazed by their distinctive music-making, their experimental approach to performing *urtyn duu*. After interviewing mostly countryside herder singers and older professional singers for my research, I realized that there was something special in the way these young *urtyn duu* singers were making their music, harmonizing, and rearranging the folk songs and that it seemed quite different from other 'folk-pop' bands I had heard from other countries.

Urtyn duu is understood as one of the representative traditional folk genres most dear to contemporary Mongolians. As a typically solo genre, it is famous for its lyrics with ornamented and elongated vowels and for the skilful manipulation/modification of consonants in relation to the vowels. Traditionally, the genre was performed with up to 32 verses (*badag*), but it is now often sung with only two verses (Sampildendev and Yatskovskaya, 1984: 8), following changes to the stage culture of folk traditions in UB made during the 1950s and 1960s under Soviet influence. Currently, the *urtyn duu* genre is still actively practised both in the countryside and in the city. Singers in the countryside are mainly herders who learned the singing from their parents and grandparents at home or sometimes from a local teacher in their town. Those in the city are either established professional singers who perform mainly in UB's theatres or younger professionals who have been trained in singing schools in UB such as the University of Culture and Art (Soyol Urlagiin Ikh Surguul', SUIS) or the Music and Dance College (Khögjim Büjgiin Surguul'). While the older generation of professional singers in UB, who are mainly the faculty of these schools, have the experience of being raised as herders and of learning the songs in the countryside, the majority of these younger professional singers, despite growing up in the countryside, have neither learned nor practised *urtyn duu* singing very much in the rural herding context.

The three members of Shurankhai — B. Nomin-Erdene, D. Üüriintuya, and G. Erdenchimeg — are among those younger professional singers, and they studied together at SUIS. Each grew up in the countryside but in different provinces (*aimag*); Bayankhongor and Arkhangai, the home provinces of Nomin-Erdene and Erdenchimeg, respectively, are not far from UB (Figure 14.1), and *urtyn duu* is common to these regions, since feasts, at which *urtyn duu* was customarily sung, are held frequently there. The style of singing in this area, which is often defined as the Central Khalkh style, is characterized by long-sustained notes and by the gentle flow and refined sound of the melodic contour. Üüriintuya's home province, Dundgov' (Figure 14.2), is famous for producing many long-song singers. The Borjigon style from this region is much more elaborated and ornamented, utilizing a nasal vocal tone. Central Khalkh style is often described as following the topography of the *khangai*, a landscape of open steppe, grassy and undulating (Figure 14.3), while the Borjigon region is well known for its lower lying and concave topography. Compared to the neighbouring *khangai* regions, the Borjigon region also has a more rocky and rugged mountainous landscape. When driving in this area, I was aware of how many rocks there were on the road leading into the nearby 'desert' area (the Mongolian word *gov'*, sometimes transliterated *gobi*, means 'desert'), creating jagged tracks. This is often reflected in the Borjigon singing style, according to local singers.

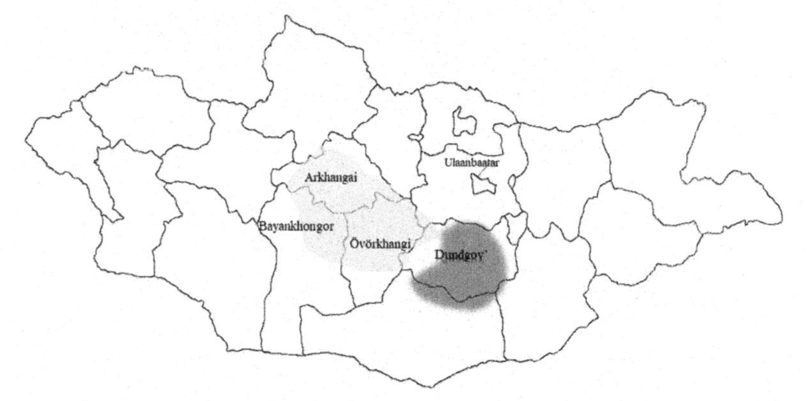

Figure 14.1 Mongolia's Central Khangai (light grey) and Gov' Regions (dark grey). Based on map © OpenStreetMap contributors, with data available under the Open Database Licence.

Figure 14.2 'Gob' region, Dundgov' province. Photo: Sunmin Yoon, 13 July 2017.

Figure 14.3 Khangai region, Arkhangai province. Photo: Sunmin Yoon, 19 June 2015.

Coming from different rural regions, these three singers have noted in interviews (28 November 2009 and 20 June 2017 with the author and in Batsaikhan (2010)) that they all began to study *urtyn duu* in UB, not in their hometowns. They learned in UB the pieces and techniques that are typically practised among professional singers, while acknowledging that they all had similar memories and experiences from their time in the countryside. As Nomin-Erdene mentioned, 'we are very close, because we all came from the countryside; together we missed our mothers and their food, and we were learning *urtyn duu* in the same school'. She also mentioned that, like their contemporaries, the members of Shurankhai also enjoyed listening to and singing pop and rock music from 'the West' while developing respect and love for *urtyn duu* (interview, 28 November 2009).[1]

Since the end of the socialist regime in Mongolia in 1990, the musical taste of the public, which had been heavily influenced by the Soviet Union for almost 70 years, has been divided into two different camps. As part of their nationalistic reinforcement of post-Soviet identity, Mongolians envisaged a strong revival of the traditional musical genres that, they believed, had thrived during the pre-revolutionary era, including *urtyn duu*, as well as *khöömii* (throat-singing), by registering these with UNESCO as their intangible cultural heritage. In contrast, there was also a great interest in embracing new sounds, in the form of a wide variety of current popular music, and in bringing the globalized world to the new Mongolia. Notwithstanding the shift of time and ideology, the spatial dichotomy of urban and rural in Mongolia also frequently aligns the popular music scene with the urban and sees traditional/folk/vernacular music as symbolic of the rural. Over the past ten years, however, my observations have indicated that Mongolia's music scenes embody a more hybridized process, rather than a separate and isolated one, between the two dissimilar sounds of traditional folk genres and synthesized and electronic popular music, mixing the two into a new style of pop or crossover music. Consequently, this spatial dichotomy has also generated a new musical sound that has emerged from both urban and rural Mongolia. As much as *urtyn duu* has traditionally been — and as it continues to be — practised by herder singers, and deeply embedded in the rural landscape

[1]It was interesting that Nomin-Erdene described Western art/pop music as 'Russian Style', while Üüriintuya referred to it as 'German Style'. I have found that the concept of 'the West' can be understood quite differently among diverse people in a non-Western country depending on their particular experiences.

and a nostalgia based on nomadic lifestyles, this traditional sound is also prevalent in the contemporary cityscape. Particular vocal techniques, such as the chest pressed voice (*kharkhiraa*) of throat-singing, or *urtyn duu's* glottal ornamentation and exclamatory vocal timbres, were adapted to genres popular with Mongolians, including electronic dance music (EDM), hip-hop, and rap, and played as background music in UB restaurants and other public spaces. Musical groups such as Zagasan Shireet Tamga, Ethnic Zorigo, and Shurankhai are typical examples of this experiment in mixing and arranging the tradition. In this way, the countryside is vivid and constant in the sound found throughout the cityscape.

14.2 Interviews: 'Tsombon Tuuraitai Khüren' (Brown Horse with Plate-Shaped Hooves)

The song that launched Shurankhai as stars in Mongolia was 'Tsombon Tuuraitai Khüren' (Brown Horse with Plate-Shaped Hooves), a rearranged traditional *urtyn duu* song in which the group's three members sing together 'in harmony'. 'Tsombon Tuuraitai Khüren' is a simple style of long-song (*besreg urtyn duu*) that is sung by many singers, both in the countryside and in UB, as a solo tradition. This song is about the protagonist's lover Egiimaa (or Igiimaa) who has moved away from him. Although I have heard it throughout the Mongolian countryside, the origins of this song are found in the eastern regions of Mongolia, including Inner Mongolia, where the Üzemchin ethnic group is located and where a more grassy steppe-plain topology is prevalent. The lyrics of the song, given below, are followed by musical examples, showing the original version of 'Tsombon Tuuraitai Khüren' and Shurankhai's version of the song.

Tsombon tuuraitai khüren maan'	My plate-hooved brown horse
Tsokhilson khar alkhaatai	beats out rough paces
Tsovoolog yanzin Egiimaa *n'*	and my lovely Egiimaa appears
Tsotsood serkhed sanagdana	when I suddenly awake
Oonyn khoyor everiig	A gazelle's two horns
Orooj zangidaj bolokhgüi	cannot be twisted together
Orchlon khorvoogiin jamig	We must not forget
Ogoogch martaj bolokhgüi	the way of this world

Musical Example 14.1 Ts. Delger 'Tsombon Tuuraitai Khüren' (traditional style):
https://www.youtube.com/watch?v=SAur1PW-T7I

Musical Example 14.2 Shurankhai's 'Tsombon Tuuraitai Khüren':
https://www.youtube.com/watch?v=micqFajC3Eo

Following a long piano introduction in Shurankhai's version of 'Tsombon Tuuraitai Khüren', the three vocalists simultaneously 'harmonize' the original tune, yet in a way clearly different from the traditional harmony typically understood in Western art music.[2] I used the English word 'harmony' in my interviews,[3] since I was not sure what the equivalent term would be, and the word became at least a channel for communication with the singers, to suggest this concept of 'singing together'. This simultaneous 'harmonic' singing was mixed with the echoing projection of each singer's exclamatory and timbrally distinctive voices in 'Tsombon Tuuraitai Khüren', but the quality of an echoing sound was also clearly presented in other songs on the album, even in the pieces where the three singers did not sing together. The more I heard their singing through the recording, the more the relationship between the music-making process and Mongolia's endless rural landscape, through which I was constantly travelling, became clearer to me. I then had the chance to meet with two members of Shurankhai — Nomin-Erdene (28 November 2009, in UB) and Üüriintuya (20 June 2017, in UB) — to talk in detail about the music-making (harmony-making) process, directly after the release of the first album (with Nomin-Erdene) (Shurankhai, 2009)

[2]Mongolian composers of Western art music use several terms for harmony. For example, Batchuluun (2008) uses two terms, *nairaldakh* ('to bring together') and *zövshööröltsökh* ('to come to agreement'), while Jantsannorov (2009) generally uses the term *zokhirol* ('musical chord', 'concord') (Bawden, 1997). Additionally, *khamsraa zokhiokh* ('to harmonize'), and *aya khamsrakh* ('put together a melody') appear in the *Oxford English-Mongolian Dictionary* (Luvsandorj, 2004). *Öngö niilüülekh* can also be used, according to Nomindar' Shagdarsüren, the former UNESCO Mongolia branch officer (email communication, 25 May 2017). I learned later in an email communication with a music teacher in Khovd, Erkhembayar Khalkh, that another Mongolian term *nairuulga khiikh* ('doing together') was used. Also, the composer Sh. Erdenebat used *duur'sakhui uhaan*, literally meaning 'resounding embodiment' (email communication with Erkhembayar Khalkh 24 April 2020). Overall, I found Mongolians' use of the term to suggest the concept of 'putting together' or 'joining' the tones in a pleasant way.

[3]When I mentioned 'harmony' in English, they often confirmed using the Russian word *garmonia*.

and after a further two albums (with Üüriintuya) (Shurankhai, 2014a, 2014b).

Both singers indicated the involvement of D. Tüvshinsaikhan, a composer and conductor of the National Morin Khuur [horse-head fiddle] Ensemble, in this rearrangement of the song from the original folk song. Üüriintuya in particular mentioned that they had needed a more 'professional' way of making 'harmony' and that Tüvshinsaikhan had arranged each singer's parts to improve on their performance when singing together. The three singers' distinctive voices had come together in a more effective way thanks to Tüvshinsaikhan organizing the relative movement and emphasis of articulation among the three separate vocalists' lines. According to Üüriintuya, they often engaged in vibrant discussions with each other, as well as with Tüvshinsaikhan, and she added that Tüvshinsaikhan also frequently experimented with, and demonstrated, possible melodic progressions on the horse-head fiddle (*morin khuur*) as they discussed. In the interview with Nomin-Erdene, I asked how the decision regarding their vocal ranges had been made — which one of them would take a high vocal tone and which a lower tone — using the Mongolian musical term *öngö* to indicate the 'vocal range' of a singer.[4] Nomin-Erdene provided an interesting answer: 'We three have very different *öngö*. I have a rather strong [*khüchtai*] voice, and Chimgee [Erdenechimeg] has skilful articulation, while Üüriintuya has beautiful glottal ornaments [*tsokhilgo, tsokhilt*]. We tried to keep everyone's vocal characteristic in our process of putting everything together'. Clearly, for them, the idea of 'harmonizing' was not about producing similar timbral expression or uniting their articulations, and this raised the question of how Mongolian singers understand the process of 'putting together' folk musical language in their music-making.

Üüriintuya and Nomin-Erdene both mentioned that when they first tried singing and making 'harmony' together, most long-song singers did not especially welcome the idea, since *urtyn duu* had originally been a solo tradition. Nomin-Erdene mentioned that arranging the traditional *urtyn duu* as an ensemble piece was not an easy process, attested to by some groups that came after them

[4]*Öngö* can be translated as 'colour, timbre (quality), tone' (Bawden, 1997). However, I also heard that Mongolian musicians use this term for the Western musical term 'pitch'.

not having such success. Nomin-Erdene felt that one reason for their success was that they were always thinking about how to combine their voices and that they experimented whenever they were travelling together:

> We think about this all the time, and sing together when we travel to the countryside, on the open steppe, inside the *ger*, and anywhere we go. We listen to each other, not just each other's voice, but also our breathing and so forth...
>
> (Interview with B. Nomin-Erdene, 28 November 2009)

Just as their minds were always tuned to one another, when Üüriintuya talked in my interview about recording sessions in the studio, she revealed a similar experience to that indicated by Nomin-Erdene:

> We stood together in the studio, facing each other so that we could see and feel our bodies together... so that we could listen to each other at the same time, and breathe at the same time...
>
> (Interview with D. Üüriintuya, 20 June 2017)

Thinking about the Western concept of harmony, whose aim I see as to create a similar timbral texture integrating melodic progress, and different ranges within the defined chord progression, Shurankhai's 'harmonizing process' seems to emphasize finding a way by which each singer can communicate with the others' musical voices, while keeping their distinctive individual timbral and ornamental qualities. Nor, though, did Shurankhai's attempts to 'harmonize' their voices using distinctive vocal timbres align with the ideas of heterophony, which is created through a particular instrumental medium's limits and capacity and necessitates that the unique techniques be considered in relation to the medium. Rather, Shurankhai's music-making was based on the layering of complex sonic and performative features. In turn, they are strongly focused on feeling each other's breathing to create a 'harmonized' sensibility — in this case, the physical place in which they are present — and their sense of that space is a vital part of their 'harmonization'.

The processes involving instrumentation and orchestration of the accompaniment also reflected the Mongolian idea of music-making. I discovered in the interviews that after they completed the 'harmonization' process of the vocal melodic parts, they added piano to their singing and later inserted the synthesized beatbox starting from the second verse. Üüriintuya emphasized that Mongolian *urtyn duu* is traditionally based on improvisation with individual ornamentations: the singing should not be restricted by the instrumental accompaniment, it should be primary, and the piano and dance beat controlled by the computer program should enter only at the end.

This seems similar to the process by which the group originally formed and worked together. Nomin-Erdene and Üüriintuya both explained that they had not intended to create the ensemble group (*khamtlag*), since the long-song is originally a solo tradition that emphasizes an individual singer's improvisational skills and singing ability. They had simply organized a concert together because, as students, they did not have enough financial sponsorship to hold separate concerts, and they thought it would cost less to arrange a concert and CD release if they performed together. For this reason, both the concert and the CD included only a couple of songs that they actually sang together as 'harmonized' pieces, while the rest of the performance and recording consisted of solos with some 'modern' and 'pop' touches. They had become a *khamtlag*, however, because people started to refer to them as such, especially when their 'harmonized' songs, such as 'Tsombon Tuuraitai Khüren', became so popular. Despite the fact that they had not intended to create a *khamtlag*, it was interesting to me that even if the songs had not at first been deliberately arranged and sung together, the feeling of having been 'harmonically' arranged was present throughout their three albums. This was precisely that same sensibility of sound projection I often heard when sitting with countryside singers on the open steppe, but now it was performed in an urban studio through their innovative rendition of *urtyn duu*.

The interview with Nomin-Erdene from 2009 dozed in my ethnography diary for a while, and it was not until, at one point in my field research, I started *feeling* the space and topography of rural Mongolia that I began thinking about it once more. As I extended my research to cover most of Mongolia's provinces, I started to understand how important what these two singers had mentioned to me had been, that they needed to feel or sense one another in order to make their 'harmony' and sing together. Regardless of the subtle differences of regional topography and ecology — particularly with the mountains and forests of western Mongolia — the Mongolian countryside is generally a continuation of open land, a massive landscape that connects with the sky on the horizon. As I learned from what the herder singers in the countryside mentioned to me, the open steppe where they herd their livestock is the practice room and the concert hall for their singing. They memorize the lyrics first, a practice which relates primarily to how they understand their place in the landscape, and then continually practise some of the difficult long-song techniques and phrases by singing out loud in the open space, while listening to the sound of their livestock and to the projection of their singing and the songs in the air and in space.

14.3 Reflection: Spatial Echoes of Singing

The word *shurankhai*, used as the group's name, refers to one of the most important *urtyn duu* techniques, one that even lay people would know. In the seven or eight words of one *urtyn duu* verse (*badag*), this technique generally appears only once. Also, depending on different categories of *urtyn duu*, as well as on the singers' ability, *shurankhai* might not appear in a singer's performance. The only category of *urtyn duu* that utilizes this technique is *aizam duu*, which Pegg translates as 'extended long-song' (2001: 44) and which is considered most important and difficult to sing. I have learned throughout my fieldwork that this extended long-song is considered musically difficult because it has a range wider than that which singers are generally able to cover and also because it uses longer sustained notes to make more ornamentations, requiring longer breaths. Most importantly, however, the lyrics in *aizam duu* were traditionally more rooted in philosophical themes than those in more musically simpler categories of long-song, which describe mundane nomadic lives. For this reason, the extended long-songs were given more respect in Mongolia's traditional feasts and among singers. Certain *aizam urtyn duu* were songs that were performed in a specific order during a wedding feast, in a ceremony for erecting/moving into a new *ger*, in family customs, and so on in order to remember the people's past and celebrate their present. Thus, *shurankhai* is a technique that *urtyn duu* singers wanted to master in their singing; mastering *shurankhai* symbolized a singer's advanced ability, and such a singer was placed in the most important seat among the participants at the feast.

The technique of *shurankhai* is often compared with Western vocal tradition's falsetto, which produces an unforced and subtle vocal tone. Most of the *urtyn duu* singing techniques use strongly pressed vocal cords and resonance in the face, particularly close to the cheeks and nose. With these methods of vocalization, a singer's power and volume often surprise listeners. In the structure of the melodic contour, the *shurankhai* appears at the highest note of the entire melodic line, signalling the climax of the song. The melodic line rises towards the *shurankhai* like mountain ridges lead upwards, as shown in Shin Nakagawa's (1980) analysis of the structure of *urtyn duu*. Just as the *shurankhai* connotes the high point of the mountain ridge in the rural landscape, where it meets the sky, which Mongolians believe is the highest spiritual entity (Tenger), so the techniques of long-song singing have developed from a singer's traditional relationship with the place in which they reside.

Shurankhai's music-making, as I have discovered from interviews with members, notwithstanding the fact that these took place in UB, highlights a deep physical, emotional, and sociopolitical connection with the rural nomadic Mongolian landscape. As Tuan says at the opening of *Space and Place*, 'place is security, space is freedom' (1977: 3), and the Mongolian countryside is both security and freedom to the people of contemporary Mongolia. The countryside (*khödöö*) is their secure home, the place where Mongolia's nomads have traditionally dwelt and created their nomadic lifeways and their heritage, while for urban dwellers the countryside is not only a physical but also a conceptual space where they can explore their identity, emotion, and nostalgia, particularly during the post-Soviet era of nation-building. The boundary of home for Mongolians, as Lewicka points out (2011: 211–213), can represent a variety of 'place scale', from the province (*aimag*) to the town (*sum*) to a specific spot located among the ridges and mountains. Szynkiewicz (1982) analyzes three levels of 'local communities' that form the conceptual boundary of the homeland (*nutag*). He sees the *nutag* as starting where Mongolians' nomadic movements are repeated and limited to a close distance. But the boundary of *nutag* can be extended as herders' nomadic movements grow broader, to satisfy their livestock's and their own needs, and it can grow to encompass a greater distance. However, as is common for those who have moved to the city, it was clear that most long-song singers, including Shurankhai's members, understand the *nutag* beyond even the boundary of the 'countryside' and as being more conceptually embedded in their memories. This encompasses wherever the traditional herding practice is alive and where Mongolians spend their nomadic lives and practise the *urtyn duu* tradition in relation to the natural environment. Thus, the concept of place/space in their homeland becomes much more fluid. And as the physical boundary becomes fluid, so the idea of place becomes more linked to individual, regional, or national communities.

The concept of 'place attachment' and 'sense of place' has been discussed by human geographers, and also by environmental psychologists, who have suggested that place produces sociopolitical and emotional relations that connect the place and those who live within it (Relph, 1976; Tuan, 1977), as well as those who travel to and from it (Giuliani, 2003; Lewicka, 2011). In the case of the Mongolians, there is often a strong 'place attachment' to rural landscape and to their *nutag* within constructions of identity. There are also nostalgic emotions for the urban mobile populations who still have memories and family members in the countryside, and they provide a sense of security, however fragile that might be. In contrast, once they are in the countryside, the countryside becomes a space free from these conceptual affects, a place in which the more realistic aspects of surviving in the environment are present. This concept of 'place attachment' certainly appeared in similar ways in my interviews with Shurankhai, not as a discourse about being Mongol, but about being an *urtyn duu* singer, in that they emphasized through this genre the importance of the connection to the rural landscape.

The close relationship between melodic lines of *urtyn duu* and rural topographies has previously been discussed by others (Desjacques, 1990: 97; Pegg, 2001: 45–47, 105–107), and I also learned that this was a common discourse through my interviews with other *urtyn duu* singers, both in UB and the countryside. However, as Levin also argued, the vocality of *urtyn duu* is not only melodic but also timbral (2011: 57), and the relation between the singing and the melodic line is not only linear but also multidimensional. One of the *urtyn duu* singers' main vocal techniques, *tsokhilgo* (*tsokhilt*, glottal trilling), has been understood as simply 'trilling' due to the quick movement between two different pitches (Nakagawa, 1980). However, it is clear that this is an alternation of two different glottal timbres, and *urtyn duu* certainly utilizes the timbral aspects. This way of using timbral manipulation in their ornamented improvisation can further be explained by my observations on the relationship between *urtyn duu* and the landscape (Yoon, 2019). This relationship is expressed not only in the melodic contour in the horizontal and linear sense, and in imitation of topography, but also in Mongolian musicians' sense of the spatiality of their sound, in particular that it projects vertically.

Curtet (2013) provides extensive discussion regarding the timbral aspects and vocal techniques surrounding the concept of verticality in the case of *khöömii* (throat-singing) and its relations to the ecological environment. He illustrates two vocal spaces, inner and outer. The inner space is the singer's body and the outer space is the singer's environment — where the singer's body is when singing. This can also be applied to *urtyn duu* singers. In the case of *khöömii*, the inner space rests in the three important points of vocal verticality — the belly, throat, and mouth (the tongue and lips) (*ibid.*: 79). I see it operating slightly differently in long-song, with the belly equivalent to the breathing, the throat as glottal ornaments, and the chest, face, and mouth as the resonating cavity to project into the space of their singing. While *khöömii* emphasizes the inner body, taking natural sounds such as those of water and wind as imitations and manifestations of the universe within the body, the long-song's technique projects from the inner body towards outer space to connect the singer with their environment and the sky (i.e., Tenger, the sky god that symbolizes their spirituality). As I often encountered in discussions during fieldwork, Mongols see themselves as entities between land and sky, humans being one of the elements of nature existing in that space. Mongols pursue spiritual harmony as Tibetans also do, mediated by Buddhist tradition (Studley, 2012). Illustrating this through mountains, *ovoo* (a cairn of stones), and the nomad's tent (*ger*), Humphrey (1995) explains the notion of verticality in the Mongolian landscape:

> If we understand the landscape, as the Mongols do, as everything around us, then the landscape includes the sky and its phenomena, such as blueness, clouds, rain, lightning, stars, and rays of light... With this in mind, the idea of 'the centre' is not so much a point in a horizontal

> disc on the earth, as a notion of verticality, for which position on the earth does not matter. The aim is reaching upwards, the making of a link between earth and sky, as with the column of smoke from a fire. 'The centre' is established anew when people make a halt. It is, in other words, not a place but an action.
>
> (Humphrey, 1995: 142)

Therefore, the *shurankhai* technique is also a part of the manifestation of the highest vertical point existing in a singer's surroundings. The wide vocal range of *urtyn duu* singers and their echoing sounds are an active response to the space; they are participating directly in the creation of the Mongolian landscape. The musical 'harmonization' of the folk song, discussed earlier through the example of 'Tsombon Tuuraitai Khüren', is, after all, the reinforcement and strengthening of how Mongols have always related to and practised with their land, ecologically harmonizing themselves between the celestial roof and the earth. In dwelling nomadically, Mongol herders move not only daily, but also seasonally from winter home to summer home, and so their movement becomes circular, repetitive, and habitual. In winter, they need to be sheltered from the strong wind, while in summer they should be close to water, and so they come back to the place where they were the previous year (Humphrey, 1995; Szynkiewicz, 1982).

> ... This begins with the basic fact that nomads do not nomadize freely, but live in mobile settlements. Hence their relation to space is regulated and follows some precise lines of social, economic and ritual obligations, not to mention ecological conditioning.
>
> (Szynkiewicz, 1982: 39)

Through their repetitive movement, however, Mongol nomads have established not only the map of the place — where to move and where to come back to — but also the sense of the place. This is a spatial awareness along with their local environment and knowledge of it, so that the experience of countryside to them becomes not mastering geography, but mastering environmental and spatial interaction — 'something with energies far greater than the human' (Humphrey, 1995: 135). These 'spatial skills' are 'what we can accomplish with our body' (Tuan, 1977: 74–75). Thus, as *urtyn duu* singers have developed awareness of the relationship in their music-making between their body as a medium on the land and their echoes projected into the atmosphere, and so the 'harmony' I heard from Shurankhai was a more extended version of these sonic relationships with the singers' specific surroundings.

14.4 Conclusion

The American philosopher Edward S. Casey states, 'body is what links this self to lived place in its sensible and perceptible features; and landscape is the presented layout of a set of places, their sensuous self-presentation as it were' (2001: 405). Mongols have made relationships with their environment in a way that 'do[es] not take over any terrain in the vicinity and transform it into something that is their own. Instead, they move within a space and environment where some kind of pastoral life is possible and "in-habit" it' (Humphrey, 1995: 135). Mongolian *urtyn duu* singers, then, take their body as a condensed universe to connect with their outer space of the rural landscape, where they interact with each other and with the natural world and make relationships in their lives. Thus, singing, and the making of music, is not simply a skilful articulation of the human body but, as a part of the natural world, the resonances and echoes of the sounding body become the channel to their survival in the Mongolian landscape.

The act of standing together to make a 'harmonized sounding' by listening, not only to one another's voice but also to one another's breathing, and also by feeling each other as well as sensing each other's presence within the surroundings, is the habitual way of living for nomadic Mongols, and something *urtyn duu* singers habitually practise as part of their daily lives. The soundscape and the echoes that provide the sense of Mongolia's rural space are becoming more and more the spirit of music-making in today's UB, and that has certainly been seen in the hip-hop, rap, folk-pop, and other popular music scenes in the city (see Zagasan Shireet Tamga (2015) and Ethnic Zorigoo (2015)). Because the rural place is where they came from, it is their secured past, but it is to a more phenomenological experience of Mongolia's rural space through new musical experiments that they would like, in their unforeseen future, to go.

References

Batchuluun, Ts. 2008. *Khögjmiin Urlagiin Tügeemel Ner Tomiyoo, Oilgolt* [General Terminology and Understanding of Music]. Ulaanbaatar: Admon.

Bawden, C. 1997. *Mongolian-English Dictionary*. London: Kegan Paul.

Casey, E. 2001. 'Body, Self, and Landscape: A Geophilosophical Inquiry into the Place-World'. In *Textures of Place: Exploring Humanist Geographies*, edited by P. Adams, S. Hoelscher, and K. Till, 403–425. Minneapolis: University of Minnesota Press.

Curtet, J. 2013. *La Transmission du Höömij, un Art du Timbre Vocal: Ethnomusicologie et Histoire du Chant Diphonique Mongol*. PhD thesis, University of Rennes 2.

Desjacques, A. 1990. 'La Dimension Orphique de la Musique Mongole', *Cahiers de Musiques Traditionnelles* 3: 97–107.

Giuliani, M. 2003. 'Theory of Attachment and Place Attachment'. In *Psychological Theories for Environmental Issues*, edited by M. Bonnes, T. Lee, and M. Boniuto, 135–170. Aldershot: Ashgate.

Humphrey, C. 1995. 'Chiefly and Shamanist Landscapes in Mongolia'. In *The Anthropology of Landscape: Perspectives on Place and Space*, edited by E. Hirsch, and M. O'Hanlon, 135–162. Oxford: Oxford University Press.

Jantsannorov, N. 2009. *Mongol Khogjimiin Egshig-Setgelgeenii Onolyn Utga Tailal* [Research in Tonal Concepts of Mongolian Music]. Ulaanbaatar: Admon.

Levin, T. 2011. *Where Rivers and Mountains Sing: Sound, Music and Nomadism in Tuva and Beyond*. Bloomington: Indiana University Press.

Lewicka, M. 2011. 'Place Attachment: How Far Have We Come in the Last 40 Years?', *Journal of Environmental Psychology* 31(3): 207–230.

Luvsandorj, A., ed. 2004. *Oxford English-Mongolian Dictionary*. Oxford: Oxford University Press.

Nakagawa, S. 1980. 'A Study of Urtiin Duu — Its Melismatic Elements and Musical Form'. In *Musical Voices of Asia, Report of Asian Traditional Performing Arts 1978*, edited by R. Emmert, and Y. Minegishi, 149–161. Tokyo: Heibonsha.

Pegg, C. 2001. *Mongolian Music, Dance, and Oral Narratives: Performing Diverse Identities*. Seattle: University of Washington Press.

Relph, E. 1976. *Place and Placelessness*. London: Pion.

Sampildendev, Kh., and K. Yatskovskaya. 1984. 'Mongol Ardyn Urtyn Duuny Övörmöts Ontslogiin Asuuldald' [Studies on the Special Characteristics of Mongolian Long-Song]. In *Mongol Ardyn Urtyn Duu*. Ulaanbaatar: State Publishing.

Studley, J. 2012. 'Territorial Cults as a Paradigm of Place in Tibet'. In *Making Sense of Place: Multidisciplinary Perspectives*, edited by I. Convery, G. Corsane, and P. Davis, 219–233. Woodbridge: The Boydell Press.

Szynkiewicz, S. 1982. 'Settlement and Community among the Mongolian Nomads', *East Asian Civilizations* 1(1): 10–44.

Tuan, Y-F. 1977. *Space and Place*. Minneapolis: University of Minnesota Press.

Yoon, S. 2019. Mobilities, Experienced and Performed, in Mongolia's *Urtyn duu* Tradition. *Asian Music* 50 (1): 47-77.

Other Media

Batsaikhan, T. 4 February 2010. 'G. Erdenechimeg: Urtyn Duuny Khamtlag Mash Olon Töröösei' [Please Produce More Urtyn Duu Groups] (news report, GoGo News Agency). Available online at http://news.gogo.mn/r/65742

Ethnic Zorigoo, featuring A Cool, Zaya, Frankseal. 2015. 'Tengeriin Huch' (music video). Available online at https://www.youtube.com/watch?v=6RqjgJwsEMo&index=7&list=RDLuVLjAhsw-w

Shurankhai. 2009. *Shurankhai* (CD). Produced by D. Tüvshinsaikhan, and P. Sükhdamba. Ulaanbaatar: CHUBA Design.

_____. 2014a. Mongolyn Ikh Khatdyn Duuts Tovchoo (CD). Produced by D. Tüvshinsaikhan, and P. Sükhdamba. Ulaanbaatar: Studio 6464.

_____. 2014b. Even Chingis Khaan (CD). Produced by D. Tüvshinsaikhan, and P. Sükhdamba. Ulaanbaatar: Studio 6464.

Zagasan Shireet Tamga. 2015. 'Mongol' (music video). Available online at https://www.youtube.com/watch?v=M7mGciZ6S2s

Designing Creative Spaces

Chapter 15

Staging *Ariodante*: Cultural Cartographies and Dialogical Performance

Benjamin Davis*

bendavisopera@icloud.com

> ... it is entirely oriented towards an experiment with the real... open and connectable in all of its dimensions; it is detachable, reversible, susceptible to constant modification. It can be torn, reversed, adapted to any kind of mounting, reworked by an individual, group, or social formation. It can be drawn on a wall, conceived as a work of art, constructed as a political action or as a meditation... The map has to do with performance...
>
> (Deleuze and Guattari, 1976: 12)

*Benjamin Davis is an independent scholar.

Musical Spaces: Place, Performance, and Power
Edited by James Williams and Samuel Horlor

ISBN 978-981-4877-85-5 (Hardcover), 978-1-003-18041-8 (eBook)
www.jennystanford.com

15.1 Introduction

Staging Handel's eighteenth-century opera *Ariodante* involves the making and navigation of various maps that are expedient to both process and performance. In this chapter, I shall employ the idea of the map to interrogate my own creative practice, as the associate director of Richard Jones's 2014 Aix-en-Provence Festival production for co-producers Dutch National Opera in 2016 and as revival director for Canadian Opera Company in 2016 and Lyric Opera of Chicago in 2019. In charting the development of this production across time and through its various incarnations with different companies and in different cities, I shall consider Christophe Den Tandt's recent theoretical 'blueprint' for the production of cultural forms, *On Virtual Grounds* (2016), zooming in on an analysis of the opera's production staging and broadening out to examine how contemporary 'operatic hermeneutics' produce spaces.

The map, as Deleuze and Guattari suggest, takes many forms and has long been associated with many different notions of realism.[1] Maps are performative in both their form and purpose, asserting a particular 'experiment with the real' in why they are conceived and how they are then navigated and employed. I wish to qualify different maps as informing different phases in the process of staging opera. These perform distinct functions, such as focusing on detail relevant to certain roles within complementary disciplines and discourses, or '*socially constructed ways of knowing some aspect of reality* which can be drawn upon when that aspect of reality has to be represented, or, to put it another way, *context-specific frameworks for making sense of things*' (Van Leeuwen, 2009: 144, emphasis in the original). Many kinds of interdisciplinary maps of meaning fall under this definition of discourse and require particular knowledge and skill to decipher and interpret,[2] for example, orchestral parts of

[1]My preferred definition of realism as a mode (rather than a genre or historical movement) was conceived of literature but is relevant to other cultural forms: 'An imaginative extension of experience along lines laid down by knowledge: referring to, reporting on, doing justice to, celebrating, analysing and being constrained by reality, not [merely] replicating, mirroring, reproducing or copying it' (Tallis, 1988: 195).

[2]'Discourse' as a field of study is more broadly related to 'culture' and has been defined as: 'distinctive ways of speaking/listening and often, too, writing/reading *coupled* with distinctive ways of acting, interacting, valuing, feeling, dressing, thinking, believing, with other people and with various objects, tools, and technologies, so as to enact specific socially recognizable identities engaged in specifically recognised activities' (Gee, 2005: 155).

an operatic score, a piano reduction vocal score, the *libretto* (or sung text), rehearsal schedule, design model box, technical drawings, rehearsal mark-out, costume and lighting plots, production book, video recording, and even a review of a performance — all of which privilege certain information, from a particular perspective and for a particular purpose, which can in turn be interpreted in a number of ways.

Auslander argues that music and its performance are inextricably imbricated with one another, intending his MAP to stand for music *as* performance, not an alloy of constituent disciplines (musicology and performance studies), but rather an 'elemental, irreducible fusion of expressive means' (2013: 354–355). I agree with this assertion, particularly when it comes to the heightened, multimodal reality of opera as a form which sets out to communicate in performance through its own array of 'expressive means'.[3] Tonal analysis, historical contextualization, and reception studies have indeed been slow to embrace the multimodal nature of opera as a visual and spatial as well as musical form of theatre.

The process of staging opera involves the collaborative and reflexive mapping of various discourses: from deciphering and interpreting the score to the realization of a design world and negotiations with company members and casts who embody and navigate these musical, physical, and conceptual spaces. I propose that in relation to the making of opera productions it is useful to consider how dialogue and dialogical theories, developed from the writings of Mikhail Bakhtin (1981), might be understood as 'thinking together' about performance from different perspectives (Isaacs, 1999), in order to broaden and deepen understanding of the subject material (in this case opera) and reach some consensus (in this case, a psychologically congruent, and hopefully coherent, production narrative). Den Tandt's performative repositioning of realism 'on virtual grounds' has proffered a blueprint for the production of

[3]Opera is described on the Opera Europa website in the following way: 'Opera is a total art form which joins music, singing, drama, poetry, plastic arts and sometimes dance. In each work, all the components of opera combine their expressiveness and their beauty. This complex alchemy makes an opera performance an extraordinary show, monopolising the sight, hearing, imagination and sensibility of the audience, where all human passions are at work' (Opera Europa, n.d.).

cultural works in a contemporary, or dialogical realist mode.[4] Whilst developed as a literary and cultural theory, I apply Den Tandt's theoretical blueprint to my discussion of the form of opera in this case study, 'staging *Ariodante*'. In so doing, I postulate that opera productions are increasingly dialogical cartographies of culture and can themselves be read as idiomatic maps. From the perspective of creative practice, I wish to advocate a dialogical model for the making and performing of opera, as well as asserting the richness of similarly dialogical and interdisciplinary approaches to scholarship and praxis.

Thus, with reference to the way the set design for *Ariodante* combines actual and imaginary elements, I examine how visual and choreographic disciplines function in a dialogical realist mode. With reference to a technical video recording, I explore how audio and visual discourses in opera can be mutually shaping within a production in order to intensify its affective experience, as well as jarring or going against one another to disrupt our perceptual flow of time and the production's metanarrative. This can be intentionally deployed for narrative effect, as breaking the imagined 'reality contract' in a Brechtian device of epic theatre,[5] or more frequently it betrays an unravelling of the immense artistic task of knitting all

---

[4]According to Christophe Den Tandt's developing theoretical framework for cultural forms, 'dialogical realism' as a mode of discourse is firstly heuristic (fact-finding): investigative, researching, and inter-medial in relation to our perceptual experience of the world. Secondly, it is reflexive (meta-discursive): for example, through self-embedment (which in the context of opera I understand to refer to the paradox of the simultaneous presence of the performer and their persona, as well as to a musical work and its interpretation); other examples of reflexivity include impersonation and pastiche. Thirdly, dialogical realism is contractual: implicit transactions make up the 'referential contract'; performative negotiations make up the 'reality contract'. Both must function simultaneously to support a psychologically congruent or plausible lifeworld. This is messy and inherently problematic to dialogical approaches; however, it is often because of these discrepancies that realism claims to open up 'an authentic search for truth'. Fourthly, dialogical realism is a mode in praxis (action-oriented): testing the limits of physical feasibility and the practical contingencies of situatedness. Following the logic of the 'reality bet' (the optimistic belief that consensus can be reached), dialogical realism is a performative practice of looking more closely at what we know, reminding us of it and transforming that knowledge into experience (Den Tandt, 2016).

[5]A familiar conventional example of this is the breaking of the 'fourth wall', where a performer steps out of a scene, or out of character, in order to address the audience directly.

of opera's component discourses and expressive means seamlessly together. None of this mapping, however, should be seen to diminish the considerable investment individuals make in navigating the live performance of a production's metanarrative, which must be uniquely embodied and can be seen as an act of orienteering. The staging of operatic events, meanwhile, can be said to extend beyond their performances in venues to engage individuals, communities, and heterotopias or 'other spaces'. This leads me to a consideration of opera's meta-stages, 'paratexts',[6] avatars, reception, and where any 'truth' in opera might exist.

15.2 'Performing' Space

The set design, by ULTZ,[7] for the 2014 Aix-en-Provence production of *Ariodante* is an interior cross section of a building inspired by photographs of seventeenth-century buildings on Scottish islands.[8] The entire set is framed within a white border, delineating the 'fourth wall'. White lines on the floor continue along the back wall and ceiling, marking out the absent interior walls of the kitchen and bedroom. The kitchen door, which opens onto the main room, is signified by an off-white door handle mounted on a hinged bar fixed to the floor, use of which effectively simulates the action of opening and closing a door without the physical presence of an actual door or door frame. This device is repeated on the other side of the main room in place of a door to the bedroom. In the screenshot from DNO's technical fixed-camera recording, taken at the beginning of Act II (Figure 15.1), the imagined kitchen door is open and the one to the bedroom is closed.

---

[6]'Paratexts' surround and extend the 'text', in this case an opera production, 'precisely in order to present it, in the usual sense of this verb but also in the strongest sense: to make present, to ensure the text's presence in the world, its "reception"' (Genette, 1997: 1).

[7]ULTZ is a well-known British designer (and director). The name, which he has legally adopted, is an acronym standing for Unity, Love, Truth, and Zeal.

[8]Photographs of some of the sources of inspiration for the design were included in the original accompanying production programme in Aix. When the production was revived in Amsterdam and Toronto, the programme images varied and the set then elicited additional associated references in press reviews. The design process involves what Den Tandt calls 'heuristic', inter-medial investigation.

Figure 15.1 Screenshot from the beginning of Act II, DNO technical wide-shot video.

The delineation of interior solid features, as an integrated design element, is intended to function dialogically with actual physical walls and doors and enables better sightlines across the set. The lines themselves perform as a haunting presence of objects removed from the actual building and also set up a readable convention for the imaginative space, providing the contractual terms of its integrity are observed by those inhabiting it (i.e. by not breaching imaginary walls or putting arms through imaginary doors and by 'playing the space' as if defined by walls and doors that are not actually present).

The design of imaginary architectural features enables the production to construct a level of simultaneous action across several contained spaces, where choreographed movement can be synchronized visually as well as musically, and it is a development from relying solely on musical cues in closed-box sets for opera. Therefore, the conventions and potential for the *mise en scène* to construct a kind of 'choreographed realism' are built into the design.[9]

The mechanism of the door without a door asks for both the cast's and an audience's complicity in the imagined reality of the world being portrayed, or in Strindberg's words from his preface

[9]This design feature became a point of interest for the marketing department at Canadian Opera Company (COC) in 2016, when I directed their revival of the production. A promotional 'insights' video for the production illustrates my rehearsal room interview with selected video clips from the staged production (COC, 2016).

to *Miss Julie*: 'Because the whole room and all its contents are not shown, there is a chance to guess at things — that is our imagination is stirred into complementing the vision' ([1888] 2008: 66). This is, of course, not unique to opera, but one of many spatial design features in theatre tailored to invite an audience to participate in the imagining of the metanarrative, or world of the production: an example of how visual and conceptual interest is hopefully created, drawing curiosity into a space. I would also suggest that the multiple divisions in the spatial design of the set play to our twenty-first century literacy of the semiotics of the screen and our capacity to read significance into simultaneity across multiple spaces. This resonates with the perception of simultaneous 'voices' and spaces possible in music through rhythm, melody, harmony, and dissonance, and in opera in particular, where distinct simultaneous voices and text are frequently discernible.

Transposing the setting for the opera to a 1970s island in the Hebrides allowed for a number of artistic licenses to be imagined, forming some of its contractual terms of reference and reality, not least the notion of the community outsider. The spoiler of that community is Polinesso, a Duke in the libretto, re-imagined as a travelling charlatan priest in the production. The supposed saviour is Ariodante, Ginevra's fiancé from a neighbouring island and as such nominated to succeed Il Re as king. This offered a given context in which to explore the nature of deception, judgment, sacrifice, and the resulting psychological fallout, rather than the traditional paradigm of redemption left open by the opera's ending. Polinesso capitalizes on the good-natured suggestibility of the Calvinist island community and its adherence to a rigid ideology of morality and gender, which is exposed and harnessed for ill. In one respect, the setting of the narrative provides a map for the dynamics of competing discourses and ideologies as well as the subsequent psychological investigation of those characters that navigate them.

15.3 Towards 'Dialogical Realism' in Opera Staging

I shall consider the production staging of Act II, scenes 9 and 10, with reference to my own vocal score annotated with production blocking (Figure 15.2) and a fixed-camera video recording of a performance of the production in Amsterdam, made as a technical record of the show.[10]

Figure 15.2 Part of scene 10 from the Bärenreiter vocal score with production blocking.

The video was not made for broadcast purposes, although it is of better quality than most technical records of productions (Figure 15.3).[11] Here I regard it as a record and map of how actual, physical elements of the set and props in the production design become vessels of imagined and symbolic meaning through a combination of heightening shifts in style and register in the libretto, scoring, lighting, and production choreography; all function dialogically to construct and propel the narrative towards

---

[10]Vocal score reproduced with the kind permission of the publisher from the Halle Handel Edition of *Ariodante* (BA4079-90) © 2007 Bärenreiter-Verlag Kassel and Aix-en-Provence Festival.

[11]Permission gratefully received for its reproduction with the courtesy of Dutch National Opera.

psychological congruence or dialogical realism.[12] In my reading of this staging, informed by my own creative practice in reviving the production, I shall show how design and choreographic features of visual discourses, such as the use of puppetry, framing, stage pictures, and the punctuation of movement in stillness, inform how we hear and experience the music of a staged opera, particularly in the treatment of repeats and Baroque musical structures, such as the *da capo* aria.[13] Audio and visual discourses, therefore, are mutually shaping and inform emotional and psychological engagement with operatic situations.

Figure 15.3 Excerpt from Act II, scene 9.
https://player.vimeo.com/video/275385557 (password: Musicology Research).

The scene begins at the pivotal moment towards the end of Act II when Il Re (the King) wrongly accuses his daughter Ginevra of immorality, and thus of being responsible for the death of her fiancé Ariodante, disgracing herself and the community: '*Non e mia figlia, una impudica*' ('Not my daughter, an immodest').[14] Il Re traverses

[12]A question persists about how much set is required in opera for there to be a believable environment in support of the metanarrative of the production, and perhaps distract enough from the technical mechanics of singing (increasingly exposed through camera close-ups), but this is beyond the scope of this chapter.

[13]A *da capo* aria is an accompanied song for solo voice with a musical structure of ABA, where the repeated A section is often embellished, ornamented, and sometimes even improvised by the singer in a display of their virtuosity. The ornamentation of *da capo* arias together with any motivation and staging are, in my experience, a question of taste and negotiated in rehearsal between performers, conductor, and director.

[14]My own literal translations.

the imaginary threshold to Ginevra's bedroom (now established as real to the world on stage). The main room is left populated with a community council still in shock from the news of Ariodante's sudden death in despair, as well as from his brother Lurcanio's call for justice in accusing Ginevra of being to blame. A secretly guilty Dalinda (Ginevra's confidante) and the misguidedly self-righteous Lurcanio are foregrounded in the main room. The audience crucially knows of Dalinda's guilt and Lurcanio's misdirected anger. The absent Polinesso has deceived Lurcanio, Dalinda, Il Re, and Ariodante, precipitating these events.

Ginevra stands on the bed. Disturbed out of unconsciousness by Lurcanio's denunciation of her, she has made it to her feet. Her innocence and moral purity are culturally signified through the vulnerability and whiteness of her costuming, skin tone, and red hair; she is the only person in the production seen in undergarments and she exposes more skin than anyone else. From an initially elevated position on the bed, Ginevra retreats to the wall when accused of being '*una impudica*' by her father, the attack endorsed by Lurcanio in the doorway. She challenges the accusation at arm's length to her father and king, who brandishes the damning evidence with an outstretched upstage arm that keeps the stage picture demonstrative, dynamic, and open to an audience (on stage and in the auditorium). This becomes a more intimate scene as Ginevra falls to her knees in an incredulous, vulnerable appeal at the foot end of the bed, interpreted by Lurcanio as 'proof' and testimony of witness to her 'immoral' conduct. The repetition and questioning of '*A me impudica?*' ('To me immodest?') turns to disbelief in the reality to which she has awoken in her text: '*Chi sei tu? Chi fu quegli? E chi son io?*' ('Who are you? Who was that? And who am I?'). The reality and promise of love, a morally pure and happy family life, indeed the whole atmosphere of joy and celebration at her engagement to Ariodante at the end of Act I, has turned into a waking nightmare for Ginevra. Dynamically, this loaded scene between father and daughter is staged in the bedroom in profile to the fourth wall and, as such, references painterly treatments of mythical and religious subjects in its spatial composition and lighting, intensifying the dramatic moment of Il Re's misguided denunciation of his daughter as a turning point in the opera.

The staging foregrounds Dalinda's interjections of '*O ciel! Che intesi?*' ('Oh Heavens! What intended?', an aside), '*Misera figlia!*' ('Miserable daughter!'), and '*Oh Dei!*' ('Oh Gods!'), positioning her downstage centre in the adjacent room. Whilst Dalinda's gaze is directed out to the audience and pierces the fourth wall, she remains anchored to the scene with the rest of the community by physical contact with the table in the main room. As a consequence of this dynamic tension in the direction of focus on stage, Dalinda comments dialogically and simultaneously on Ginevra's misfortune, the implications for Il Re and the community, as well as her own shameful predicament: having impersonated Ginevra at her lover's behest, she has precipitated Ginevra's denunciation and accounts of Ariodante's death. Dalinda's text '*Ohimè, delira!*' ('Oh delirious!') is ambiguously bracketed as an aside in the libretto. In the staging it is delivered to Il Re, in an impotent attempt to deflect his fury, and is loaded with dramatic irony; Dalinda was herself complicit in Ginevra's drugging in a silent play at the beginning of Act II.

Dalinda's text also sets up a transition that follows the brief image of innocent sacrifice, dismantled as Ginevra steps off the bed and passes into the main room, up onto the table for the heavily accented chords in the orchestra that signal a shift in the poetic and dramatic register of her text as she summons the Furies from Hades: '*Uscite dalla reggia di Dite! Furie, che più tardate?*' ('Get out of the palace of say! Rage, what more delay?'). The erupting orchestral fervour of the *accompagnato* (a form of *recitativo* with full orchestral accompaniment) is liberally embellished in the pit in support of the staging of the text that follows: '*Su, precipitate nell'Erebo profondo quanto d'amor voi retrovate al mondo*' ('Up, hurl into Hell's depths all of love you can find in the world').

The register of the staging shifts with the move from *recitativo*,[15] which has been mainly naturalistic in performance style, to *accompagnato* and a flurry of more expressionistic stage action. The

[15] *Recitativo*, or 'recitative' in English, is a style of delivery in which a singer adopts the rhythms of ordinary speech and ranges from *secco*, or 'dry', at one end of a spectrum where the singer is accompanied with minimal plucked or fretted instruments (standardized at the time of Handel as a harpsichord and viol or violoncello) through to *accompagnato, obbligato*, or *stromentato* at the other where the full orchestra is employed as an accompanying body and where sung ordinary speech bridges into something more song-like. This latter form is often used, as in this case, to underscore a particularly dramatic text.

chorus members, who throughout the production form a community that bears witness to, and endorses the wielding of moral judgment, move the table with Ginevra on it downstage, remove chairs from the main room, and furiously tear down the wedding decorations that were joyfully erected across the four spaces of the set in Act I. The whole sequence completes in just moments (five bars of music) disrupting our sense of naturalistic time. The chorus assembles upstage in a line across the back of the main room. Ginevra runs out onto the porch pursued by Dalinda over her appeal '*Principessa!*' ('Princess!'), followed by Il Re who stops in the kitchen to watch the women, focusing the stage picture for a punctuating extended *fermata* in the orchestra (at the end of bar 15).[16] There is stillness and silence across the stage and in the pit — a picture — a moment in which to take in what has just happened. Out of this comes a chord, embellished with some artistic license in performance by the lute and strings to serve the dramatic pacing of the scene and change of mood. The chord signals Ginevra's private questioning of space and meaning on the porch in the wake of Ariodante's death and accusation that she is to blame for it: '*Dov'é? Ch'il sa me'l dica! Che importa a me, se il mio bel sole è morto?*' ('Where is he? Who knows how to tell me! What does it matter to me, if my beautiful sun is dead?). The meaning of this text and Ginevra's position in the stage picture echo the staging of the ending of Ariodante's famous aria 'Scherza Infida' (Enjoy Yourself Unfaithful Woman), in which he bewails his betrothed's (supposed) infidelity, the last an audience has seen or heard from him.

The violence in the text and orchestral *accompagnato* finds metaphorical expression through a disruption to the temporal flow of the narrative, whilst remaining within the aesthetic and space of the production's own realist terms: the community scatters to tear down the heart-shaped decorations and reconfigure the furniture in the main room. This spatial reconfiguring shifts the aesthetic from the relative naturalism of the recitatives to a heightened

[16]A *fermata,* or pause, is a symbol of musical notation indicating that a note or rest should be prolonged beyond what the note value would normally indicate. Here, it is not printed in the score. I added it in pencil (as would the conductor in his score) during production rehearsals. As it occurs during *recitativo*, there is more artistic licence taken over pauses and the orchestra is held until the desired stage action is completed.

expressionistic space in which to stage Ginevra's aria 'Il Mio Crudel Martoro' (This Cruel Anguish). The video excerpt finishes there.

Although the transition between sequences described above occurs quickly in performance, detailed maps of the flight paths of cast and props are developed in rehearsal to achieve these scenic modulations safely and at the desired temperature and speed for dramatic momentum. The community's gaze and the staging of the aria that follows focus on Ginevra, whose movements quietly resonate with visual references to the Crucifixion story in part A, reminiscent of the choreographic language of Pina Bausch; in the B section, the staging echoes Ariodante's own position of despair on the floor against the front door in 'Scherza Infida', further emphasizing their spiritual connection, and in the *da capo* section of the aria,[17] she is dragged by Il Re from the kitchen floor across the main room and onto the table, where she is metaphorically sacrificed.

This heightened idiomatic staging has been progressively signalled and operates dialogically with Ginevra's text, music, and the sub-textual world that has been constructed to support it. The culmination of the stylistic modulation is a metatheatrical puppet show, titled 'Sin City' in rehearsal (Figure 15.4), in which the community morally passes judgment and denounces Ginevra at the end of Act II, during the opera's orchestral 'dances'.[18]

How the story of *Ariodante* is told in the production, how an audience is invited to be imaginatively complicit and emotionally involved in the telling, forms a significant part of the contractual basis of reference and reality, or imaginative truth of the world of the

[17] *Da capo*, meaning 'from the head, or beginning' here refers to the repeated section of a *da capo* aria, with a musical structure of ABA.

[18] Puppetry, as a metatheatrical narrative device, is also used during the dances at the end of Act I where the community presents a surprise puppet show finale to the celebrations that are staged to mark Ariodante and Ginevra's engagement, with puppet-size replicas of the singers and significant props featured during the action of Act I. The community performs an endearing backstory of how Ariodante and Ginevra met, projecting an innocent vision of a happy family life to come, under the banner of the Good Book. This first celebratory puppet show is reprised, under the blundering attempts of Il Re at the end of Act III to smooth over the atrocities that have happened and reunite Ariodante and Ginevra in marriage, becoming an expression of ideological intransigence in Il Re and community, and illuminating the oppression and suffering caused by religious dogma, among other things. Ginevra abandons the charade, rejecting both Ariodante and the island community in the production's significantly 'modern' solution to staging the opera's 'happy' ending.

production. The level of willingness of an audience's suspension of disbelief can only be 'bet' upon (to use Den Tandt's terminology) by stakeholders in a production through committing to what *they* can believe in and by participating in dialogical processes in both visual and musical languages.

Figure 15.4 'Sin City' puppet sequence — screenshot from the 'zoom' technical video, DNO.

Deleuze and Guattari conceive of music as being rhizomatic in the way it 'has always sent out lines of flight' (1976: 11).[19] Conflating the two analogies of rhizome and map, they conceive of connecting strata and dimensions of space that we experience as real but that are not always actual; space includes that experienced in the mind suggested, unfolded, or created by what we perceive, feel, and imagine: the augmented actual, the virtual, and the transcendent.[20] Modulations in the musical and visual register culminate in the puppetry sequences as 'virtual' to the reality of the onstage community in this production of *Ariodante*. They are staged as the imaginative expression of that community's joys, hopes, and fears at various points in the opera's story. Similarly, as for the onstage

[19]A rhizome is a continuously growing horizontal underground stem which puts out lateral shoots and adventitious roots at intervals. In *A Thousand Plateaus*, it is explored as a metaphor for networks of connections and meaning in contrast to the structural and metaphorical hierarchy of the tree.

[20]Two fascinating volumes inform my own understanding of Deleuze and Guattari's analogy here: Buchanan and Lambert (2005) and Buchanan and Swiboda (2004).

world, visual and musical 'lines of flight' traverse our own personal narratives as makers and audiences. It is at these intersections that we experience something meaningful, something real to us in our encounter with the performance of music. Visual information and contexts have been shown to influence what we hear in music at both cognitive and perceptual levels (Schutz, 2008: 91). Indeed, our other senses influence our experience and interpretation of sound and vice versa.[21] Whilst music is shaped and spaced in time, it arguably shapes our sense of time and space, or temporality (Kramer, 1988), connecting our sense of the present place to sounds and ideas from elsewhere or the past and even to notions of the future.

A significant operatic feature is the possibility of a congruent 'assemblage' of multiple, simultaneous voices and spaces within which cultural values are more or less cogently expressed and nuanced. Within this production, the metadiscursive puppet shows during the 'dances' at the end of each act are examples of Den Tandt's performative dialogical realism that imaginatively express and shape the reality of the community that creates them. We can see this relationship reach beyond the world of the production to reference actual spaces in the cities where the production was performed. The Act II 'Sin City' puppet sequence, for example, took on particular resonance when the production was performed in Amsterdam at DNO, not far from the city's famous red-light district.

Positioning opera productions within a discourse of staged 'events' that necessarily take in time, place, and the demographic of performers and audiences allows for a compelling expression of culture. Such territory in the twenty-first century is arguably becoming increasingly virtual, reaching well beyond the confines of actual space through the way contemporary performances enact affective interrogations. Such events, of course, do not exist in a void.

15.4 Towards 'Dialogical Performance'

Singers, music staff, conductors, and directors ideally arrive at production rehearsals 'knowing' the score to the best of their ability: what everyone is voicing in the libretto, in whatever language they are

[21]See the work of Charles Spence and Crossmodal Research Laboratory in experimental psychology. For a more light-hearted article, see Barton (2012).

singing, as well as having a personal understanding of what it means to them. Increasingly important are the skills necessary to explore the group's understanding and expression of character, phrasing, and subtext, together with others in their roles through rehearsals. For singers, this becomes about embodiment: assimilating, through a process of layering of information, the discursively mapped role into the voice, thought, and emotional architecture, movement, and relationships in scenes. This is an extraordinary feat of orienteering when accomplished successfully and considering the climate and constraints of international opera productions.[22]

Crucially, mapping and navigation also exist on the level of programming repertoire, detailing and deploying the producing company's resources, such as the availability of casts and staff members, rehearsal spaces, orchestra and chorus sessions, costume fittings, stage time, and so forth, in the form of the technical and rehearsal schedule. It is beyond the scope and expedience of this study to unpack just how important, unseen, and political the power struggles over this territory can be; however, in my experience, many of the most critical and diplomatic negotiations occupy these planning spaces and determine the relative dialogical success of opera productions.

On the level of the metanarrative (world of the production), the director hopefully mediates the various maps towards some consensus (generally in the interest of coherence, but sometimes intentionally not). This includes character motivations for flight paths through actual and imagined elements of the set, recorded as 'blocking'. Meanwhile, the conductor and director exert their influence over other stakeholders in the production, hopefully, but by no means consistently, towards the same end. This can be messy, difficult, combative, and wonderful. Diplomatically speaking, practitioners have a certain amount of agency within their defined professional and character roles, whereas personalities understandably manifest variously under pressure.

Opera rehearsals, in my opinion, should be about human beings navigating the musical and associated territories about negotiation and discovery. These are led at various phases in the schedule by

[22]In accord with my own view, another form of recent sector mapping has endorsed the holistic training of opera singers in the UK (where acting, physical training, and command of languages are not merely add-ons to vocal performance) to meet the growing demands of the profession (see GDA, 2016).

different roles. Crucially, they involve an act of faith in the process to generate a set of choices for staging opera that are plausible to the group of people assembled, resulting in the creation of a cohesive metanarrative, mapped in turn in the form of the production book. I would therefore agree, in theory, with Den Tandt's insistence on an open mind and faith in dialogical processes to reach consensus, although I must stress that, in practice, this manifests in participants in varying degrees under the pressures of time and within the constraints of available resources. Investigating the given material of the score and libretto from a number of different perspectives dialogically, when entered into in this way can, in my experience, lead to greater communion with the music, oneself, and others whilst encouraging a sense of purpose and commitment to a shared 'imaginative truth' for those involved. Therefore, counter to what one might instinctively believe about the heightened form of opera, staged productions can be made following the blueprint of a dialogical realist discourse, but they crucially require an act of faith in the process, and a clarity and acceptance of contractual terms, whether they hope to reference a naturalistic, expressionistic, and/or other aesthetic.

This latter point, regarding a production's artistic prerogative to create an imagined or virtual element of its 'reality' contract, is a major point of contention not only for new cast members in revivals, but with regards to audiences and in an opera production's reception in the media. Whether it is a new production or a revival, how the metanarrative is embodied or 'bought into' by cast members and how it is 'received' by audiences are always a gamble, relating to Den Tandt's 'reality bet'. Nevertheless, from the perspective of those performing in and producing opera productions, it is my view that live performances of opera should be seen as acts of cultural orienteering and faith in production maps to hold value and interpretive meaning.

15.5 Return as a Musical and Production Conceit

The style and directorial approach of this production of *Ariodante* is deliberately melancholic and intense, in the vein of the psychological realism of a play by Strindberg or Ibsen. The Baroque formal structure of *recitativo* and *da capo* aria under this approach becomes a framework for the psychological development of characters, so that each musical section emerges in narrative progression, allowing for a motivational excavation of tonal features, *ritornelli,* repeats, *coloratura* ornamentations, and *cadenzas.*[23] In the co-production revivals, staging 'real time thinking'[24] across musical repeats in an existing production presented a challenging task for new cast members. We needed to work together on 'rediscovering' reference points and character associations that cast members could find plausible in this particular configuration, anchoring thoughts and more abstract ideas within the physical space of the set for each musical figure and section.

The combination of thoughts 'in real time' and choreographed sequences, based on the production map, featured visual reminiscences, rhymes, and 'ghosts' or hauntings of scenic space. This production idea unfolds from the potentialities within Handel's eighteenth-century musical structures and the design of the set, already mapped out in my discussion of the 'performing space'. In the opera, the character Ariodante in particular reflects on the events of Act I, experiencing them differently during the course of Acts II and III; in this production, the same stage blocking is echoed at different times, generated and understood anew, as the contexts change in the light of the unfolding plot. With different knowledge at such junctures, Ariodante must repeatedly reinterpret the events in which he took part or to which he bore witness. Thus, the same blocking is invested with different meaning at different times in the opera through choreographic rhyme (as Ariodante retraces his steps) and visual reminiscence (from the audience's perspective). The visual grammar and choreographic use of rhymes in the

[23]*Ritornelli,* in this context, are orchestral passages between verses and sung phrases of an *aria* or song; *coloratura* literally means 'colouring' and in Baroque music refers to a technique of singing that requires florid musical runs, trills, and wide interval leaps; *cadenzas* in Baroque singing are improvised or written out ornamental passages allowing virtuosic display, generally occurring over the penultimate note of an important subsection or end of an aria, or duet.

[24]By this I mean to refer to 'units of thought' that translate into staged action, following Stanislavskian techniques, where the amount of time it takes for a character to believably think and act on thoughts as they might occur in a naturalistic way, and in response to a situation, need to be plotted alongside the score.

production were conceived to work alongside the Baroque musical features already listed above — as well as phrase structure, rhythms, repeats, and vocal embellishments — to invite a meditation on the opera's themes and their relevance to performers and audiences. This production convention is set up during the overture in a contemporary Brechtian interface with the public: the surtitles are used to project passages of Old Testament text (deliberately chosen for their 'misogyny' from the perspective of contemporary gender politics) that accompany a choreographed sermon, thus establishing the onstage community and setting on a remote Scottish island in the 1970s. The font and colour of the surtitles were chosen to suggest Old Testament quotations, later 'echoed' on stage in the banner that is unfurled for the culminating picture of the puppet show finale to Act I (Figure 15.5). The infiltration into the island community of Polinesso, cast as an ill-intentioned visiting preacher in this production, threatens the seeming innocence and order brought about by an ideologically unquestioned religious framework to the island community.

Figure 15.5 Screenshot of Act I finale: 'happy' puppet show.

Dancing and the heartfelt celebratory puppet show follow Ariodante's recitative '*Pare, ovunque mi aggiri, che incontri'l gaudio e'brio*' ('It seems wherever I turn, that I find joy and animation') and the event of Ginevra's appearance in her wedding dress for the finale of Act I. In Act II, the joy and truth of Ariodante's experience is then brought into question, through the insinuation that Ginevra is Polinesso's lover, in the recitative at the beginning of scene 2 that launches Ariodante's aria 'Tu Preparati a Morire' (Prepare to Die). Ariodante is then witness to a duplicity in the following recitative where Polinesso greets Dalinda dressed as Ginevra, as part of an unwitting sexual game, in a choreographic echo of the moment in Act I when Ariodante greeted Ginevra in her wedding dress. Lurcanio's intervention prevents Ariodante from impetuously taking his own life and then from confronting Ginevra. The awful irony highlighted in the staging is that during Lurcanio's aria 'Tu Vivi' (You Live) Polinesso rapes Dalinda in the bedroom (while she wears Ginevra's wedding dress). Ariodante, believing his bride to be unfaithful, progressively retreats from the site of these events; his psychological devastation is staged progressively across the main room, kitchen, and porch areas through the three respective musical sections (AB *da capo*) of the opera's central aria 'Scherza Infida'.

Whilst the set performs acoustically extremely well, emptied of people and action, it also serves as a visual echo chamber and signals the loss of joy and undoing of innocence, as Ariodante remembers and locates in the space where happier but now sullied memories were created in Act I. The staging of the aria is built around thought changes that trigger emotions occurring with harmonic shifts in the orchestration as well as sung text, *melisma*, and vocal ornamentations.[25] With no other visual 'distractions', the audience's focus is solely on the singer and their ability to tell a story through incorporation or embodying the music.[26] The significance of the staging, however, draws on the memory of what the audience has already witnessed occupying and animating the space for its dramatic context, supported by shifts in how the set is lit.

On his miraculous but lamentable return to the island in Act III, Ariodante revisits, decodes, and reinterprets the deception event with fresh knowledge garnered from Dalinda (that it was her, dressed as Ginevra he witnessed with the imposter Polinesso) in

[25]A *melisma* is an expressive vocal phrase or passage consisting of several notes sung to one syllable.

[26]I am mindful of using the word 'embodying' here in Varela, Thompson, and Rosch's (1993) organismic sense of the term, which may appear at odds with the use of the word in a materialistic sense; however, the point here is that both phenomenological experience and the representation of ideas are relevant to creative practice and staging performances of opera.

their recitative before his next big aria 'Cieca Notte' (Blind Night). In this complex scene, which is staged with Ariodante in the central room as a physical duet with Dalinda (Figure 15.6), Ariodante's experience of the 'ghosts' of the events in Acts I and II becomes increasingly investigative and heuristic; throughout sections A and B, Ariodante's outrage, suffering, and incredulity lead to uncovering the metonymic significance of the Bible in the main room as the means of deceiving the community, signalled as the point of focus throughout the *da capo* section of the aria. Dalinda also embodies echoes of the physical blocking from earlier scenes, such as Polinesso's duplicitous comforting of Ariodante in the B section of the latter's aria 'Tu Preparati a Morire'; however, now Dalinda's sympathy for Ariodante is genuine and full of remorse.

Figure 15.6 Screenshot of 'Cieca Notte'.

Meanwhile, juxtaposed in the adjacent bedroom, a simultaneous mock-exorcism of Ginevra is staged as the embodiment of Polinesso's exploitative charlatanism that succeeds in brazenly manipulating a community's fears. The community is both blind to and complicit in the actual damage being inflicted; recasting Ginevra as hysterical and possessed by the devil legitimizes Polinesso's opportunistic groping of her on the bed, as Il Re and the rest of the community clutch their Bibles.

Polinesso eventually gets his comeuppance, but the critique of a dour Calvinist discourse around the Fall, sacrifice, and redemption permeates the production. In Toronto, the staging of Ariodante's aria 'Dopo Notte' (After Night) evolved from Sarah Connolly's 'ecstatic' characterization through rehearsals with Alice Coote (for whom the production was new and somewhat alien); Alice wanted to 'anchor' Ariodante's psychological progression in the space, communicating his decoding of the meaning of objects within it to the community of onlookers. We took this path through the scene possibly because Alice was struggling to embody certain aspects of the production's very particular reading of the piece in the absence of the original director (for whom she has great affection), along with dealing with huge emotional stress in her personal life; however, the results were compelling and gave Alice a visceral framework for what at this point in the evening is a monumental task vocally and emotionally for any singer of the role. The additional blocking created is a practical example of how ideas *constellate* around objects to form a realist narrative and, in our situated telling of the story, how Ariodante makes sense of his surroundings and feelings in the moment. They were added to the production book and adapted for the revival in Chicago, which developed the idea further.[27]

In the opera, following 'Dopo Notte', Ginevra is confronted all at once with the miraculous return of her 'dead' fiancé, apologies from her father Il Re, Lurcanio her accuser, and Dalinda, her duplicitous

[27]During the staging of 'Dopo Notte' in Toronto, the photograph of Ginevra on the wall as an innocent child — which Polinesso had corrupted in Ariodante's mind and Il Re had taken off the wall to caress with lamenting nostalgia after the public disgrace of his daughter in his aria (no. 38) 'Al Sen ti Stringo, e Parto' (I Embrace You and Depart) — is taken up anew; the purity of the libelled image is now restored in Ariodante's mind, whilst the community is reminded of what they have put Ginevra through (unbeknownst to Ariodante) as they and the rest of the principal characters experience shame. The stack of Bibles upon which Polinesso rests his chin during the same aria of Il Re (no. 38) is toppled and scattered in the *da capo* of 'Dopo Notte', in Ariodante's attempt to communicate the folly of religious dogma to the community. Working together, Alice and I also 'discovered' space for a final 'transcendent' moment of forgiveness between Ariodante and Dalinda that ties up an otherwise loose narrative thread in their relationship. During the 'B' section of the aria, Dalinda burns the dead Polinesso's clothes in the kitchen, whilst during the playout of the aria, in the Chicago version of the production, we restaged the scene so that Ariodante unites Dalinda and Lurcanio before following the community into Ginevra's bedroom, tying up the tentatively redemptive subplot, as their duet was one of the casualties of cuts that needed to be made in Chicago. The tragedy of this trajectory for Ariodante is precisely that he has not witnessed and has no knowledge of just how shamed and exploited symbolically Ginevra has been.

confidante. The production plays out to its psychological conclusion within a broader social narrative that details devastating individual casualties and denial within the patriarchy, with a final act of female liberation and rejection of the atrocities caused by men, forbidden desire, and unyielding religious dogma. In the end, the community attempts to recreate the celebrations of Act I (Figure 15.7) as if all can simply be forgotten because the two 'innocent' lovers are reunited. Ginevra's dismay, objection, and final packing of her suitcase to leave the island — by breaking the fourth wall and stepping out of the set — is a Brechtian theatrical coup, recognized in the press as a 'postmodern' solution to resolving a morally 'problematic' end to the opera. Ariodante is left numbed, bereft, and impotent in the wake of Ginevra breaching the frame of island kingdom.

Figure 15.7 Excerpt from the reprise of the Act I puppet show in Act III. https://vimeo.com/408037259 (password: Musicology Research).

15.6 Beyond the Physical Stage

Performances of opera reverberate beyond the space of the physical stage to occupy other virtual territories through their paratexts in programme notes and avatars in digital broadcasts, DVDs, marketing show reels, 'insight' promotional material, on social and other recorded media, such as YouTube, or the technical video made by DNO used in this case study. The reception of these performances and their virtual reincarnations continues the process of interpretation and intersection of individual narratives (that we each bring to the event) with the metanarrative of the opera production, igniting further interplay between actual and virtual worlds, be it by an audience in a venue on the night, in the press, on television, YouTube, Twitter, Facebook, or other virtual presence. The more intersections of personal narrative with the metanarrative there are, the greater the chance of connection between them, following the 'reality bet'. The physical touring and international co-productions of opera, often 'revived' with new company and cast members, create further cultural ripples in their spatio-temporal conjunctions with new audiences.

In the midst of the well-documented and debated circumstances of the opening night of *Ariodante* in Aix-en-Provence in 2014,[28] where it was targeted for disturbance and sabotage amidst national demonstrations on behalf of the *intermittents du spectacle* in France, a large portion of the metanarrative, such as the use of puppetry and Polinesso's disguise of religious authority in the production, was picked up and commented upon in the press. The publicity and marketing departments' involvement, wherever a production is presented, adds another level of discourse that clearly steers, frames, and inflects the space of any production's reception.[29] Marketing strategies open up other debates and 'lines of flight' around engagement and impact.[30]

Reviewers that were sympathetic towards Richard Jones' production of *Ariodante* in Toronto, when I directed it for Canadian Opera Company in 2016, focused on local resonances with the island setting and *how* the story was told, citing its use of puppetry, choreography, 'postmodern' treatment, and ending (informed through direct relationships with the company's public and media relations department, interviews with the production team, the

[28]Whilst there is much to say on this subject, it is unfortunately outside the scope of this chapter. Sarah Connolly, who sang the title role in both Aix and Amsterdam, wrote her own vivid account of the evening of the first performance (Connolly, 2014).

[29]See the *Globe and Mail*'s preview article by Robert Harris (2016a) subsequent to separate interviews with me and the conductor, Johannes Debus, as an example of an informed and sympathetic framing of the production in Toronto.

[30]For relevant recent marketing material on 'culture segments' and the mapping of audiences' receptiveness to and engagement with arts and heritage sectors, see MHM (n.d.).

conductor, and members of the cast). For example, where the 'Sin City' puppet sequence at the end of Act II was particularly noted and enjoyed by company members in the Netherlands as referencing Amsterdam's neighbouring red-light district, in Toronto the same was described salaciously in the press to generate controversy, or an overall parallel was drawn between the production's metanarrative and the communities and culture of the Canadian Maritimes.[31] Just how much resonated with the thousands of audience members in each of the co-producing venues forms part of Den Tandt's 'reality bet' and remains resistant to measurement and codification, beyond the scale of the applause, polling audience's immediate reactions, or studies that have attempted to measure the longer term impact of the arts in general.[32]

Negative, reactionary reviews typically aimed at the production rather than the music, conductor, or singers (who were almost universally praised for the quality and suitability of their voices in

---

[31]For example, Michael Vincent's review for the Toronto *Star*: 'A puppet on a stripper's pole. A wall of knives and invisible doors. A tattoo-covered priest in a "Canadian tuxedo" and combat boots with a kink for sniffing women's knickers... Opinions will surely be divided, but with a show as dramatically complex as this was, it is also exactly why it is so deliciously interesting' (Vincent, 2016). Contrast with the *Globe and Mail*'s considered endorsement by Robert Harris in Toronto, 'The decisions Jones has made to update and deepen the resonances of the opera work beautifully both to preserve the integrity of the original and add to it touches and textures that only a modern audience can appreciate. Those resonances begin with the setting. Jones has taken the story of Ariodante, a simple Renaissance tale of jealousy, deception and eventual reconciliation, and placed it in the suffocating world of the 1970s Outer Hebrides of Scotland. Intentionally or not, a Canadian audience cannot fail to see in these homespun clothes, immediate passions and narrow, dogmatic, community life echoes of a Newfoundland outport or isolated Maritime village... Then there's the set for *Ariodante*, a clever tripartite arrangement, three rooms divided by the simplest of gates that allow action to proceed in all three places simultaneously, allowing for a wealth of psychological suggestions and counter-suggestions that immediately modernize the original. Those gates are immensely symbolic in Jones's productions, which is all about doors opening, and more often closing, in the community's life — closing on people, ideas, forgiveness, faith... If you needed one example to demonstrate why modern staging and perfectly realized music from the past need each other, this was it' (Harris, 2016b).

[32]There have been many studies into these areas using both quantitative and qualitative measurement, often deployed politically to endorse the cultural value of the arts beyond the economistic value-for-money argument made in response to devastating cuts in state funding. It has also been acknowledged, however, that there will always be an element of experiencing the arts that remains subjective and elusive to measurement (see Shishkova, 2015).

Toronto) largely betray the writers' own aesthetic tastes (which tend to be rather literal when it comes to assessing visual components of the performance), exhibiting self-aggrandizing prejudices, and a pejorative, if amusing writing style.[33] Rarely, in this case, are reviews balanced and neutral; however, several reviewers provide caveats to temper their frustrations with the production in phrases such as 'taken on its own terms' and 'within its dramatic suppositions' (Sohre, 2016). These, at least, acknowledge the contractual parameters of an approach that licenses artistic prerogatives of the production, but to my mind fall back on preconceived or pre-drawn maps of the operatic work and a vaguely envisioned spectacle, while citing some rather disembodied notion of 'the music'.

Audio recordings and radio broadcasts of opera performances allow listeners to imagine their own visual fantasies, arguably appealing further to an audience's own narrative experience, sensibilities, and listening environments to engage with them. As a staging practitioner, my own frustrations with reviewers (and collaborators) arise at the lack of self-proclaimed bias and political agenda that so often colours a sense of 'ownership' (informed or otherwise) of the material. This, to my mind, remains an obstacle to accepting the contractual terms of dialogical realism (on both sides of the pit), influencing the odds of the 'reality bet'.

15.7 Re-Conception, Reception, and Controversy in Chicago

It is worth reporting in detail some of my observations and experiences of directing this production's most recent revival, as

[33]One example was James Sohre's review for *Opera Today*: 'the updated "realm" consists of a massive, unattractive setting that is one large part community meeting hall and one small part private residence. Well, "residence" in the sense that Ginerva's bedroom and a cramped, ill-used "foyer" were all that were seen other than the rather primitive common room. From the numbingly ugly bedroom wallpaper, to the floating doorknobs that open/close non-existent "doors," to the confusing configuration of entrances, this was a depressing, intentionally dull atmosphere, meant to convey a mandated routine and an oppressive societal structure. Set designer ULTZ was also responsible for the drab, purposefully provincial costumes. The attire was at times confusing (chorus women were dressed as men but danced as women), at best functional (such as the wedding gown that gets passed around), and at worst, defeating (the titular prince looks like a hang-dog village simpleton)' (Sohre, 2016).

its reception caused significant controversy. The running time of the production (with the necessarily longer interval breaks at Lyric Opera of Chicago (LOC)) was going to come dangerously close to the contractual limits of the orchestra's hours for performances scheduled for that season. Amidst contractual renegotiations with AGMA and the orchestra,[34] and the unconscionable risk of triggering overtime payments, the management decided that twenty minutes was needed to be cut from the opera. Inevitably, this had a significant impact, but after some discussion via email, the conductor (Harry Bickett) and the production's original director (Richard Jones) approved the cuts I was asked to propose and we were able to calculate bringing the show in, with two 25-minute intervals, safely under four hours. This was to be the first time LOC had ever performed the opera but the production's fourth outing for its final co-producing venue.[35]

The venue itself, one of the largest in North America, presented other challenges that needed addressing, such as in the scale and legibility of 'naturalistic' details. Physical gestures, props, puppetry sequences, sightlines, and written text (the gothic font used for the Bible quotations on the surtitle screen during the overture and on the banner unfurled as part of the puppetry sequences in Acts I and III) were all subsequently reconceived.

In anticipation of the challenging nature of the production (based on its previous incarnations), General Director Anthony

[34]AGMA is the unionized American Guild of Musical Artists.

[35]With a seating capacity of 3563, the Lyric Opera of Chicago is the second largest opera auditorium in North America. Dramaturgically speaking, the cuts served to advance the action of the story. The Gavotte in the overture was cut, as was scene 4 in its entirety (Polinesso's recitative and aria 'Coperta la Frode' (If Deception is Covered)). The middle and *da capo* sections of Il Re's Act I aria ('Voli colla sua Tromba' (Sound the Trumpets)) were cut, meaning the celebratory decorations erected across the set in the production needed to happen more quickly to transform the space. The orchestral ballets that preceded the puppet shows (and were staged as Scottish dances) were cut, as were a couple of other orchestral sinfonias, Ginevra's short 'resurrection' recitative and arioso, the whole of the Dalinda-Lurcanio duet, and middle and *da capo* sections of the Ginevra-Ariodante duet in Act III. The silent plays at the beginning of Acts II and III of the production were also substantially contracted. Musical casualties and transitions aside, the overall impact of the re-conception of the show did make it flow more like a play and 'feel' less long to us as modern-day storytellers. In that respect, the cuts worked in the production's favour. The lighting designer preferred this version and, as a production team, we agreed that the show had never felt more polished in spite of regrettable musical omissions.

Freud confided in a conversation with me that, in his opinion, audiences at LOC follow a path to shutting down new productions if they are not clearly signposted: 'I don't understand it' leads quickly to 'it makes me feel stupid' and 'I don't like to feel stupid', which converts to 'it is stupid; reject and condemn it as stupid'. With this in mind, I was invited to write an accessible synopsis and 'Director's Note' explaining the production's approach to staging *Ariodante*, to be published in the programme (as I had done for COC). The programme also featured a more scholarly article on the history and musical features of the opera by Lyric's resident dramaturg, Roger Pines, as well as an article by Meg Huskins filling the programme's 'Modern Match' slot (an attempt to reach out to audiences through popular cultural references) in which structural and thematic comparisons were drawn between *Ariodante* and modern American television shows such as *Glee*, *Smash*, and *Crazy Ex-Girlfriend*. This particular attempt to engage American audiences was new to me, may seem rather crass to European eyes, 'dumbing down' to the *cognoscenti* or indeed laudable to those in public relations. In any case, I read it as an attempt to increase accessibility to the art form by providing relatable cultural context, orienting potential audiences, or signposting Den Tandt's so-called 'referential contract'; however, further research on the effectiveness of such strategies to engage (or inadvertently alienate) audiences of opera would be valuable.

At the post-performance event, along with customary congratulatory feedback, I was somewhat press-ganged by a group of sponsors who had many questions they wanted answering. Whilst some were clearly disturbed by the production choices (the casting of Polinesso as a charlatan priest and the staged sexual violence against Dalinda were predictably contentious), all those with whom I had quite heated discussions by the end expressed how much they had appreciated the evening. Questions included the perennial 'Why update the setting?' as well as 'Is the message anti-church?' and 'Why did Polinesso have to be a priest?' I came away with a sense that the production had generated strong reactions and heated debate; nonetheless, I was thankful for the somewhat gruelling opportunity to defend the production, whilst championing the cast, and heartened by the power of post-show dialogues with the cast and creative team to meaningfully engage with audiences and sponsors.

The press notices in Chicago were polarized, in a similar vein to those in Toronto, although more extreme. Those reviewers that were positive about the performances and production were wholehearted in their praise, awarding maximum stars in the *Chicago Tribune*: 'Vocally, visually and dramatically arresting' (Reich, 2019),[36] and *Bachtrack.com*: 'A vibrant, urgent *Ariodante* at Lyric Opera of Chicago' employing phrases such as: 'platinum quality in every sense' and 'top tier musical-dramatic figure' (Syer, 2019). Meanwhile elsewhere, vitriolic criticism was levelled at the production's 'overrated English directors who foist their solipsistic outrages on great operas' (Johnson, 2019) and British General Director of LOC (Anthony Freud) for bringing 'Eurotrash' to Chicago, resonating ironically with the production's conceit of an outsider (Polinesso) who causes havoc to the moral compass of a community.[37]

One consistent feature of the published criticisms levelled at the production in Chicago (and Toronto) is a relish in descriptive details of 'offensive' or 'crass' sexual content amongst the crimes against Handel's 'beautiful' music. To my mind, fixation on these details (in sequences totalling less than ten minutes of the four-hour evening) betrays an author's political and aesthetic bias, in this instance towards conservatism. Moreover, what such moral outrage reveals under critical scrutiny is an underpinning set of values that sound alarm bells in the ears of those that have taken them to heart: deafening and blinding them to other perspectives, colouring their whole experience. The use of language condemning production choices as 'adolescent' versus the 'adult' audiences of Chicago (in a blanket reaction to sexual content and an emotional rejection of violence on stage in opera) is a noticeable difference to the production's press reception in more liberal societies, such as Amsterdam's. One irony of this is that Chicago is a city renowned for its liberal theatre scene and audiences, within the same vicinity

[36]The full review, 'A Startling Update of Handel's *Ariodante*', by Howard Reich (2019) is available to those in the USA on the *Chicago Tribune* website; however, content from this website is unavailable in most European countries.

[37]Lawrence A. Johnson's (2019) well-observed and highly polemical review 'Worthy Handel Singing Buried by Scabrous Excess in Lyric Opera's *Ariodante*' for online publication *Chicago Classical Review* also features a comments page with readers' posts mostly concurring with Johnson's view of the production.

as the Lyric Opera.[38] Another aspect of this somewhat entrenched political divide (around the relative (im)maturity of opera directors and audiences) is that (unsurprisingly) only positive reviews of productions are quoted on the LOC website and Facebook page; marketing is performative and good notices (rather than controversy or debate) are still relied upon to sell tickets to diverse audiences.

The more considered reviews situate the production within its wider political contexts, as well as detailing external factors that contribute to its realization, such as singers' individual circumstances, as distinct from production ideas about character.[39] They consider a production's scheduling within a season of operas, and they voice the values they expect to be honoured by an opera-producing company in their city.[40] These are less common and cut to the chase: in directly addressing the values of politically divided nations, art made according to a contemporary realist blueprint is going to divide opinion.

As a last, ethically sensitive detail, I include my reflections on a telephone conversation I had with Anthony Freud on my return to the UK after the second performance (where Alice Coote performed the role of Ariodante with aplomb, by most accounts, in spite of suffering from late-diagnosed pneumonia). Two details of the production were thought (by Anthony Freud) to be 'the straw that broke the camel's back' for a reactionary segment of the audience base: the bruises on Dalinda's body after a stylized but violent

[38]That such different music, theatre, and opera cultures co-exist in Chicago and audience members overlap (to an extent) was fascinating to me, particularly in the light of an unforgettably inclusive evening I spent at Chicago Blues venue Kingston Mines (in between the dress rehearsal and opening night of *Ariodante* in 2019) where local artist Nellie Travis gave a witty and affecting performance of Blues standard 'Oil and Water Don't Mix' (see Travis (2009) for a recording of a similar performance).

[39]Examples cited in the press include Alice Coote's unfortunate ill health (and cancellation) on the opening night, and the fact that Brenda Rae (by all accounts) gave a convincing performance in the role of Ginevra whilst she was six months pregnant with her second child.

[40]For example, Hannah De Priest's (2019) positive review for *Schmopera.com* concludes: 'All in all, this *Ariodante* exemplifies what I hope to see more of from the Lyric in future seasons: creative, specific staging performed by committed and compelling singer-actors, with top-notch direction in the pit. Especially when compared with *Idomeneo* and *La Traviata* this season, in which singers with few observable dramatic instincts were set adrift in productions that felt half-baked, *Ariodante* pulsed with emotion and suspense — no small feat for a lesser-known Handel opera clocking in at around the four-hour mark.'

sexual encounter with Polinesso, and the latter's drawing of an erect penis — the sketch planted with others in Ginevra's bedroom to incriminate her as immoral.[41] In considering sponsor criticisms and sensitivities, Anthony and I took the pragmatic decision together to cut that particular drawing and dull down the make-up of Dalinda's bruises for the rest of the scheduled performances. Neither alteration in my opinion (which I know would not have been shared by everyone in the production team) would severely impact the production narrative; however, the affected singers were understandably unhappy about the changes (because they had integrated and been committed to these elements in their universally praised performances). Weighing up the merits of such an artistic intervention, whether these changes would indeed render the production more palatable to LOC audiences is difficult to claim, but I understood this as a concession that Anthony could report to sponsors who threatened to withdraw their patronage in a move to keep the conversation open (and money flowing). On reflection, I am not so sure I would have reached the same decision based on similar feedback in a European country because of my own bias towards how opera is still (at least partly) funded by the state in relatively liberal societies, where the creative remit is oriented more overtly towards challenging as well as delighting audiences. To my mind, these are among the more nuanced political and diplomatic decisions taken in running an opera company (particularly in America), balancing

[41]The idea of portraying Ginevra as an amateur artist was not in the Aix version but was added to the production late in the process in Amsterdam, during its first co-production revival, when Richard Jones joined for the stage rehearsals. Ginevra, in her bedroom, drew a charcoal portrait of Ariodante during his first aria 'Qui d'Amor nel suo Linguaggio' (Here of Love in its Language), presenting it to him at the end. Following Ariodante's celebratory aria 'Con l'Ali di Costanza' (With Wings of Constancy), Dalinda then tidied away the drawing in a scene with Polinesso, who noticed it, later planting a number of additional, lewd images in Ginevra's bag (during his Act II aria 'Se l'Inganno Sortisce Felice' (If My Deception is Successful)) as a way of discrediting her. The drawings of male nudes were then discovered by Lurcanio (her accuser) and brandished as proof of debauchery by the community to condemn Ginevra at the end of the Sin City Act II puppet show. On a metaphorical level, this staging was an attempt to illustrate how the community turns on and vilifies those with imagination (celebrated in Act I) when they appear to challenge its codes of morality. The 'pornographic' drawings, therefore, were a considered directorial choice to make an important conceptual and political point (whether in good taste is clearly debatable, but the decision was not made on an 'adolescent' whim).

stakeholder expectations and tastes, while demonstrating a desire to educate and continue dialogue between artists and audiences.

15.8 Conclusion

As a group of practitioners and makers of productions, we aim to make the work in which *we* can believe as artists gathered from an international and interdisciplinary community, albeit a relatively well educated and privileged one. When it comes to realism in opera, as perhaps any form, the question gravitates towards *whose* realism is under scrutiny. An opera production inevitably communicates cultural values that are variously affirmed, ignored, or challenged on the level of the production's reception and in the press. These discourses can, in their turn, be the subject of analysis in a continuation of the dialogical spaces generated by the production, particularly as it not only travels to but is also revived by different co-producing houses for their opera-going communities. Occasionally, changes to the production itself will be made as a result of its critical reception. It is, however, in the virtual spaces of social media and the blogosphere as much as academic circles that the critics themselves may be critiqued, although the impact of negative notices on the perceived success of opera productions, and on the careers of those involved, is rarely held to account.

Ultimately, operatic hermeneutics produce spaces from the interplay of actual and virtual elements that constellate around musical form. Opera, as a total art form, demands an open and complex engagement with the virtual as an imaginative exploration of the real or of reality as experienced by human beings — a co-created hallucination, if you will; it tests the limits of the contractual parameters of reference and reality, becoming most powerfully affecting, plausible, and 'real' when its metanarrative intersects with the personal narratives we each bring to the event. From the cartographic processes involved in creating and promoting a production to the varied off-shoot forms, paratexts, and commentary that reinterpret it, the increasingly dialogical staging of opera repositions the performer, critic, and audience (in the broadest sense) as cultural orienteers and performers in their own reality matrix.

In conclusion, I return to the quotation from *A Thousand Plateaus*, which is a definition of music, of the map, of the way art reimagines what we perceive to be real in order to glimpse at some truth about our sense of space and place in the world. Whether it is to our taste, resonant or dissonant to an audience's set of values, staged opera can be both poetic and real to the extent to which it is given meaningful context, with reference to actual forms and human experience, and is grounded by them in our understanding. In my opinion, the 'truth' of music as performance and of any MAP, therefore, is to be sought in the live encounter and interplay of actual and virtual worlds. Opera may be uniquely positioned to explore this territory because its heightened idiom continues to test the boundaries of our 'willing suspension of disbelief' in its propositions and because of the potential dialogical richness of its expressive means and staging in the twenty-first century.

References

Auslander, P. 2013. 'Music as Performance: The Disciplinary Dilemma Revisited'. In *Taking it to the Bridge: Music as Performance*, edited by N. Cook, and R. Pettengill, 349–357. Michigan: University of Michigan Press.

Bakhtin, M. 1981. *The Dialogic Imagination: Four Essays*, edited by M. Holquist. Fifth Edition. Austin: University of Texas Press.

Buchanan, I., and G. Lambert, eds. 2005. *Deleuze and Space*. Edinburgh: Edinburgh University Press.

Buchanan, I., and M. Swiboda, eds. 2004. *Deleuze and Music*. Edinburgh: Edinburgh University Press.

Deleuze, G., and F. Guattari. 1976. *A Thousand Plateaus*. London: Eulenberg.

Den Tandt, C. 2016. *On Virtual Grounds: Blueprint for a Postmimetic, Dialogical Realism*. Unpublished working paper. Available online at https://www.academia.edu/11861833/On_Virtual_Grounds_Blueprint_for_a_Postmimetic_Dialogical_Realism

Gee, J. 2005. *An Introduction to Discourse Analysis: Theory and Method*. New York: Routledge.

Genette, G. 1997. *Palimpsests: Literature in the Second Degree*, translated by C. Newman, and C. Doubinsky. Lincoln: University of Nebraska Press.

Kramer, J. 1988. *The Time of Music: New Meanings, New Temporalities, New Listening Strategies*. New York: Schirmer Books.

Isaacs, W. 1999. *Dialogue and the Art of Thinking Together: A Pioneering Approach to Communicating in Business and in Life*. New York: Currency.

Schutz, M. 2008. 'Seeing Music? What Musicians Need to Know about Vision', *Empirical Musicology Review* 3(3): 83–108.

Strindberg, A. [1888] 2008. '*Miss Julie*, Preface'. In *Miss Julie and Other Plays*, translated and edited by M. Robinson. Oxford: Oxford University Press.

Tallis, R. 1988. *In Defence of Realism*. London: Edward Arnold.

Van Leeuwen, T. 2009. 'Discourse as the Recontextualization of Social Practice: A Guide'. In *Methods of Critical Discourse Analysis*, edited by R. Wodak, and M. Myer, 144–161. Second Edition. London: Sage.

Varela, F., E. Thompson, and E. Rosch. 1993. *The Embodied Mind: Cognitive Science and Human Experience*. Cambridge, MA: The MIT Press.

Other Media

Barton, C. 22 October 2012. 'How Sound and Smell Can Create Perfect Harmony' (news report, The *Guardian*). Available online at https://www.theguardian.com/science/2012/oct/22/sound-and-smell-create-harmony

COC [Canadian Opera Company]. 2016. 'Inside Opera: Ariodante' (online video). Available online at https://www.youtube.com/watch?v=JXbIUzIS98o&list=PL7KMuQtaRtozfCqPez2Ib6U78Cq3QubPh&index=22Sarah

Connolly, S. 8 July 2014. 'Nightmare in Aix: Sarah Connolly on a Shocking First Night' (online review, *Theartsdesk.com*). Available online at http://www.theartsdesk.com/opera/nightmare-aix-sarah-connolly-shocking-first-night

De Priest, H. 6 March 2019. 'Chicago *Ariodante* Pulses with Emotion and Suspense' (online review, *Schmopera.com*). Available online at https://www.schmopera.com/chicago-ariodante-pulses-with-emotion-and-suspense/

GDA [Graham Devlin Associates]. 2016. *Opera Training for Singers in the UK: How Should It Evolve to Meet the Changing Needs of the Profession?* (commissioned by National Opera Studio). Available online at https://www.nationaloperastudio.org.uk/News/opera-training-report

Harris, R. 14 October 2016a. 'COC Brings Handel's Ariodante to 1960s Scotland' (newspaper preview, The *Globe and Mail*). Available online at https://beta.theglobeandmail.com/arts/theatre-and-performance/coc-brings-handels-ariodante-to-1960s-scotland/article32361072/?ref=http://www.theglobeandmail.com&

_____. 18 October 2016b. 'COC Beautifully Updates Handel's Ariodante for the Modern Stage' (newspaper review, The *Globe and Mail*). Available online at https://beta.theglobeandmail.com/arts/theatre-and-performance/theatre-reviews/coc-beautifully-updates-handels-ariodante-for-the-modern-stage/article32409878/?ref=http://www.theglobeandmail.com&

Johnson, L. 3 March 2019. 'Worthy Handel Singing Buried by Scabrous Excess in Lyric Opera's *Ariodante*', (online review, *Chicago Classical Review*). Available online at http://chicagoclassicalreview.com/2019/03/worthy-handel-singing-buried-by-scabrous-excess-in-lyric-operas-ariodante/

MHM [Morris Hargreaves McIntyre]. n.d. 'Culture Segments' (marketing material). Available online at http://mhminsight.com/articles/culture-segments-1179

Opera Europa. n.d. 'Opera, the Art of Emotions' (website). Available online at http://operavision.eu/en/library/stories/opera-art-emotions (last accessed 10 July 2017)

Reich, H. 3 March 2019. 'A Startling Update of Handel's *Ariodante*' (newspaper review, *Chicago Tribune*). Available online at https://www.chicagotribune.com/entertainment/music/howard-reich/ct-ent-lyric-ariodante-review-0304-story.html

Shishkova, V. 2015. *General Mapping of Types of Impact Research in the Performing Arts Sector (2005–2015)* (report, International Network for Contemporary Performing Arts). Available online at https://www.ietm.org/en/publications/mapping-of-types-of-impact-research-in-the-performing-arts-sector-2005-2015

Sohre, J. 26 October 2016. 'COC'd Up *Ariodante*' (online review, *Opera Today*). Available online at http://www.operatoday.com/content/2016/10/cocd_up_ariodan.php

Syer, K. 6 March 2019. 'A Vibrant, Urgent Ariodante at Lyric Opera of Chicago' (online review, *Bachtrack.com*). Available online at https://bachtrack.com/review-handel-ariodante-davies-rae-bicket-lyric-opera-chicago-march-2019

Travis, N. 2009. 'Oil and Water Don't Mix' performance at Kingston Mines (online video). Available online at https://www.youtube.com/watch?v=BsDX5t5nGP0

Vincent, M. 17 October 2016. 'COC's Ariodante Offers Audiences Plenty to Ponder' (newspaper review, The *Star*). Available online at https://www.thestar.com/entertainment/stage/2016/10/17/cocs-ariodante-offers-audiences-plenty-to-ponder.html (last accessed 10 July 2017)

Chapter 16

Musicians in Place and Space: Impact of a Spatialized Model of Improvised Music Performance

David Leahy
Centre for Research and Education in Arts and Media (CREAM), University of Westminster, Harrow Campus, Watford Road, Northwick Park, London, HA1 3TP, UK
dafmusic@gmail.com

16.1 Introduction

Free improvisation is often explained as a non-hierarchical musical process that emerges out of the precise acoustic, emotional, environmental, psychological, and social conditions in existence at the time of the music's creation. But given that free improvisers continue to perform in conditions that involve the static positioning and formal separation of the performer and the audience, the extent to which these claims can be realized is questioned by this chapter. I report on a practice-based research project, 'Musicians in Space' (MiS), which aims to provide insights into how the free improviser

Musical Spaces: Place, Performance, and Power
Edited by James Williams and Samuel Horlor

ISBN 978-981-4877-85-5 (Hardcover), 978-1-003-18041-8 (eBook)
www.jennystanford.com

can develop an all-encompassing and non-hierarchical musicking practice by offering the performers and audience the opportunity to shape their personal listening experience by modifying their spatial location during the performance. An outline of the research process and the findings from the first stage of the research are given, before the discussion is extended to look at the connection between spatialized free improvisation and deep ecology.[1]

16.2 Free Improvised Music

Free improvised music emerged in Europe during the mid-1960s, with musicians eager to strip away the performative expectations and restrictions that they were used to (Morris, 2012: 100). Inspired by developments in free jazz and contemporary Western classical music, the music is constructed through a dialogical process where 'everybody's actions and ideas impact everybody else's' (Vargas, 2013: 25). The roles of soloist and accompanist are shared fluidly amongst the improvising participants, with the musicians utilizing any musical, technological, or structural means at their disposal to maintain the flow of the music. This is achieved with little or no reference to any identifiable melody, rhythm, or harmonic structure. Instead, improvisers engage in a process of music-making that pushes at the boundaries of the known and recognizable, and this regularly involves subverting what is expected, led by the 'desire to make something important happen' (Wachsmann, 2012: 20).

In 1975, free improvising saxophonist Evan Parker suggested that his 'music of the future' would be played by groups of musicians 'who improvise freely in relation to the precise emotional, acoustic, psychological and other less tangible atmospheric conditions in effect at the time the music is played' (1975: 12–13). This potential for free improvisation to respond to any aspect of the performative experience has become a prominent and widely acknowledged characteristic of the music. The all-inclusive nature of the musical process is regularly associated with the idea that free improvisation involves a largely egalitarian approach to music-making. As Hargreaves, MacDonald, and Miell (2012: 6) point out, a range

[1] I respectfully acknowledge the contribution made to this research project by all the participants of the performances.

of theories to situate the practice of musical improvisation has blossomed in recent years. Many have noted that the implications of the practice go well beyond the boundaries of musical performance and practice, but encroach on wider cultural, educational, and political contexts also. This idea was championed by, amongst others, Spontaneous Music Ensemble's founder, John Stevens, who referred to free improvisation as 'free group music'. Stevens advocated a music constructed by individuals focused on creating an immersive group sound.

I question the extent to which all-inclusiveness and egalitarianism are realistically possible within free improvised music, given the largely unquestioned and continued adherence to the static and stratified positioning of the musicians and the audience. I argue that while the improvising musicians continue to perform fixed in one position, the extent to which they can experience all the possible aspects of the performative environment is left in doubt. Additionally, the stratified positioning of the participants, I suggest, does little to support the heterarchical aspirations of the musical form; the implied fourth wall makes the listener a mere bystander to the process that, through their continued presence, nevertheless involves them. Stevens himself grew to feel that the collective principles of the music were later sidelined by an increased emphasis on 'individualism, personal instrumental virtuosity, and musical elitism, which he believed detracted from the more profound musical, spiritual and political implications of free group music' (Scott, 2014: 104).

16.3 Musicians in Space

'Musicians in Space' (MiS) is a practice-based research project investigating the ways in which free improvising musicians relate to both the listener and their spatial embeddedness within the performance environment. It introduces a spatialized approach to the performance of free improvisation that allows all the participants the opportunity to move during the improvisation. Worded as an invitation, the participants are given the option to shape and regulate their individual involvement in the performative process without being prescribed a set behaviour or mode of action.

Considering the intersubjective and unfixed nature of improvisation, the task of structuring the research was approached with a clear desire to protect the integrity of the improvisational process in all its complexity. To that end, an emphasis was placed on identifying a research design and method that could accommodate and faithfully celebrate the inherent polysemic nature of the practice of free improvisation without restricting the improvisational process.

These considerations led to a design utilizing a pragmatic approach to practice-based research in combination with the phenomenology-based heuristic research method developed by Clarke Moustakas (1990). The heuristic research method places the researcher at the centre of the process and emphasizes the importance of the 'tacit dimension' that Michael Polanyi ([1966] 2009) believes underlies all knowledge. The pragmatic approach to the research reflects the inherent nature of free improvisation and supports an open structuring that allows for the selection of methods based on what is considered most appropriate and 'fit-for-purpose' (Kupers, 2011).

To contextualize the research, a theoretical framework that emphasizes the social and interconnected nature of musical activity was used. This consisted of Christopher Small's 'musicking' (1998) and a broad range of ideas and themes related to the link between musical performance and ecology.[2] The referents converge to support the idea that free improvisation can exemplify an individually centred, but collectively based means of self-expression and co-creation, which balances individuality and selfishness, and collectivism and totalitarianism (Fischlin, Heble, and Lipsitz, 2013).

The research was structured in two stages, with the first stage involving a series of public performances that occurred during May and June 2017 (see Figure 16.1). The second stage builds on these performances with a series of participant interviews. The five performances, in the British cities of Canterbury, Liverpool, Oxford, and two in London, all involved improvisers experienced in large group improvisation, with most of them affiliated to improvising orchestras such as the London Improvisers Orchestra, the Merseyside Improvisers Orchestra, and the Oxford Improvisers.

---

[2]For example, Borgo (2002, 2006, 2007), Clarke (2005), Cobussen (2014), Davis (2008), Di Scipio (2003, 2015), Nelson (2011), and Waters (2007).

This ensured that levels of experience and skill from across the improvising community were broadly represented.

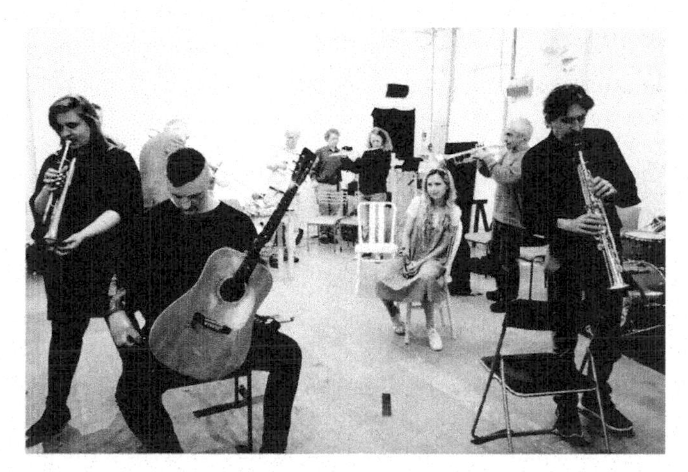

Figure 16.1 London Improvisers Orchestra (LIO) performance, 21 May 2017. Photo: Séverine Bailleux.

During three of the performances, the improvisers were initially split into playing in smaller improvising ensembles before the large group improvisation began. This gave the improvisers the opportunity to experience the spatialized improvisations as both performers and audience members. The instructions offered to the 49 improvisers simply involved an invitation to move — but only if they felt motivated to do so — to enhance their listening and playing practice. The public audiences were also invited to modify their spatial relationships to the musicians just as they would within a sound art installation that incorporates an array of loudspeakers. The intention was to allow the participants the freedom to interpret the invitation as they saw fit, and it was emphasized that it was in no way obligatory that they moved at all.

All the improvisations, and the post-show discussions that followed each performance, were filmed and recorded binaurally from various positions in each venue to provide a range of listening and viewing perspectives for future reflection. The post-show discussions, which involved both musicians and audience members, were all transcribed and, along with the footage, served as the source of data for the reflections reported in this chapter. A subsequent set of interviews, with a small number of improvisers, have since taken place. These will develop the themes that emerged through the initial process and will constitute the second stage of the research. Footage of all the performances is available for viewing via www.dafmusic.com and on an associated YouTube channel.

16.4 Observations and Experiences

The various approaches to the spatialized performance process employed by the musicians can be seen to fall under two headings. Firstly, there were 'added affordances' which resulted in a greater level of diversity of interactions, activity, and ways of listening, all contributing to more visible relationships and connections within the improvisational process. Secondly, there was an increased level of 'inclusivity' amongst the improvisers, the audience, and the physical and sonic elements present within the performance environment. Encouragingly, a definitive set of actions and responses from the musicians were not observed during the MiS performances. Had this occurred, it may have indicated that the improvisers felt inhibited by the process. Instead, the actions of the musicians were seen to exist on a continuum of possible responses to the complexity of the process. This demonstrated that they were taking the opportunity to move as just another improvisational parameter to utilize, subvert, or ignore. It meant that while some musicians chose to remain in the same place, or moved just once during an improvisation, others embraced the idea of engaging with the ensemble from different locations, making the shifting topography of the performance space a central focus for their improvisation.

16.4.1 Added Affordances

Despite an established tradition of improvising in a close spatial arrangement, no specific distance between improvisers was observed as being preferred over any other. Musical connections were made both distally — from opposite sides of the room — and proximally — with musicians right next to each other (Figure 16.2). The movement of the musicians was generally guided by their personal interests, coupled with an intention to maintain a clear perspective on the whole ensemble. This meant that musicians moved towards or away from what they wanted to engage with, while facing into the room and with an open stance in relation to the group.

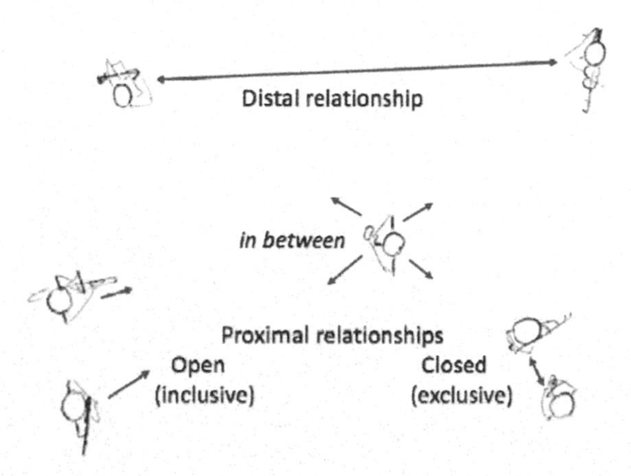

Figure 16.2 Possible playing positions observed during the performances.

Only rarely was it observed that musicians chose to block out the other participants by playing proximally close together and facing inwards. The other musicians chose to respond to this exclusivity either by stopping and waiting or by encroaching musically and/or physically on the closed dialogue.

Musicians readily took advantage of small pauses in the music to move and reconfigure their spatial relationships. This had the effect of making some of the improvisations feel episodic in structure. Musicians were regularly observed coming together to form subsets of the larger ensemble.

Video Example 16.1 Oxford Improvisers, 20 June 2017:
https://youtu.be/0bHfbO4aW70?t=4m15s

These subsets continually formed and reformed as improvisers moved towards or away from one another, with musicians taking the moments when they were not playing to listen or move. Thus, a cascade of musical activities was sometimes observed, with one group leading the ensemble while others took time to listen or regroup in preparation for establishing something new. This grouping strategy was also noticed at times of adversity when, for instance, the volume of the music increased and the quieter instrumentalists would come together for solidarity (Figure 16.3).

Video Example 16.2 London Improvisers Orchestra, 21 May 2017: https://youtu.be/QvtyJ57Tilo?t=6m55s

Figure 16.3 String solidarity — London Improvisers Orchestra (LIO) performance, 21 May 2017. Photo: Séverine Bailleux.

But just as much as the improvisers chose to group together, at other times they were also seen to remain separate from each other. This afforded a greater ability to hear the entire ensemble and to see the various disparate subsets of musicians that made up the whole. In fact, the option to play in a spatially isolated way and 'in between' different subsets of the ensemble, listening and responding to multiple groups simultaneously, was frequently taken by musicians.

Video Example 16.3 London Improvisers Orchestra, 21 May 2017: https://youtu.be/_AQ5Uhb08Ew?t=4m31s

To maximize their ability to hear and see everyone in the ensemble, the musicians naturally formed a circular arrangement (Figure 16.4).

Video Example 16.4 I'Klectik Arts Lab, London, 11 June 2017: https://youtu.be/BCjFMv3ZQfw?t=4m55s

Figure 16.4 The musicians naturally chose to perform in the round, optimizing their ability to hear and see everyone.

This corresponds to research by Healey, Leach, and Bryan-Kinns, which found that the musicians, without any instruction, orientated themselves in a circle to support a cooperative ethos where all the participants have more or less equal 'speaking rights' (Healey, Leach, and Bryan-Kinns, 2005: 2). This positioning was then maintained throughout the improvisations, with other musicians moving into the free areas when gaps appeared. However, when a musician or musicians decided to enter the circle or cross through the central space, this action regularly resulted in the subsequent actions of that musician gaining added gravitas and poignancy.

Video Example 16.5 Sydney Cooper Gallery, Canterbury, 12 May 2017: https://youtu.be/p3eGT9UoxGs?t=14m25s

Musicians frequently reflected on having to change the way that they approached the improvisation process. In particular, they could not close their eyes whilst playing, as they would normally. This was not always seen as a negative aspect of the process, but rather as an unexpected point of departure for exploring the ensemble in a new way:

> I found it very illuminating in this environment to be encouraged to use my sight. And that also led me to different kinds of relationships with the other musicians. And the ability to move around also. But... the aspect of sight and being able to move, I thought, was really exciting because I felt that I could explore the geography of the acoustics of the space.
>
> (Improviser 1, Canterbury performance)

Others commented on needing to adapt playing techniques to account for standing up and walking, while some chose to leave parts of their musical equipment in different areas of the room.

Video Example 16.6 I'Klectik Arts Lab, London, 11 June 2017: https://www.youtube.com/watch?v=qqpt1lhdOY8&feature=yo utu.be&t=4m49s

In two instances, both voiced by trombone players, the need to remain more proprioceptively aware of the other people in the space was seen to have a detrimental effect on their playing. However, most musicians appreciated that they could always return to a static position, with eyes closed, to reduce the complexity of what could at times become a cluttered performance environment.

Where the performance involved established groups, such as the London Improvisers Orchestra, the spatialized practice led to a confounding of expectations for the more static musicians. They found that the sounds of various instruments were not coming from where they were expecting them to.

> What was surprising to me was that the sounds kept... not coming from where I expected. Although I didn't know that I was expecting anything until they stopped doing that...
>
> (Improviser 2, LIO performance)

It was regularly remarked that this approach to performance impacted on the overall nature and sound of the improvisations. The improvisations were felt to be less cluttered, possibly for two reasons. Firstly, as the musicians instinctively faced what they were listening to, the other improvisers could gain more understanding of the relational dynamics at play. Secondly, the spatially dispersed arrangement of the musicians afforded, as noted by Henry Brant, more clarity and ease in locating the various sonic contributions of the participants (Harley, 1997: 70).

Video Example 16.7 Oxford Improvisers, 20 June 2017: https://youtu.be/zqzc6nRCqA0?t=9m25s

These factors meant that the improvisers could enter and exit the music with more understanding of the unfolding relational dynamics, with many choosing also not to play and just listen more frequently.

16.4.2 Inclusivity

The MiS performances allowed many improvisers the opportunity to focus on the variability of the acoustic topography of the space. This led to an engagement with the physical and spatial qualities of the performance environment rarely seen within free improvised music. Frequently, musicians chose to play into the corners of the room, into the floor, towards different surfaces, or they moved outside of the room entirely. Sonic qualities of the rooms were also taken advantage of, with resonant or creaking floorboards and walls being played, and extraneous objects that were lying around being incorporated into, or used on, instruments.

Video Example 16.8 Oxford Improvisers, 20 June 2017: https://youtu.be/zqzc6nRCqA0?t=20m49s

This added a playfulness to the performance and made visible an inquisitive quality to the improvisers' engagement with their environment. It also highlighted the musicians' interest in the varied characteristics of the aural architecture, with the focus of the participants expanding outwards beyond the musical activity on the stage. Instead, the listening experience became spatially specific to the individual and could go from encompassing the entire space to being narrowed onto a specific location. The term 'telescoping awareness' (Koteen, Smith, and Paxton, 2008), which comes from the work of the dancer Nancy Stark Smith and her framework for contact improvisation, the Underscore, may describe this diverse range of listening experiences possible. Stark Smith uses it to refer to the ability of the dancer to focus on anything from the macro to the micro, in terms of 'personal awareness to sensation, activity and any other information or aspect of your improvising practice' (Leahy, 2014: 43). Similarly, the nature of this adaptive listening process can be compared to 'Deep Listening' (Oliveros, 2005), which emphasizes a more expansive and inclusive approach to listening to the world around us. This wider listening awareness, it is suggested, resulted in the music regularly becoming increasingly quiet, as mentioned here:

> I was really... taken by the sensitivity of everyone's listening. And the fact that we could... really hone to each other and play very, very quietly and still keep a kind of intensity to what was happening, with such space. Personally, I found that very engaging.
>
> (Improviser 3, Canterbury performance)

This greater sensitivity also developed an increased awareness of the multi-sensorial nature of the musicking process, emphasizing the link between perception and action (Clark, 2015; Clarke, 2005; Ingold, 2011; Thompson, 2007) and the physical body, as exemplified here:

> Immediately, by the sounds that were being emitted and the energy that was being constructed from almost everybody, I was immediately, [click of fingers] kind of, triggered to start dancing, and to start moving. Maybe not necessarily like... [gesture] but like start to actually allowing my body to express itself. Which is something that I hadn't really felt in a lot of spaces, so that was great.
>
> (Audience member 2, LIO performance)

Regularly, both musicians and listeners remarked on the pleasure of moving very close to players to hear from the musician's perspective:

> It was also nice for musicians with louder instruments that you can go right up to someone who has got a quieter instrument and, sort of, hear it from their point of view, and tune into that. Even against the sound of your own instrument or... the whole sound.
>
> (Improviser 4, London group performance)

While the positive aspects of these performances were widely discussed, the possible negative implications of spatialized improvising were also raised. It was suggested that the movement and the spatially divided positioning could distract the listener and performer from the focused and 'reduced' listening experience regularly associated with a free improvised performance (Chion, 2012: 50). But it was felt that this concern was outweighed by the added possibilities that the spatiality of the performance process afforded.

While free improvisers have focused, for the last 50 years, on questioning and subverting the musical forms, techniques, and meaning of musical performance, they have expected their audience to remain as polite and silent bystanders to the process. As stated initially, I argue that the objective of the improviser to arrive at a unique musical experience that incorporates all the possible aspects of the performative environment may elude them while the presentation of the music continues to support a stratified and hierarchical relationship between the musicking partners. The MiS performances effectively demonstrated that it is possible to afford a level of freedom to the audience without adversely affecting the musical process. This results in an interesting bridge between the realms of participatory and presentational music-making as outlined by Thomas Turino (2008). As the performances progressed, the audience gained more confidence in exploring the varying acoustics and the spatial relationships present within the performances. Like the musicians, some listeners enthusiastically explored the 'psycho-sonic' nature of the performance (Improviser 13, LIO performance): lying on the floor, turning their heads, or circling on the spot to highlight the immersive listening experience. One audience member remarked that

> ... it was really nice to see how there was this organic, kind of, conversation happening that was generating a topography of sound around the space... And it made the music, and the playfulness of the music, really... characterize the space that it was occupying.
>
> (Audience member 10, LIO performance)

The audience appreciated the freedom to self-regulate their relationship to the musical process, modifying their position in the room depending on what interested them. As expressed here:

> The option to detach a little by going to the outer edge of the space made it quite relaxing. Equally, there was a choice to enter into the space a little more and in doing so, it felt like I was having a more active influence on the performers.
>
> (Audience member 9, LIO performance)

It was initially imagined that audience members, inspired by their new-found freedom to move, would decide to join in musically as they used to do during the performances of the People Band, a London-based ensemble active from 1965. But this only happened once, with this individual choosing to jingle some keys and shuffle across the performance space. This instance was not seen as distracting or a violation of the performers' space, but rather it was simply accepted into the unfolding musicking process. The MiS performances, therefore, demonstrated a way that the roles of audience member and improviser could be maintained and respected, while at the same time providing an opportunity for the social conventions associated with a musical performance to be blurred, played with, and mutually subverted. Collectively, everyone seemed to benefit from the more heterarchical dynamic that provided new possibilities for sonic and social engagement. The audience appreciated the greater freedom to shape their listening experience, while the improvisers gained an increased appreciation for the contribution made by the listener to the performances. This shared sense of responsibility for the creation and maintenance of the performance process has subsequently emerged as an important area for further investigation. In the limited space left, I will address some of the wider implications of this spatialized approach to free improvisation by relating it to deep ecology.

16.5 Deep Ecology

Deep ecology emerged in the 1970s, through the work of Arne Naess, as a way of combating the ecological destruction resulting from excessive human action and intervention. It emphasizes the intrinsic value of all life forms and environments, separate to any material or monetary value that may be placed on them by interested human parties. Built on a deeply felt understanding of the interconnectedness of everything and the right for all life to flourish, deep ecology promotes a 'process of ever-deeper questioning of ourselves, the assumptions of the dominant worldview in our culture, and the meaning and truth of our reality' (Devall and Sessions, 1985: 8). The dominant worldview referred to emphasizes an anthropocentric, hierarchical, reductionist, and rational mode of thinking and being, which leads to modes of behaviour and action based on the belief that the world is inherently cruel, competitive, and something reducible to ever smaller parts that can be engineered and manipulated by humans.

Deep ecology promotes an alternative, more holistic worldview that does not see humans as above or separate from other life. It emphasizes qualities of equality and cooperation, while attempting to balance rational reasoning with intuitive and tacit forms of knowledge and understanding. Only by targeting the underlying assumptions and values, the deep ecologist argues, it is possible to facilitate a sustainable change to the way that humans live and interact with all life on earth. Fritjof Capra encapsulates this by explaining:

> ... the connection between an ecological perception of the world and corresponding behaviour is not a logical but a *psychological* connection... if we have deep ecological awareness, or experience, of being part of the web of life, then we *will* (as opposed to *should*) be inclined to care for all of living nature.
>
> (1996: 12)

Interest, inside and outside academia, in the connection between ecology and music has greatly increased since the 1970s and the time of the World Soundscape Project (Schafer, 1969, 1977), with an ecological perspective on music seen as particularly useful in highlighting its social and interactive nature. Ecological approaches to music have emerged from a variety of musical and musicological fields. These include Steven Feld's 'acoustemology' (2015), 'acoustic ecology' (Schafer, 1994), and a 'performance ecosystem' (Di Scipio, 2015; Waters, 2007), to name a few. Each approach has appealed to different areas of musical endeavour and has been used to frame a host of composition, improvisation, sound art installation, listening, and virtual AI-instrument experiments and practices. While free improvisation can also be contextualized within an ecological perspective, as pointed out by Marcel Cobussen (2014), the expansion of the improviser's practice off the stage and into the performance space has facilitated a subtle but profound shift in the dynamics within the performative space. This shift is possibly due to the greater sense of physical and social engagement that arises out of the spatially fluid process, where audience members are now experienced more as equal partners in the performance.

As 'the way we see things is affected by what we know or believe' (Berger, [1972] 2008: 8), the way we engage with music also corresponds to the set of values and assumptions that makes up our ontological perspective and worldview. This connection is outlined by Small (1998), who suggests that the specific music an individual engages in corresponds to a particular view of the world. Moreover, playing and listening to music with others provides us with an opportunity to congregate with like-minded individuals, to affirm, explore, and celebrate our desired version of reality (*ibid.*: 183). I argue that the MiS performances bring into existence a more egalitarian and collaborative musicking experience, one where the contributions of all the participants are valued, irrespective of their particular roles within the process. This heterarchical musicking experience, when viewed from an ecological perspective, demonstrates a process more accepting, tolerant, and resilient to the complexities that exist within a musical improvisation. It, therefore, leads me to draw parallels between MiS and the interconnected and collaborative values of deep ecology.

Returning to Capra's words above, which emphasize the connection between the deep ecologist's outward behaviour and a deeply rooted valuing of equality and cooperation over competition, we can see that the elimination of the fourth wall between the audience and the improviser has produced a greater sense of collective responsibility and care for the musicking process. This, therefore, could signal to the deep ecologist that an appropriate means of cultivating an individual's deep-rooted belief structure could arise in a musicking experience that reflects and celebrates the same heterarchical and ethical beliefs. From a shared commitment to questioning the basic assumptions and behaviours of our dominant worldview, deep ecology and spatialized free improvisation entertain similar orientations towards diversity and adaptability, complexity rather than complication, and a sense of inclusivity and equality that comes from appreciating the inseparability of all things. They both aim to construct something sustainable and meaningful through a dialogical and dynamic process that includes, not excludes, dissenters. They see idiosyncrasy and difference as positive points for creative departure, rather than as adversities to overcome. The free improviser also joins the deep ecologist in seeing a healthy level of diversity and variety as positive signs of a rich and resilient system or community.

The MiS performances revealed a middle ground between participatory and presentational modes of performance that respects the separate roles of improviser and listener, but within a heterarchical environment. They provided a forum for divergent and creative excellence, while also cultivating an environment that empowers all the participants by promoting an active and self-directed engagement in the musicking process. By drawing parallels between deep ecology and the MiS performances, I have attempted to show that the subversive, resourceful, and adaptive nature of the spatialized free improviser has implications beyond just the musical. The wider implications of the MiS free improviser's skill set is something that, I would argue, begins to resemble the original vision for free improvisation from John Stevens, who proposed that 'spontaneity between human beings is a way of serving the community' (Scott, 1987). This also relates to the political dimension of music, as emphasized by Fischlin, Heble, and Lipsitz, who see improvisation as 'more than an artistic conceit, more than a spontaneous creation of notes by musicians' (2013: xii). They suggest instead that 'improvisation is the creation and development of new, unexpected, and productive cocreative relations among people' (*ibid.*).

Like any complex self-organizing system, free improvisation exists as a constantly emerging process, where ideas are shared, tested, developed, and discarded. It is then the responsibility of all the participants to make sense of the experience for themselves, given what is known from previous experience, what is unknown now but just under the surface, and what is unknowable. Improvisation keeps music pliable and able to adapt; it does the same for the individual. By inviting the audience to engage more fully in the musicking process, the complexity, diversity, and vibrancy of the improvising experience were increased. Therefore, the free improviser's primary objective in creating a musical expression that balances new with old material while avoiding habituated responses was also made easier with the invitation to move in space.

16.6 Conclusion

The 'Musicians in Space' performances made visible the inseparable link between the improviser, the performance space, and the audience. By building on the existing practices of the free improviser and inviting them to explore the performance space, a range of added affordances became available that provided further clarity to the musicking process. At the same time, in allowing the audience the opportunity to shape their own spatial relationship to the listening experience, they enjoyed an increased degree of freedom to actively engage in the musicking process. For these reasons, I argue that this spatialized approach to performance comes closer to realizing the ideal of free improvisation as an all-inclusive and heterarchical musical process. Additionally, I suggest that the deeply rooted connection universally felt towards music can provide a key, for the deep ecologist, to tap into and challenge a set of values and assumptions that can otherwise remain difficult to reach. It is hoped that given favourable conditions, such as room to move, spatialized musicking processes like 'Musicians in Space' can be used to establish more inclusive modes of performance — approaches that not only blur the boundaries between the separate roles of the musicking participants, but also exemplify a sustainable and ethically sound approach to relating to ourselves, to those we collaborate with, and to the world around us.

References

Berger, J. [1972] 2008. *Ways of Seeing*. London: Penguin.

Borgo, D. 2002. 'Negotiating Freedom: Values and Practices in Contemporary Improvised Music', *Black Music Research Journal* 22(2): 165–188.

_____. 2006. 'Sync or Swarm: Musical Improvisation and the Complex Dynamics of Group Creativity'. In *Algebra, Meaning, and Computation: Essays Dedicated to Joseph A. Goguen on the Occasion of His 65th Birthday*, edited by K. Futatsugi, J-P. Jouannaud, and J. Meseguer, 1–24. Berlin: Springer.

_____. 2007. 'Musicking on the Shores of Multiplicity and Complexity', *Parallax* 13(4): 92–107.

Capra, F. 1996. *The Web of Life: A New Synthesis of Mind and Matter*. London: HarperCollins.

Chion, M. 2012. 'The Three Listening Modes'. In *The Sound Studies Reader*, edited by J. Sterne, 48–53. New York: Routledge.

Clark, A. 2015. *Surfing Uncertainty: Prediction, Action, and the Embodied Mind*. New York: Oxford University Press.

Clarke, E. 2005. *Ways of Listening: An Ecological Approach to the Perception of Musical Meaning*. Oxford: Oxford University Press.

Cobussen, M. 2014. 'Steps to an Ecology of Improvisation'. In *Soundweaving: Writings on Improvisation*, edited by F. Schroeder, and M. Ó hAodha, 15–28. Newcastle upon Tyne: Cambridge Scholars Publishing.

Davis, T. 2008. *The Ear of the Beholder: Ecology, Embodiment and Complexity in Sound Installation*. PhD thesis, Queen's University Belfast.

Devall, B., and G. Sessions. 1985. *Deep Ecology: Living as if Nature Mattered*. Salt Lake City: Peregrine Smith Books.

Di Scipio, A. 2003. '"Sound is the Interface": From *Interactive* to *Ecosystemic* Signal Processing', *Organised Sound* 8(3): 269–277.

_____. 2015. 'The Politics of Sound and the Biopolitics of Music: Weaving together Sound-Making, Irreducible Listening, and the Physical and Cultural Environment', *Organised Sound* 20(3): 278–289.

Feld, S. 2015. 'Acoustemology'. In *Keywords in Sound*, edited by D. Novak, and M. Sakakeeny, 12–21. Durham: Duke University Press.

Fischlin, D., A. Heble, and G. Lipsitz. 2013. *The Fierce Urgency of Now: Improvisation, Rights, and the Ethics of Cocreation*. Durham: Duke University Press.

Hargreaves, D., R. MacDonald, and D. Miell. 2012. 'Explaining Musical Imaginations: Creativity, Performance, and Perception'. In *Musical Imaginations: Multidisciplinary Perspectives on Creativity, Performance and Perception*, edited by D. Hargreaves, D. Miell, and R. MacDonald, 1–14. Oxford: Oxford University Press.

Harley, M. 1997. 'An American in Space: Henry Brant's "Spatial Music"', *American Music* 15(1): 70–92.

Healey, P., J. Leach, and N. Bryan-Kinns. 2005. 'Inter-Play: Understanding Group Music Improvisation as a Form of Everyday Interaction'. *Proceedings of MSRC International Forum, Less is More: Simple Computing in an Age of Complexity, Cambridge*.

Ingold, T. 2011. *The Perception of the Environment: Essays on Livelihood, Dwelling and Skill*. London: Routledge.

Koteen, D., N. Smith, and S. Paxton. 2008. *Caught Falling: The Confluence of Contact Improvisation, Nancy Stark Smith, and Other Moving Ideas*. Northampton: Contact Editions.

Kupers, W. 2011. 'Embodied Pheno-Pragma-Practice — Phenomenological and Pragmatic Perspectives on Creative "Inter-Practice" in Organisations Between Habits and Improvisation', *Phenomenology & Practice* 5(1): 100–139.

Leahy, D. 2014. *Discovering a Music-Based Underscore: An Account of the Translation Process from the Danced Underscore to a Music Based Practice*. MA thesis, Trinity Laban.

Morris, J. 2012. *Perpetual Frontier: The Properties of Free Music*. N.p: Riti.

Moustakas, C. 1990. *Heuristic Research: Design, Methodology, and Applications*. Newbury Park, CA: Sage.

Nelson, P. 2011. 'Cohabiting in Time: Towards an Ecology of Rhythm', *Organised Sound* 16(2): 109–114.

Oliveros, P. 2005. *Deep Listening: A Composer's Sound Practice*. New York: iUniverse.

Parker, E. 1975. 'Speech to the SPNM Forum on "Music in the Future"', *Musics* 1: 12–13.

Polanyi, M. [1966] 2009. *The Tacit Dimension*. Chicago: University of Chicago Press.

Schafer, R. 1969. *The New Soundscape*. Scarborough: Berandol Music.

_____. 1977. *The Tuning of the World*. New York: Knopf.

_____. 1994. *The Soundscape: Our Sonic Environment and the Tuning of the World*. Rochester, VT: Destiny Books.

Scott, R. 2014. 'The Molecular Imagination: John Stevens, the Spontaneous Music Ensemble and Free Group Improvisation'. In *Soundweaving: Writings on Improvisation*, edited by F. Schroeder, and M. Ó hAodha, 95–109. Newcastle upon Tyne: Cambridge Scholars Publishing.

Small, C. 1998. *Musicking: The Meanings of Performing and Listening*. Middletown, CT: Wesleyan University Press.

Thompson, E. 2007. *Mind in Life: Biology, Phenomenology, and the Sciences of Mind*. Cambridge: Belknap Press.

Turino, T. 2008. *Music as Social Life: The Politics of Participation*. Chicago: University of Chicago Press.

Vargas, M. 2013. 'Notice and Contribute: Collaborative Negotiations Between Improvised Music and Dance', *Contact Quarterly* 38(1): 24–27.

Wachsmann, P. 2012. 'Klaus Wachsmann and the Changeability of Musical Experience: My Experience as a Performing Maker of Music'. In *Ethnomusicology in East Africa: Perspectives from Uganda and Beyond*, edited by S. Nannyonga-Tamusuza, and T. Solomon, 16–32. Kampala: Fountain Publishers.

Waters, S. 2007. 'Performance Ecosystems: Ecological Approaches to Musical Interaction'. In *Proceedings of the Electroacoustic Music Studies Network conference on The Languages of Electroacoustic Music, Leicester*. Available online at http://www.ems-network.org/IMG/pdf_WatersEMS07.pdf

Other Media

Scott, R. 1987. 'John Stevens' (interview on blog). Available online at http://richard-scott.net/interviews/334-2/

Chapter 17

Space, Engagement, and Immersion: From La Monte Young and Terry Riley to Contemporary Practice

Joanne Mills
Faculty of Arts, University of Wolverhampton,
Wulfruna Street, Wolverhampton, WV1 1LY, UK
j.t.mills@wlv.ac.uk

17.1 Introduction

Traditionally, the terms 'audience' and 'visitor' have been set in passive contexts: where 'audiences' attended performances — in Western art music — held in auditoriums or theatres, while 'visitors' viewed art works in exhibitions. However, Clare Bishop notes that minimalism as an artistic movement 'had radical implications for the way art had hitherto been understood', and that it 'initiated an important shift in the viewer's perception of the gallery space' (2005: 54, 66). Minimalist artistic installations became holistic narratives in constructed spaces, in a similar manner to theatrical performances with audiences rather than visitors.

Musical Spaces: Place, Performance, and Power
Edited by James Williams and Samuel Horlor

ISBN 978-981-4877-85-5 (Hardcover), 978-1-003-18041-8 (eBook)
www.jennystanford.com

La Monte Young (b. 1935) and Terry Riley (b. 1935) are known as two of the 'pioneers in the evolution of musical minimalism' (Potter, 2014); however, it has been noted that many composers 'disavowed' the term 'minimalism' (Gann, Potter, and ap Siôn, 2013: 3). Gann suggests that '[m]any of the major minimalist works of the 1960s and 1970s seemed to embody a new performance paradigm', where

> ... works were often evening-length and suited to a listening mode more ambient and less formal than that of the standard classical-music concert; audience members might lie down or sit on the floor and could come and go as they pleased.
>
> (2013: 39)

Young and Riley are here presented as occupying a space which I propose should be retrospectively recognized as immersive, with certain examples of their work situated within performance, relational art, and participatory practice. While certain works from the 1960s (e.g. Young and Marian Zazeela's *Dream House*) are considered as audiovisual, immersive, environmental installations, in this chapter, I aim to establish the extent to which the use of 'active' space within the early work of Young and Riley might contribute to audience engagement and immersion and furthermore inform a contemporary practice in the visual arts.

This chapter presents observations from a wider investigation into the relevance of both composers to contemporary immersive practice, highlighting their work during the 1960s — a period in which North America was recovering from the Second World War and in which developments in post-war technology coincided with an economic boom that saw new materials and technology (including projectors and custom-designed structures) become available to artists and composers alike, in 'an environment of excess and experimentation' (Riley, 2008: 21). Around this time, Young and Riley were collaborating with those from other disciplines to bring their compositions, together with lighting, incense, projections, and visuals, into non-traditional performance spaces, where their audiences could move freely around or engage with their compositions from the comfort of their own homes.[1]

The use of a non-traditional performance space, together with collaboration across disciplines and the move towards consideration of audience experience during the 1960s,[2] led to the creation of an expanded narrative (see Figure 17.1). The result was engaging, multiple-media works which immersed their audience or encouraged active participation.

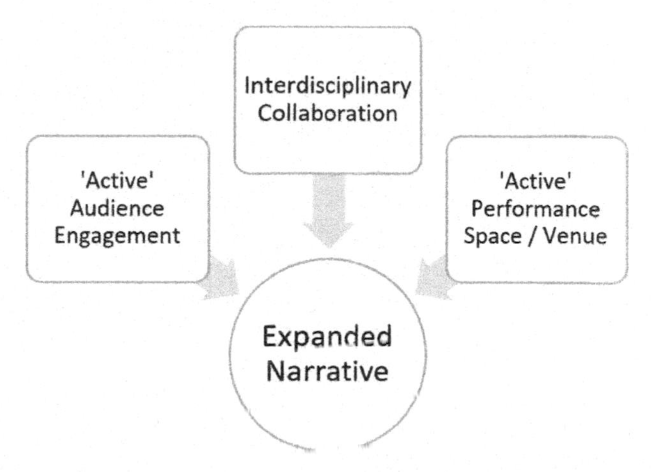

Figure 17.1 How space, collaboration, and audience engagement contribute to an expanded narrative.

Three works by Young and Riley were identified as exemplifying these features which facilitated an expanded narrative and allowed for an immersive experience:

- The *Time Lag Accumulator* (by Riley, 1968), which illustrates the engagement of an 'active' audience

[1]For example, in printed form — including Young's *An Anthology of Chance Operations* (1963), *Dream Sheet* (1965, edited by Diane Wakoski), and the ninth edition of *Aspen* (1970); or recordings circulated with periodicals — including the third and fourth editions of *Shit Must Stop* (1968).

[2]Other instances that saw the redefinition of the audience's role around this time include: 'happenings', with the audience often taking a participatory role within the work; pieces in which consideration was extended to the space in which work was situated, as in those by Donald Judd, Robert Morris, and Fred Sandback; and the adoption of the concept of cybernetics, with a two-way exchange created between audience and artwork.

- *LaMonteYoung&MarianZazeelaTerryRileyJonHassell: A Collaboration* (hereafter referred to as '*A Collaboration*') (by Young, Riley *et al.*, 1969), an example of interdisciplinary collaboration (in this case between composers and a visual artist)
- *Dream House* (by Young *et al.*, 1969—present), which exemplifies the construction of an 'active' performance space

Two new works created as part of this investigation, *The Lull* (2017) and *A-Wakening* (2018), one online and one offline, are presented as incorporating these three expanded narrative components through both the physical and the digitally mediated virtual space — the latter as a contemporary form of non-traditional 'active' space.

While the three works above by Young and Riley exemplify all of the features of the expanded narrative to varying degrees, the following sections are dedicated to one work in turn to focus separately on the three features, starting with the engagement of an 'active' audience — which both contributes to, and becomes part of, a work.

17.2 Active Audience

The growing awareness that audiences construct an individual narrative environment is demonstrated by the movement towards participatory art during the period. As clearly exemplified in early works of Young and Riley, constructed spaces allow the audience to more richly experience performed compositions, installations, and environments.

According to the composer and musicologist Wim Mertens, 'perception is an integral and creative part of the musical process since the listener no longer perceives a finished work but actively participates in its construction' (1983: 90). While Mertens was specifically reflecting on repetitive compositions, he also recognized Young and Riley's interest in the 'physiological and psychical effects of music' (*ibid.*: 36). This sentiment applies to the early works of both composers, particularly *Dream House* and the *Time Lag Accumulator*. The former enables the listening of an 'active' audience to be enhanced through the engagement of multiple senses within a constructed performative space, while in the latter the construction of the installation itself encourages the movement and sonic interaction of the audience.

17.2.1 *Time Lag Accumulator* (1968)

The *Time Lag Accumulator*, in particular, references the move towards participation and immersion during the 1960s, in that audiences at the time were beginning to recognize their own role in the interpretation of a piece, from which it was only a small step towards actively contributing to or controlling one's own experience within it. The installation formed part of the *Magic Theater* exhibition at the Nelson-Atkins Museum of Art, Kansas in May 1968, of which its curator Ralph T. Coe states:

> [t]here is evidence [at the time of the exhibition] that... audiences are becoming... less visually restricted... they are beginning to cross borders of materialism altogether in looking at art neither as illusion nor as object, but as a psychic entity, a sort of mental probe.
>
> (1970: 14)

This work is an example of how the location of audience members within specially constructed chambers (an 'active' space) — with recording equipment allowing their voice and sounds to form the basis of the installation (as an 'active' audience) — can contribute to an immersive experience. The work itself could stand alone in an exhibition (as indeed it did), but it is the presence and action of the audience which completes the installation and allows it to exist in its intended form.

The *Time Lag Accumulator* took the form of an 'octagon with eight pairs of glass doors opening into chambers' in which people could occupy the space and make utterances which would be relayed/delayed to the other chambers (Coe, 1970: 190). Riley's vision for the installation was of a machine in which 'live voices heard within a chamber... would be repeated in delayed time sequence, projecting the past into the present' (*ibid*.: 88). The relay system was set up so that while 'voices talking in one chamber could be heard faintly in the adjoining cubicle, the repeats [were] broken up, so [as not to be] heard on the same delay-relay line' (*ibid*.: 193).

In his description for the installation, Coe refers to the audience-centred dynamic of Riley's concerts, during which 'the audience [are] allowed to recline on pillows and mats and move around according to their own feelings without the usual formal concert restrictions' (*ibid.*: 190). Coe described the intention for the area around the installation to be comfortable, containing cushions and chairs to encourage a relaxed atmosphere, although this was not possible at the Kansas exhibition due to space restrictions. This method of informal audience engagement references *Dream House*, in which visitors have the choice to move or rest at their leisure within the environment (as opposed to within a traditional concert venue with fixed stage and seating). This also evidences the shift during the 1960s which, as noted earlier, broke away from the conventions of a passive audience viewing a performance, allowing for an active relationship between audience and artwork.

The *Time Lag Accumulator* exemplifies how the ideas of composers can be made into reality with the experience of engineers, while the following section examines collaboration between the composers and the visual arts.

17.3 Interdisciplinary Collaboration

While Young and Riley's early works were concerned with the manner of reception of their pieces,[3] developing innovative forms of engagement meant adopting ideas from other disciplines. This mirrored the cultural shift at the time towards mixed media and reflected John Harle's observations that minimalist music's aesthetic was 'essentially... cross-disciplinary' and that its proponents 'existed apart from the mainstream' during the 1960s (2013: 384).

A Collaboration is an example of how multi-sensory elements can combine to create an engaging and immersive environment for an audience. A continuous performance of sound and light, originally presented at the Albright Knox Gallery in Buffalo, New York on 3 May 1969, it formed part of the Evenings for New Music

[3]Other early works by both composers which exhibit similar characteristics of challenging the traditional relationship between audience and performance include Young's *Compositions 1960 #6*, which required performers to watch and listen to their audience, and Riley's *Ear Piece* (1962), where a 'performer' places an object of their choosing onto their ear before creating sounds by interacting with it.

series of concerts organized by the Center of the Creative Arts at State University of New York at Buffalo.

17.3.1 *LaMonteYoung&MarianZazeelaTerryRileyJonHassell: A Collaboration* (1969)

The performance included three compositions, one each by Young, Riley, and Jon Hassell, accompanied by a lighting installation created by Zazeela. According to Renée Levine Packer, who worked for the centre at the time,

> One heard the sound even before one entered the darkened, amplifier-strewn auditorium. Mauve and green Indian-inspired filigree projections bathed the walls on either side of the stage. Incense permeated the space.
>
> (2010: 93)

Young and Zazeela have collaborated through both performance and sound and light environments since the 1960s, including in *A Collaboration* and *Dream House* (see Section 17.4.1). Of their partnership and that between aural and visual elements, Young remembers,

> I found at a very early stage in my performing career that the lighting could make all the difference in the world... As we worked together in our studio we evolved a type of light environment we felt we could perform best in and which helped the audience to get in the mood we wanted.
>
> (in Pelinski, 1980: 8)

In works such as these, and the *Time Lag Accumulator*, Young and Riley also demonstrate the creation of, or control over, an 'active' space, in which the audience is effectively invited to 'step into' a performative environment, as a form of escapism to an alternate reality, for the duration of the performance.

17.4 Active Space

During the 1960s, minimalist composers were utilizing non-traditional performance spaces partly through necessity; as Philip

Glass observed, '... no-one else would play the music... the avenues for presenting new music were closed to us' (in Smith and Smith, 1995: 134). Together with Coe's concept for his exhibition *The Magic Theater*, which brought together a collection of works that 'personally involve[d] the viewer' through sensorial engagement and saw the audience itself make the exhibition 'come alive' by becoming 'immersed in [its] mystery' (*ibid.*: 14-15), this suggests that the newly engaged and activated audience within these new spaces may have developed different expectations and that open or 'active' spaces have allowed for and informed advances in artistic practice.

This 'active' space contributes to the immersive experience of the 'active' audience discussed previously in this chapter by allowing for the construction of, or control over, environmental factors and sensory stimuli, and it has parallels with what Lucy Bullivant (2007) terms the 'transformative effect' of interactive installations and environments. According to Bullivant, such environments are both 'porous' and 'responsive', inviting the audience 'to spontaneously perform and thereby construct alternative... meanings', and she suggests that it is 'the behavioural aspects — the unpredictable, "live" quality of installations — that is compelling and with the active involvement of visitors "completes" the identity of the work' (*ibid.*: 7).

Dream House is selected here to demonstrate the purposeful construction of an 'active' space; Young's compositions are presented against a backdrop of Zazeela's lighting installations, and audience members are free to self-navigate around the space through movement or repose in order to experience the work. As Mertens observes, 'For Young, the positioning and the spatial mobility of the listener are an integral part of the experience of the composition' (1983: 30).

17.4.1 *Dream House* (1969–Present)

The first public exhibition of the work took place at the Galerie Heiner Friedrich in 1969, with each iteration of *Dream House* being site specific, allowing Young and Zazeela full control over the lighting, audio, and the set-up of the environment.[4] The continual nature of the work lends itself to being visited at various times of day, in different seasons, and with small or large audiences, all of which affect engagement with and experience of the work. Young's biography on the MELA Foundation website describes the concept of *Dream House* as 'a permanent space with sound and light environments in which a work would be played continuously' (MELA Foundation, n.d.). Against a backdrop of Young's compositions,

> Zazeela's mobile forms are arrayed in symmetrical patterns with lights placed in precisely symmetrical positions creating symmetrical colored shadows; the wall-mounted light sculpture and the neon are both symmetrical forms.
>
> (MELA Foundation, 2016)

Describing his experience of *Dream House* during the exhibition *Seven+Eight Years of Sound and Light*, which opened at the MELA foundation in 1993, Ted Krueger remarks on the scent of incense and the availability of 'several pillows on the floor [which] invite repose', while '[t]he white walls, ceilings, woodwork and carpet are bathed in an amazing magenta light, and an extraordinary sound pervades the space'. At the same time, 'the frequency and intensity of the tones vary in each ear and... the changes correlate with even the slightest movement' (2008: 13). Together with the sound, in Zazeela's *Imagic Light* suspended aluminium spirals' 'ultra-slow spin is induced by air currents from a viewer's movements or thermal differences in the room', which 'creates a slowly changing composition of shadows and objects in varying intensities of contrasting hues' (*ibid.*: 14). Krueger suggests that such work requires a 'commitment' from its audience:

> It cannot be taken in briefly or casually, nor is it intended to be experienced passively. This is not entertainment. It requires an investment from the participant and rewards them in proportion to their degree of engagement.
>
> (*ibid.*: 15)

While parallels can be drawn between *Dream House* and *A Collaboration* in their engagement of three senses, and through the

[4]It has since been exhibited in different forms and locations until its current incarnation at the MELA Foundation in New York.

use of sound, coloured lighting, and scent, the *Time Lag Accumulator* was a physical structure with which the audience could actively interact, thus participating in both the content and the creation of meaning for the work.

Architectural and spatial works such as *Dream House* and the *Time Lag Accumulator* combine multi-sensory elements to create an immersive environment for the audience. While each is controlled by their creators to a certain extent, the experience of the works varies with different lighting, temperature, and audience numbers. It is precisely this consideration of the environment in which a work is sited, and the experience of the constructed space as a whole (almost as if setting a scene for a theatrical performance), which places the audience in a central role. The visitor is thus able to share a space with the work and to experience it in a more engaging way than, for example, framed works or small intricate sculptures, which are traditionally placed on plinths in a space shared with a passive beholder.

All three of Young and Riley's works under discussion in this chapter draw the audience into an 'active' space as separate from 'everyday' experience. This engagement can allow for 'active' audience movement or involvement. This chapter recognizes the potential of the venue itself to be opened up for the audience to undertake a performative role and ultimately to be combined with a new technology (such as virtual environments) to create new forms of 'active' spaces. In 2017–18, I created two works to examine how the three concepts of the 'active' space, cross-disciplinary collaboration (leading to multi-sensory engagement), and the 'active' audience together form an expanded narrative to promote a more meaningful encounter with artworks.

17.5 Exploring Space, Collaboration, and Audience Through Practice: *A-Wakening* and *The Lull*

Young and Riley's use of non-traditional spaces, and their embracing of other disciplines, suggests that their works became holistic narratives in constructed spaces in a similar manner to theatrical performances, placing the audience *within* the environmental 'scene' itself. To explore this proposition further, two works were developed, with one performed in a constructed real-world space and the other in the virtual environment *Second Life*. Thus, the task also drew comparisons between the creative processes of producing works within physical and virtual environments.

A-Wakening, developed in collaboration with the composer Chris Foster, concerned live practice of interdisciplinary collaboration, as with the work of Young and Zazeela, while *The Lull* incorporated a shorter clip of the audio from *A-Wakening*, which was then looped.[5]

17.5.1 *A-Wakening* (2018)

The primary consideration in the creation of this work was whether the listening experience could be enhanced through additional senses and movement. This emulates the collaborative work of Young and Zazeela, where Zazeela's lighting installations (e.g. in *Dream House*) are intended to contribute to the overall experience. Consideration was given both to audience perception and to the concept of the 'extended conversation', which Brandon LaBelle describes as being placed between the 'presence of a viewer or listener, and object or sound, and the spatial situation' (2010: 81).

An octophonic playback system was used to play eight separate sound channels to the audience through eight speakers arranged around the edge of a black box theatre space, with the aim of creating an immersive soundscape that surrounded the listener as they moved around, allowing the audience to experience different combinations of sounds with each step. The position of the speakers, together with the direction that the audience entered the space, is shown in Figure 17.2, where 'projection' indicates the position of the projected visuals.

Visual and scent elements were intended to augment the audience's experience of the sound composition. Animated shapes were projected to the back of the space, through mist, to give an

[5]The difference in the length of audio used in both works arose in part due to the restrictions on imported files within *Second Life*.

almost tangible quality to the light itself (see Figure 17.3, illustrating both the visuals as projected, and the light source itself, appearing to 'hang' in the air). Small red and blue plastic cubes containing scent were available to be selected by the visitor as handheld objects to carry around within the installation. These cubes were intended to promote the concept of an 'active' audience, with the ability to select an object at will connecting them with the work. The engagement of the olfactory sense was also intended as a reference to both *A Collaboration* and *Dream House*.

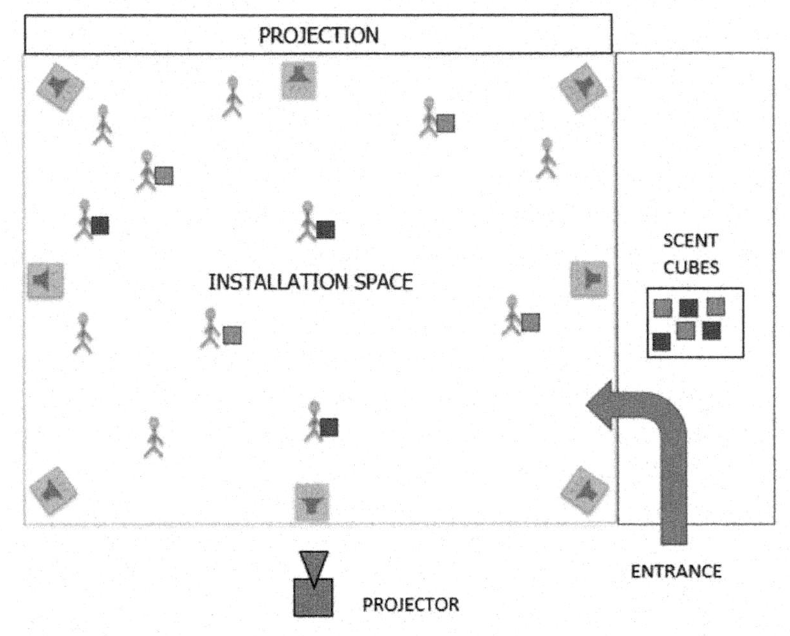

Figure 17.2 Diagram showing the placement of audio, visual, and scent within *A-Wakening*.

While *A-Wakening* was located in a physical venue, the second work, *The Lull*, was sited in the virtual environment of *Second Life*. In both of these immersive, dream-like works, the feature of free navigation within constructed spaces was intended to evoke both *Dream House* and the *Time Lag Accumulator*.

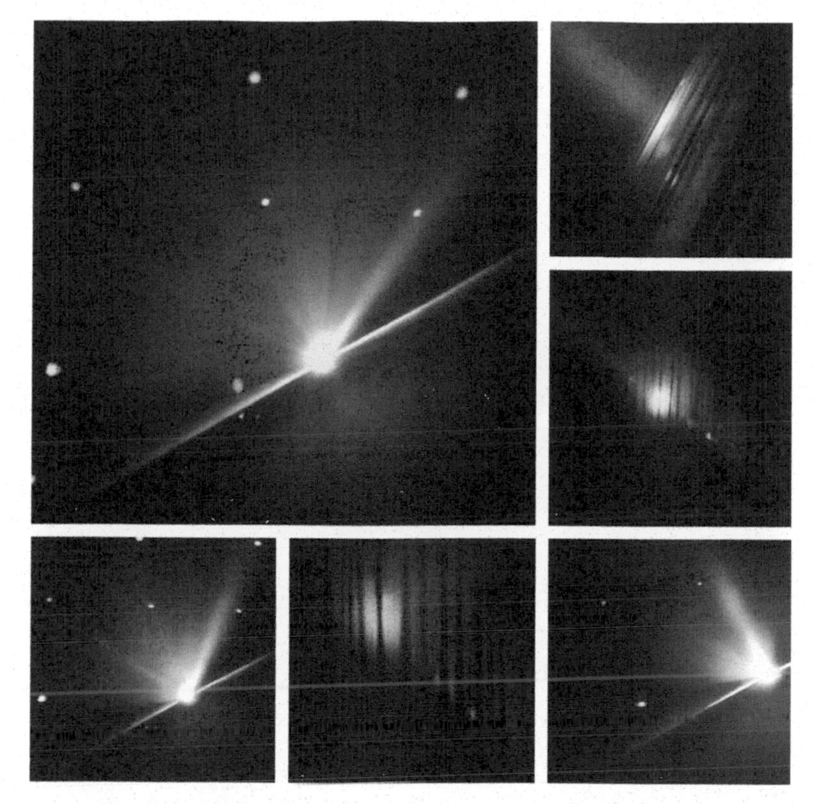

Figure 17.3 Composite of images from *A-Wakening*. Photos by Joanne Mills.

17.5.2 *The Lull* (2017)

In contrast to *A-Wakening's* real space, *The Lull* was a temporary virtual installation created within *Second Life*, officially opened for visitors in December 2017.[6] It was intended as a re-creation, or re-imagining, of *A-Wakening*, with an avatar used in place of the audience's physical presence (see Figure 17.4). It used the sound piece composed for *A-Wakening* as the starting point, and as an audio element within the installation, where the volume of the audio depended on the presence and placement of the avatar within the space.

---

[6]While *The Lull* was realized prior to the public showing of *A-Wakening*, both were developed simultaneously.

Figure 17.4 Stills from *The Lull*. Images by Joanne Mills.

The Lull highlights the virtual as a new non-traditional 'active' space and its potential for new forms of artistic practice, both for visual artists and for musicians and composers. The *Second Life* platform was chosen because it was known to be used as a virtual space by other composers such as Pauline Oliveros, a contemporary of both Young and Riley, whose own practice was informed by the use of sound within the virtual environment. Examples are her 2008 version of *The Heart of Tones*, for which she recreated her 1999 work of the same name, and her association with the Avatar Orchestra Metaverse (who performed the new work). Of the virtual work, which was performed again in 2017 shortly after Oliveros' passing, it was noted:

> The resultant beats, timbre shifts and audio illusions create rhythms, transformations and textures that are precisely mirrored in visual colour spectrum shifts on activated capes worn by the avatar performers.
>
> (Avatar Orchestra Metaverse, 2017)

Through the *Second Life* platform, which offers visitors the experience of defying gravity (in that they can fly) and a novel sense of embodiment in a form outside their own (through the avatar), I explored the possibilities (and limitations) of transposing a digitally supported physical work into the virtual environment.

Digital technology can be used to create a perception of virtual immersion, by extending the space of the spectator beyond the screen, leading to a state in which the action is shaped by the collaboration between audience and artwork. This shift from physical spaces to virtual worlds, as enabled through advances in technology, allows for a new form of artistic and performance space which might open up situations for co-creation and collaboration, facilitated *with* instead of *for* the audience.

Both *A-Wakening* and *The Lull* consider the creative potential in joining the two discrete areas of music composition and visual art.[7] The former, in particular, was influenced by the use of space and collaborations within the three works of Young and Riley discussed in this chapter — where multi-sensory environments were created to allow for 'active' audience engagement, movement, and open exploration of 'active' spaces.

17.6 Conclusion

This research into the early works of Young and Riley led to an understanding of how the use of 'active' performance spaces and cross-disciplinary collaboration generated new forms of artistic practice and, together with the consideration of the 'active' role of the audience, informed the creation of an expanded narrative, which

[7]This also has parallels with Dick Higgins' (1965) concept of 'intermedia', which concerned interdisciplinary artistic activities that 'fall between' traditional (single) mediums and coincided with the growth of groups championing this form (e.g. the San Francisco Tape Music Center and Experiments in Art and Technology (EAT)) and of intermedia programmes at institutions including New York University.

in turn provided an immersive experience for deeper engagement. By moving beyond traditional forms of performance, three of Young and Riley's works first performed during the 1960s — the *Time Lag Accumulator, A Collaboration*, and *Dream House* — are positioned within the context of the changing role of the spectator during the 1960s from passive viewer to active participant, a relationship which places the audience as central within the constructed space of the performance or artwork. They present very different forms of immersion — the latter two relying on an almost transcendent, meditative experience, while the former requires concentration on the moment-by-moment experience of the installation.

Both *Dream House* and *Time Lag Accumulator* have continued to be presented since their first public appearances in 1969 and 1968 respectively, albeit discontinuously. *Dream House* has been on show since 1993 at the MELA Foundation, New York, with differing combinations of lighting and sound works, while Riley's *Time Lag Accumulator* has been reinvented twice since its original form in the *Magic Theater* exhibition: the *Time Lag Accumulator II* (2003) and the *Time Lag Accumulator III* (2019).

This chapter shows how the use of 'active' space contributes both to the blurring of boundaries between music and visual arts and to the creation of engaging and immersive experiences for audiences, while recognizing the continued relevance of disrupting the traditional concert hall set up in the development of such works.

References

Bishop, C. 2005. *Installation Art: A Critical History*. Ann Arbor: University of Michigan Press.

Bullivant, L. 2007. 'Alice in Technoland', *Architectural Design* 77(4): 6–13.

Coe, R. T. 1970. *The Magic Theater: Art Technology Spectacular*. Kansas City: The Circle Press.

Gann, K. 2013. 'A Technically Definable Stream of Postminimalism, Its Characteristics and Its Meaning'. In *The Ashgate Research Companion to Minimalist and Postminimalist Music*, edited by K. Potter, K. Gann, and P. ap Siôn, 39–60. Abingdon: Routledge.

Gann, K., K. Potter, and P. ap Siôn. 2013. 'Introduction: Experimental, Minimalist, Postminimalist? Origins, Definitions, Communities'. In *The Ashgate Research Companion to Minimalist and Postminimalist Music*, edited by K. Potter, K. Gann, and P. ap Siôn, 1–16. Abingdon: Routledge.

Harle, J. 2013. 'Performing Minimalist Music'. In *The Ashgate Research Companion to Minimalist and Postminimalist Music*, edited by K. Potter, K. Gann, and P. ap Siôn, 381–384. Abingdon: Routledge.

Krueger, T. 2008. 'This is Not Entertainment: Experiencing the Dream House', *Architectural Design* 78(3): 12–15.

LaBelle, B. 2010. *Background Noise: Perspectives on Sound Art*. New York and London: Continuum.

Levine Packer, R. 2010. *This Life of Sounds: Evenings for New Music in Buffalo*. New York: Oxford University Press.

Mertens, W. 1983. *American Minimalist Music: La Monte Young, Terry Riley, Steve Reich, Philip Glass*, translated by J. Hautekeit. London: Kahn and Averill.

Pelinski, R. 1980. 'Upon Hearing a Performance of the Well-tuned Piano', *Parachute: Contemporary Art* 9: 4–12.

Potter, K. 2014. 'Minimalism'. *Grove Music Online*. Available online at https://doi.org/10.1093/gmo/9781561592630.article.40603

Riley, R. R. 2008. 'Liquid to Light: The Evolution of the Projected Image Light Show in San Francisco'. In *The San Francisco Tape Music Center: 1960s Counterculture and the Avant-garde*, edited by D. W. Bernstein, 21–23. California: Berkeley.

Smith, G., and N. W. Smith. 1995. *New Voices: American Composers Talk About Their Music*. Portland: Amadeus Press.

Other Media

Avatar Orchestra Metaverse. 29 October 2017. 'Still Listening: For Pauline Oliveros AOM Performs *Heart of Tones*' (online article, *Avatar Orchestra Metaverse*). Available online at http://avatarorchestra.blogspot.com/2017/10/blog-post_29.html

Higgins, D. 1965. *'Synesthesia and Intersenses: Intermedia'* (online article, *Arte* Sonoro). Available online at http://www.artesonoro.net/artesonoroglobal/intermedia.html

MELA Foundation. n.d. 'La Monte Young' (online article, *MELA Foundation*). Available online at http://www.melafoundation.org/ly1para8.htm

_____. 2016. 'Dream House Opens for the 2016–2017 Season — Our 24th Year' (online press release, *MELA Foundation*). Available online at http://www.melafoundation.org/DHpressFY17.html

Musical Spaces and Power

Chapter 18

Micronational Spaces: Rethinking Politics in Contemporary Music Festivals

Jelena Gligorijević
Department of Musicology,
FI-20014 University of Turku, Finland
jelgli@utu.fi

18.1 Introduction

In this chapter, I scrutinize the current politics of music festivals along two binary dimensions — left versus right and corporate versus independent. I aim to accomplish three interrelated goals. The first is to argue that music festivals today are an integral part of wider branding practices and as such are primarily focused on creating sign values through the commodification of festival experience in its totality, from its material expressions to its affective attachments (cf. Blackett, 2003; Klein, 2000; Lash and Lury, 2007). The second is to critically assess the underlying contradictions and

Musical Spaces: Place, Performance, and Power
Edited by James Williams and Samuel Horlor

ISBN 978-981-4877-85-5 (Hardcover), 978-1-003-18041-8 (eBook)
www.jennystanford.com

political implications arising from the production of music festivals as *brandscapes* (Carah, 2010). The third and last aim of the chapter is to instigate discussion on politics with a capital 'P' in contemporary music festivals by exploring the political potential of a concept I developed in my doctoral study on (Serbian) national identity and music festivals: the idea of music festivals as *micronational spaces* (Gligorijević, 2019). To be more precise, I turn to a derivative of this concept — to the idea of *microcitizenship* as a form of membership to music festival collectivities — to provide a politically invigorating corrective to existing theorizations of such collectivities (e.g. Turner's (1969) *communitas*, Maffesoli's (1996) *neo-tribes*, or Anderton's (2006) *meta-sociality*).

Before I proceed with a critical review of major political trends in the transnational music festival industry, it is important to note that the focus on popular music festivals (as events with popular music as one of the focal points in their programming) is indispensable to the discussion at hand. There are arguably two main reasons that can account for this. First, most music festival scholars agree that the familiar historical link between countercultural musical practices and emancipatory politics surrounding the hippie movement of the 1960s was articulated and fortified through the institution of the music festival (Bennett, 2004; Bennett and Woodward, 2014; McKay, 2015). Such historical baggage explains why music festivals of the present day strive so often to keep up with the countercultural image and rhetoric of their forerunners. And second, what renders music festivals additionally suitable for all kinds of political engagement and identity work are not only certain discourses about the music itself and its relation to the outer world (such as the popular myth of music's universality), but perhaps more importantly 'music's potential for sociality and community' (Hesmondhalgh, 2014: 85).

18.2 Politics in Music Festival Brandscapes

Either international or local in their content and scope — or indeed both, as it is most often the case — contemporary music festivals are overall progressive and cosmopolitan in their outlook (Bennett and Woodward, 2014: 18). They tend to celebrate the ideas of cultural diversity and mutual tolerance, that is, different identity groups and lifestyles such as those of: (1) indigenous people,[1] (2) diasporic communities,[2] (3) neighbourhoods/local communities,[3] (4) minorities, be they ethnic-racial,[4] sexual,[5] or others, (5) women,[6] and (6) various sorts of music 'neo-tribes', to borrow Maffesoli's term (see e.g. Dowd, 2014; Luckman, 2014).

Arguably, the celebration of difference and liberal values in contemporary music festivals conforms to the prevailing logic of branding under conditions of advanced globalization. In a book on branding in popular music that is particularly relevant to the aims of the present chapter, Carah suggests — following Sherry — that music festivals have indeed long since transformed into *brandscapes*, that is, into 'experiential social space[s] where marketers engage consumers in the co-creation of brand meaning' (2010: 8; see also Klein, 2000; Lash and Lury, 2007). In the Lefebvrian terminology of space production, festival brandscapes manage, profile, and lay claim to 'the spaces of consumption' (both physical and mediated) and 'the consumption of spaces' (offering such spatial qualities as festivity, sun, sea, fantasy, or nostalgia). Within such spaces, they provide festival consumers with resources from which to build their identities, lifestyles, taste cultures, and social experiences. It is precisely through social actions and cultural practices of festival consumers that corporate brands (music festivals included) can generally become so powerful, meaningful, and enjoyable. Or put differently, brands generate value for corporations from the meaning-making potential of cultural practices they accommodate. In this regard, music festivals, just like many other corporate brands, use methods of the so-called *experiential branding*, which is exactly 'about acquiring and deploying cultural capital' (Carah, 2010: 71).

---

[1]For example, Stylin' Up, in Inala, Queensland, which is Australia's largest indigenous hip-hop and R&B music and dance event (Bartleet, 2014).

[2]For example, London's Notting Hill Carnival, which is led by members of the British Afro-Caribbean community, or the Mela events across Western European countries celebrating the culture of the Indian diaspora (Carnegie and Smith, 2006).

[3]For example, Sardinia's La Cavalcata Sarda (Azara and Crouch, 2006) or Canada's Hillside Festival (Sharpe, 2008).

[4]For example, Africa Oyé, Britain's largest annual celebration of African music and culture taking place in Liverpool; or WOMAD Festival (Chalcraft and Magaudda, 2011).

[5]For example, Gay Pride festivals (Hughes, 2006; Taylor, 2014).

[6]For example, Ladyfest all around the world, or the South African Women's Arts Festival (Marschall, 2006).

Inventing ever-new extensions of festival brands is central to the successful reproduction of festival brandscapes — a phenomenon also known as a *brand canopy*. As Klein explains, 'th[is] concept is key to understanding not only synergy but also the related blurring of boundaries between sectors and industries' (2000: 148). Examples of such tendencies in contemporary music festivals may include: merging the event and tourism industries by opening the festival's own Tourist Office; founding the festival's own music label; facilitating on-site and online purchase of festival merchandise; making special product offers in synergy with other enterprises such as banks, mobile phone providers, or oil and gas companies; expanding the scope and repertoire of cultural activities by organizing other types of events, by exchanging artists with other festivals, by establishing sister festivals across countries and continents, by boosting cooperation with a great variety of music labels and clubs, digital media and music companies, and so on. Another crucial part of festival branding is, of course, *advertising*. The latter involves such activities as making special ticket offers (the festival loyalty programme included), organizing promo campaigns in major urban centres both within and across countries, and disseminating promotional messages all year round (e.g., through the festival live feed, trailers, aftermovies, interviews, and concerts with selected festival performers from previous years) to target populations using both old and new means of mass communication (including the festival website, social media, mobile apps, and email address databases of all registered festival visitors).

One more commonality among contemporary music festivals is the discourse of *corporate social responsibility* (CSR henceforth) with all its underlying paradoxes. Apparently, many festivals today tend to take a socially responsible role in such domains as substance abuse, ecology, and humanitarian aid. But what these and similar initiatives point to are certain internal contradictions arising from the very production of contemporary music festivals as brandscapes. The major paradox here lies specifically in the irreconcilable tension between the festival brand's primary drive for capital accumulation, on the one hand, and the discourse of CSR it adopts, on the other hand. Thus, when endorsing the socially responsible language of anti-drug, anti-smoking, and safe-driving campaigns, as well as that of sustainable living, music festivals facilitate at the same time 'spaces which promote the causes of those very social problems' (Carah, 2010: 119). As Carah convincingly shows in his study of pop brands, the primary goal of the latter is not social problem solving, but profit maximization. The discourse of CSR that music festival brands fetishize should therefore be understood primarily as 'a mode of capital accumulation' (*ibid.*: 125), which does not aim at restructuring existing 'social relations but at educating the "ignorant few"' (*ibid.*: 120).

Two main political implications follow from this underlying attitude of corporate brands, music festivals included. The first is that social problems are deemed less structural than individual in their nature and that their solving is considered simply a matter of self-policing (Carah, 2010: 115). The second implication of 'socially responsible' branding practices is a belief that their profit-making activities 'actually serve as positive forces for good in society' (Hilton, 2003: 47). As I illustrate elsewhere using Serbia's Exit Festival as a case study (Gligorijević, forthcoming 2021), brands tend to offer ready-made solutions pertaining to such socio-economic, political, and ethical domains as ecology, charity, youth support, or restoration of cultural monuments. In doing so, they divert attention from the complexity and contradictions of real social life and therefore from the possibility of coming up with alternative solutions to real social antagonisms. In fact, '[t]he social, ethical and political discourses brands construct relieve us of the duty to think so that we can continue to enjoy' our participation in consumer society (Dean, cited in Carah, 2010: 112).

Ideologically speaking, a majority of present-day music festivals arguably vacillate between what Malpas calls 'a consumerist form of cosmopolitanism' (2009: n.p.) — pertaining to 'the conception of the individual as having no independent affiliation to any place in particular beyond the financial and lifestyle affordances of that place' — and a seemingly more serious cosmopolitan mode of political engagement (as in New York's Global Citizen Festival or London's (Up)Rise Festival). Their common aspiration is apparently to project a progressive and cosmopolitan image using discourses of globalization, libertarianism, cultural diversity, sustainable living, creativity, and technological progress. However, as Carah rightly points out following Goldman and Papson, '[t]hese discourses obscure the frictions of "class, race, gender and global inequalities" inherent in capital' (Carah, 2010: 61).

It should also be emphasized that examples of music festivals with conservative or openly nationalist leanings are less common in this branch of the cultural industry. Lifest (pronounced 'life fest') is an annual Christian youth music festival in Oshkosh, Wisconsin, characterized by its affiliation with conservative evangelical Christianity (Caton *et al.*, 2013). In October 2016, there was a news report on a 'neo-Nazi' music festival in Unterwasser in eastern Switzerland, featuring far-right bands, including several German groups and the Bern-based Amok, who had previously been convicted of racial discrimination and incitement to violence (The Local, 2016). Similar controversies are time and again echoed in music festivals featuring notorious nationalist or pagan black metal acts. For instance, the Montreal black metal festival Messe des Morts was cancelled after protests against Graveland, a Polish pagan/black metal band with white supremacist and neo-Nazi beliefs, which was set to headline on one of the festival days (Pasbani, 2016).

Either way, politics in music festival brandscapes should not be considered solely through the lens of such dichotomies as cosmopolitan-nationalist, progressive-conservative, or left-right. Perhaps even more revealing in this regard are academic and media discourses placing festivals along the major-minor, mainstream-alternative, and corporate-independent axes. Within this discursive framework, critical views of the ever-increasing commercialization of festivals are pitted against what is seen as an alternative music festival model (of which more below). Contemporary music festivals are accordingly described as 'a sanitized version of the past' (Anderton, 2006: 348), or as 'the simulacrum of festival counterculture', whereby the earlier search for alternative lifestyles and forms of social organization became 'neutralized into safe, common and expected forms of leisure' (Robinson, 2015: n.p.). Some commentators associate increasing commercialization trends in the music festival industry with 'essentially a massive change in the kind of person you see attending music festivals in this day and age' (Spencer, n.d.). Among new kinds of festival attendees are, for example, hipsters and *bobos* (short for 'bourgeois bohemians') (Delistraty, 2014), or 'many youths' buying into the 'cool' image that music festivals sell (Morris, cited in Simonsen, 2015). This line of reasoning resonates well with a commonly held view in academia that the increasingly commercial character of postmillennial music festivals owes to growing consumerism and higher disposable incomes within the context of neoliberal globalization. This, so the argument continues, brought about changes in the composition of the festival audience (in terms of their broadening and mainstreaming), many segments of which 'would refuse to tolerate the amateurish event management and poor living conditions that prevailed at some of the pioneer pop festivals of the late 1960s and 1970s' (Stone, 2009: 213; see also Anderton, 2006 and Robinson, 2015). The paradoxical result of such an attitude is what Žižek (2004) calls *decaffeinated empowerment* — an explicit request for the experience of 'authentic' music culture, but within a safe and comfortable environment.

Offered as an alternative to overtly commercialized and corporate music festivals is what has been dubbed, at least in British public discourse, *boutique festivals*. According to Robinson, boutique festivals are 'small, "arty" and relatively unknown' (2015: n.p.), offering to their audiences that which their massified and overly commercialized counterparts seem to lack: 'intimacy, uniqueness and responsiveness to nuanced demand'. Arguably, boutique festivals seek to combat corporate influence in several ways: (1) by selecting local, independent businesses as sponsors and food vendors rather than global, corporate entities,[7] (2) through conspicuous 'anti-sponsorship' self-promotion that allows them to 'innovate their own profitable tactics... oriented towards maximizing festival-goer expenditure', for instance, through premium food and luxury camping options (Robinson, 2015: n.p.),[8] and (3) by moving away from the concert-model festivals towards immersive environments and direct audience participation.[9] The latter is achieved through festival theming and attendant audience theatricality (part of which is costume-wearing and role-playing), 'action camps' (theme camps conceived and led by festival volunteers), interactive art installations, art cars/boats (vehicles transformed into pieces of art by festivalgoers), a variety of both on- and off-site competitions

[7]Examples include Hillside Festival (Guelph, ON), Eastlake Music Festival (Oakland, CA), Blissfields (Winchester, UK), or Beautiful Days (Ottery St Mary, UK).

[8]For example, Nevada's Burning Man, Wales's Green Man, or England's Hop Farm.

[9]For example, Winchester's BoomTown Fair or Abbots Ripton's Secret Garden Party.

in which audience members take the spotlight (e.g. dance-off competitions, or open calls all year round in domains of creative and tourism industries), and so on. The conceptual model of boutique festivals is apparently consistent with the logic of self-branding, where such values as authenticity, creativity, and (self-) reflexivity come to the fore (see Carah, 2010: 93, 100). At a more general level, the boutique festival ideology and practice seem to produce ambivalent political effects, given that its audience members end up being simultaneously empowered and exploited as a free workforce. Festivalgoers, in other words, extract a great deal of pleasure and enjoyment from participating in the meaning- and identity-making processes that the festival's various cultural programmes afford. Then again, the festival's participatory model raises issues of unpaid labour and exploitation for the profit of music festival brands and their corporate partners (see Robinson, 2015).

From all the foregoing, it seems to me that contemporary music festivals — be they constituted as brandscapes or as anti-corporate and democratized sites focused on audience participation — do not have sufficient capacity for visionary projections of society. Apart from profit-making, their primary concern is arguably with creating opportunities for festivalgoers to reaffirm, explore, or reinvent their cultural identities and alliances within the discursive framework of localized and/or postnational imaginings of community. By implication, the dominant form of politics on the global music festival scene today is that of *life politics* (also known under the labels 'identity politics' and 'post-politics'). In contrast to emancipatory politics, which tackles socio-political issues of domination and exploitation in different spheres of human life, life politics is occupied with 'a reflexive relation to the self' (Lury, 2011: 198). There has arguably been a general move away from emancipatory to life politics, at least across Western liberal democracies, coinciding with the period of transition from 'culture' in classical culture industry as a site of power struggles, mediated in and through representation, to 'culture' in global culture industry as a ubiquitous and *thing*ified entity dominating the economic and the everyday. Lash and Lury describe this change also in terms of a shift away from identity to difference; that is, from 'determinacy of objects of culture industry' (resulting in the construction of identities) to 'indeterminacy of objects of global culture industry' (resulting in the construction of difference, with no serious hints of resistance) (2007: 4–5).

But more to the point, life politics, according to Carah, revolves around 'identity and meaning-making processes, and those... [are in turn] located in the social spaces and practices of consumption' (2010: 158). Or put simply, life politics suggests that we are what we consume. Understood this way, life politics has no power to subvert, let alone to fight the current form of capitalism. Life politics emerges, rather, as a successful mode of capital accumulation, reducing even marginal and minoritarian identities to a set of product choices. It is accordingly symptomatic of what Klein calls 'the politics of image, not action' (2000: 124). She specifically claims that the prior focus on structural inequalities, and the use of concrete political and legal remedies to counter them, came to be superseded in the 1990s by the lasting obsession with issues of representation and political correctness. Klein notes in addition:

> ... while it may be true that real gains have emerged from this process, it is also true that Dennis Rodman wears dresses and Disney World celebrates Gay Day less because of political progress than financial expediency. The market has seized upon multiculturalism and gender-bending in the same ways that it has seized upon youth culture in general — not just as a market niche but as a source of new carnivalesque imagery.
>
> (2000: 115)

Moreover, what 'the politics of image' keeps failing to address is the larger question of how the rising power of corporations — both in terms of their size (stemming from continuing consolidation in various industries) and in terms of their political influence — has affected our social and life-worlds. The politics of image is specifically not concerned with discussing a general sense of social insecurity and widening income inequality across the globe (Milanović, 2016), brought about by a shift to a multinational, informational, and consumerist form of capitalism and its corollary, neoliberal ideology and practice, resulting in loss of jobs, a diminished role of organized labour and the welfare state, and fewer market opportunities for small businesses and minority and community groups (Harvey,

1989). In short, the politics of image neglects such crucial issues as 'the corporate hijacking of political power... [and] the brands' cultural looting of public and mental space' (Klein, 2000: 340).

While I am inclined to acknowledge some political value in our consumption choices and habits, since it is through them that we build a sense of who we are and what we stand for, I must at the same time agree with Klein, Žižek, and other like-minded leftist thinkers that political ramifications of such choices are limited in their scope. The latter assertion can be defended on a number of accounts. First, there is a conspicuous lack of transparency in business operations, which makes it virtually impossible for citizen-consumers to determine what goods and services are produced in an ethical way. Second, a strict focus on life politics in (festival) brandscapes detracts attention from the problematic role of corporations in the reproduction of increasing structural inequalities around the globe. And if there is no reflection on such matters, there is not even the possibility of imagining fundamentally different socio-political realities (cf. Carah, 2010). Branding and life politics instead shape contemporary (music festival) culture in a way which enables consumers to suture over basic social antagonisms ensuing from an asymmetrical distribution of resources and power. Besides, people are willing to draw on 'the symbolic fictions... [produced by their social] reality, even if they know them to be false, [not only] because their subjectivity and enjoyment are dependent on those very fictions' (Carah, 2010: 117), but also because they feel that they have no other choice but to partake in brand-building practices. Lastly, life politics dwindles political action into a series of discrete, particularistic demands, thereby preventing (festival) consumers from uniting and acting politically around such universal categories as class and capitalism.

While class and capitalism are indeed rarely part of the political agenda in music festivals today (The Left Field as a travelling stage at several British festivals stands out as a noble exception), they do take a central place in transnational anti-capitalist and anti-war movements that St John (2008) calls *protestivals*. Examples here include Global Day of Action, Carnivals Against Capitalism, For Global

Justice, and Occupy Wall Street. Born out of causes that cut across large segments of the world population (namely, local autonomy, global distributive justice, and peace), these carnivalized and globally orchestrated street gatherings have at least succeeded in putting income inequality back at the centre of political and popular discourse. The reason they have not achieved more is, in Chibber's view, their weak connection to labour, as manifested also in the type of places that underwent 'occupations' (i.e. streets and parks instead of factories) (Farbman, 2017). Chibber's point is clear here: not until social protests begin to disrupt the processes of production and profit-making will the ruling elite acknowledge them.

Considering all above, it is safe to conclude that the (re) production of music festivals as brandscapes ultimately promotes a 'fight for "global capitalism with a human face"' (Žižek, 2008: 459). According to this agenda, as Žižek explains, the reasons for all real antagonisms and problems we confront today are not sought in the system as such. Rather, or so we are told, strategic ends should be achieved by devising ways of making the existing system work more efficiently. Even if this is so, the question still remains whether we should agree to the system which lets corporations dictate and shape our political views and ethical norms. Also, do we feel comfortable with the system which will never allow us to do away with the basic antagonism between capital and labour, simply because this flaw is already inscribed into the system itself? As Chibber warns us, using a vivid analogy between capitalism and cancer:

> This is why socialists have said that you can have a more civilized capitalism, and you should fight for that more civilized capitalism, but understand that it's like a cancer: you can keep giving it chemo, you can fight back the growth of the cancer cells, but they always keep coming back.
>
> (Farbman, 2017)

I turn now to consider whether and how contemporary music festivals can reclaim politics with a capital 'P'. But to accomplish this properly requires a few preliminary words about the nature of political struggles today, as well as about how to address them, both in general and in relation to music festivals.

18.3 Reclaiming Politics in Music Festivals Today

Today, as has always been the case, people's struggles are directed against different forms of oppression and injustice. People specifically fight for parliamentary democracy over autocracy, for the welfare state over neoliberalism, for new forms of democracy over corruption in politics and economy, against sexism and racism, especially against demonization of refugees and immigrants, and against the global capitalist system (see Žižek, in Belinski, 2017). It is within this context that I raise the big questions of what role music festivals have in these struggles, and whether they can make any difference.

In answer to these questions, I argue that contemporary music festivals are typically perceived as politically meaningful when organized in oppressive societies. Serbia's Exit Festival, which was launched in 2000 in the northern city of Novi Sad as a lengthy youth protest against the Milošević regime, provides an excellent example of how music festivals can oppose authoritarian rule and right-wing populism to the point of an actual political change (see Gligorijević, 2019). Another comparable example is Mali's Festival au Désert (Festival in the Desert) (since 2001), which went into exile in 2012 due to threats from Al Qaida-linked extremists. In 2013, the festival was given the Freemuse Award for the continual efforts to 'defend... freedom of musical expression and... keep music alive in the region in spite of extreme Islamists' attempts to silence all music in Mali' (see Freemuse, 2013). There is, in addition, Kubana (since 2009), Russia's biggest open-air festival of international rock music, which moved too (in 2014) from a Black Sea venue to the Kaliningrad region, a Russian exclave on the Baltic Sea, because it did not sit well with the rising political right in the country. But even at its new location, Russian Orthodox activists have protested against the event because of its alleged promotion of 'decadent' behaviour (see Kozlov, 2015).

At the same time, however, there are good reasons to claim that music festivals can produce only limited political effects. First, and as highlighted above, capitalism feeds off carnivalesque imagery, especially in Western liberal democracies whose citizens generally enjoy many civil liberties and political rights. Second, I am sympathetic to Žižek's (2012) view that joyful and transgressive moments of festivity do not really disrupt the realm of everyday life when things get back to normal. As he puts it: '[c]arnivals come cheap — the true test of their worth is what remains the day after'. One might think indeed about the people's disappointment and disillusion after the Exit Festival-related protests leading to the overthrow of Milošević in Serbia or after Greek protestivals against austerity measures at Sentaga Square in Athens. Third and finally, for true social change to happen, it is necessary to escape the trap of what Žižek calls 'false gradualism' (in Belinski, 2017) and what Brah defines as a common 'tendency to assert the primacy of one set of social relations [whether they pertain to class, gender, or race] as against another' (1996: 216). With that said, I contend that the question of left politics in and through music festivals is an important one. While it is true that the new millennium has witnessed the emergence of many protest movements, both nationally and transnationally, none of them seems to have offered a coherent programme. Unlike Klein (2000), in her hope that the anti-corporate activism of the late 1990s would evolve into a large political movement, other leftist thinkers have not been optimistic. According to Harris (2016), the surest sign that the Western left is in crisis is its increasing incapacity to cope with 'three urgent problems: the disruptive force of globalisation, the rise of populist nationalism, and the decline of traditional work'.

If we agree with Žižek (2012) that carnivals can only be the announcement of hard and committed work towards social change and not the end in itself, then they could perhaps function as a means of mobilizing the masses, possibly in a way suggested by Chibber (Farbman, 2017) — namely, by having the left operate outside academia and 'implant itself within labor' as it did in the past. Žižek, for his part, renounces a course of action based on making abstract demands for the abolition of neoliberal global capitalism (in Belinski, 2017). He suggests, rather, following French philosopher Badiou, that left-wing supporters should centre their politics around the so-called *points of impossibility* within the system. They should, in other words, make 'small specific demands' that seem realistic but are simultaneously sensitive for the society in question. In Žižek's view, the point of impossibility for, say, the United States amounts to the idea of universal health care, and for Turkey it is the idea of multiculturalism and minority rights. The main premise here, so Žižek's argument goes, resembles that in sci-fi movies: if you press the right button, the entire system collapses.

Coming close to this suggestion is perhaps also Fabiani's (2014) theorization of art festivals as platforms best suited for *critical interventions* — that is, for tackling pertinent political issues using art and critical discussions. Following McGuigan's revised notion of the public sphere, Fabiani specifically makes a case for the capacity of art festivals to operate in a space conveniently situated between *uncritical populism* (referring here to festivals' uncritical approach to consumerism as a form of citizenship) and *radical subversion* (as articulated in the discursive and performative repertoires of the festival countercultural heritage). It is therefore between these two extremes that possible 'critical interventions' may take place. This is clearly a position which acknowledges the limiting effects of a *transgression model*, whereby 'power' and 'resistance' are said to stand in a relation of binary opposition. By implication, power and resistance are inevitably caught up in a circular struggle, whereby one set of oppositions undergoes the reversal of the status quo soon to be succeeded by another set of oppositions. In contrast to that, the potentially critical space of (music) festivals should rather follow 'a model of articulation as "transformative practice"' (Grossberg, 1996: 88). Within this model, as Grossberg clarifies, the question of identity is rearticulated into an approach to subjects as historical agents, capable of forming alliances in their joint struggle for social change.

Furthermore, since music festivals operate in the micropolitical sphere of society's political practice, they also might assist in the creation of utopias, defined in the Deleuzian terms as the now-here (rather than no-where) places. In such utopias, imaginations of new socio-political realities are thus no longer placed in the future but in the now-and-here timespaces. According to Pisters (2001), of crucial importance here is a critical stance towards society rather than fixed and long-term projections of a perfect society to be reached through revolution. Or in her words, it is critical interventions that might enable 'a "becoming-revolutionary of a people"... available to everybody at any moment in the passing present' (2001: 16). The fact that present-day music festivals are part of the capitalist machinery, which constantly reproduces itself by co-opting its oppositional fringes, does not automatically mean that all their critical interventions are doomed to failure. On the contrary, perhaps the only way to politically engage with the outside world in a meaningful way is to perform critical interventions from within the system; that is, by schizophrenically 'producing and "anti-producing" at the same time' (Deleuze and Guattari, cited in Pisters, 2001: 25). Besides, the idea of total recuperation within the immanent system of capitalism is untenable when attained from a spatial perspective, simply because space itself, even when dominated by the homogenizing images of the spectacle, can never be subjected to a closure. As Massey notes, 'there are always cracks in the carapace' (2005: 116).

To contribute to the discussion at hand, I turn now to the concept of music festivals as *micronational spaces*. After highlighting some of the key assumptions behind this concept, I go on to discuss its derivative — the idea of *microcitizenship* as a possibly emergent form of music festival collectivities — and its political potential for social change in today's world.

18.4 Contemporary Music Festivals as Micronational Spaces

The theorization of music festivals as micronational spaces relies upon a fruitful dialogue between a humanist account of space, as articulated in the work of Lefebvre ([1974] 2009), Massey (2005), and Soja (1996), and the revisited notion of national identity, as emphasized through the use of the 'micronational' terminology. First, I briefly discuss the core assumptions underpinning each aspect of this two-part concept.

The adoption of a spatial theoretical perspective can be justified on multiple grounds. To begin with, Lefebvre's ([1974] 2009) theory of socially produced space allows for a context-sensitive and multidimensional analysis of national identity articulations in light of 'the multiplicity of spaces' that contemporary music festivals may instantiate. On the broadest level, music festivals can be said to perform the function of what Lefebvre calls *consumed spaces*. In his definition, these are unproductive forms of 'the consumption of space' that reflect people's nostalgic search 'for a certain "quality of space"', incorporating such elements as sun, snow, sea, festivity, fantasy, antiquity, and the like ([1974] 2009: 353). However, the consumption of festival spaces is not univocal in its meaning, insofar as such spaces have potential to be transformed into *counter-spaces* by means of 'diversion' (i.e. by having the original space's function put to an alternative use), or into *utopian spaces* by means of domination of the symbolic and the imaginary (i.e. by having the original space appropriated by the work of symbols), or into *organic spaces* by 'looking upon [themselves] and presenting [themselves] as a body' (*ibid.*: 274), or into *masculine spaces* by means of demonstration of phallic power, and so on. Clearly, the multifaceted uses of space, as demonstrated in Lefebvre's analytical insights into the workings of spatial practice in the modern world, resonate profoundly with the ways in which music festival spaces, too, are typically constituted, experienced, and interpreted.

In my considerations of national identities and music festivals, I assert that both of these entities have a profound grounding in space. I also argue that it is through their joint examination from a spatial perspective that a number of crucial insights can be gained and enhanced. The first arises from the possibility of tackling some widely exploited misconceptions associated with the global-local and modern-traditional dichotomies, as well as with the related notions of cultural heritage and cultural difference (see Massey, 2005). The second gain from a spatially guided analysis of national identity amounts to a potentially deeper understanding of the complexity of social relations and material practices coming from both endogenous and exogenous sources and intersecting in music *festivalscapes* (Chalcraft and Magaudda, 2011: 174). And finally, spatial theories may serve as an invaluable discursive source for the perspective that challenges the current form of neoliberal global capitalism — a contextual framework which appears to be taken for granted in a majority of recent festival studies. In Gligorijević (2019), I show in detail how theories of socially produced space can be employed and developed further to assist in a more extensive, fruitful, and politically engaged analysis of the national dimension in contemporary music festivals.

The use of the term 'micronational' is likewise driven by several factors. The first is to underscore awareness that contemporary music festivals are just a tiny piece within the larger system of national identity representation. The second is to emphasize the centrality of the micropolitical sphere (a sphere pertaining to society's 'beliefs' and 'desires') in national identity formations, which precisely incorporates such cultural phenomena as popular music festivals. The third is to show that the micronational discourse that emanates from the idea of music festivals as symbolic microstates seems to help the latter enhance their profile, commercial value, and thus their chances of survival in the increasingly competitive festival market, both nationally and transnationally. And the fourth reason for using the idea of micronationality is to stress the relative autonomy and creative capacity of music festivals to project alternative worlds.

This brings me to the main point: the theorization of music festivals as micronational spaces is based on a dialectic of fixity and fluidity. I specifically argue that music festivals are conceived and staged as symbolic microstates operating in their own right, while simultaneously adhering to the existing policies and dominant regimes of truth within actual nation-states that host them. Or put more elaborately, music festivals are *real places* embedded in the geography of pre-existing locations and their wider networks, national, transnational, and otherwise. Therefore, they invariably draw on the experience of a given locality and actively partake in the (re)construction of *space-based identities*, including nation-building projects. At the same time, a majority of music festivals attempt to surpass the constraints of locality and the given conditions of global 'power-geometries' (Massey, 2005). They function as self-contained worlds, very often envisioned as one-of-a-kind fantasy worlds, or as 'sacred' places that 'festival-pilgrims' around the globe prepare themselves to visit each year. In this aspect of their management and experience, music festivals act perhaps most strikingly as symbolic micronations, largely promoting the ideals of egalitarianism, universality, love, peace, and happiness. As such, they recommend themselves as *utopian places*, predominantly defined by a certain type of attitude, feeling, spirituality, and state of mind. They typically seek to annihilate a sense of space in favour of the experience that foregrounds 'the "time based" identity (contemporary-ness) of cosmopolitanism' (Massey, cited in Simić, 2009: 144). Or as Fabiani puts it, they aspire 'to develop a post-national form of cultural citizenship' (2011: 93).

In the concluding section of the chapter, I likewise draw on the corresponding idea of *microcitizenship* to propose new theoretical terms for imagining music festival collectivities. I do so under the assumption that it is through the invention of new concepts that the world and societies come to be re-described and set in motion towards new futures. To work towards this end, the concept of music festival collectivities is revisited in a way that envisions alternative formations of political identities and alliances alike. Existing theorizations, such as Turner's (1969) *communitas*, Maffesoli's (1996) *neo-tribes*, or Anderton's (2006) *meta-sociality*, are apparently all too apolitical in their implications. Instead, I put forward the idea of *microcitizenship* as a specific form of belonging to festival collectivities.

18.5 Instead of Conclusion: Concepts of Microcitizenship and Festival Coming Communities

The concept of festival microcitizenship is predictably analogous to that of citizenship. They both draw on the same principle of universality, focusing thereby on one's 'position in the set of formal relations defined by democratic sovereignty' rather than on inscriptions of one's identity in cultural terms (cf. Donald, 1996: 174). This way, the political sovereignty of music festival participants and their 'rights of microcitizenship' in festival 'microstates' are guaranteed on equal terms, rather than compromised by divisions between festival community members along cultural lines, which, if drawn, would inevitably include some members but exclude others. As a concept emptied out of cultural meaning, the term *(micro) citizen* is therefore used to 'denote... an empty place... [which] can be occupied by anyone — occupied in the sense of being spoken from, not in the sense of being given a substantial identity' (Donald, 1996: 174).

By extension, the festival microcitizenship calls to mind another concept of a belonging without substantial identity — that of Agamben's (1993) *singularity*. Singularity is a being which is inessential in its nature; that is, a being which is not discernible by

> ... its having this or that property [being red, being French, being Muslim], which identifies it as belonging to this or that set, to this or that class (the reds, the French, the Muslims) — [nor is] it reclaimed... for another class nor for the simple generic absence of any belonging, but for its being-*such*, for belonging itself.
>
> (Agamben, 1993: 1; emphasis in the original)

The terms under which a singularity lays claim of belonging to a wider whole, or to what Agamben calls *the coming community*, are comparable to the metonymical character of the relation that 'the example' holds to a set of items which it is said to exemplify. As Agamben explains, '[n]either particular nor universal, the example is a singular object that presents itself as such, that *shows* its singularity... [by] hold[ing] for all cases of the same type, and, at the same time... [by being] included among these' (1993: 2). A singular becoming of a community is an empty, exterior space of infinite ideational possibilities to which a singularity relates only by means of bordering. In other words, '[b]elonging, being-*such*, is here only the relation to an empty and indeterminate totality' (*ibid.*: 15–16). When applied to music festivals, a politics of singularity would lay the foundations for people's belonging to a common collective a festival coming community — on the grounds of their singularity rather than on a single definition of their cultural identity. Put differently, it is through a politics based on the 'coming community' that various factions of the festival crowd could be pulled together into a political struggle for change.

Rethinking music festival collectivities in political terms (with a capital 'P') apparently opens up the possibility of constructing collective agency across a broad spectrum of the political field, letting music festivals come close to what Soja (1996) calls *thirdspace*. He formulates the latter as 'a space of collective resistance', 'a meeting place for all peripheralized or marginalized "subjects"', and thus a 'politically charged space, [in which] a radically new and different form of *citizenship* (*citoyenneté*) can be defined and realized' (Soja, 1996: 35; emphasis in the original). The festival coming community clearly diverges from Soja's thirdspace in its revisited approach to the notion of resistance (see above) and therefore in its focus on a singularity as the ground of alternative political action (rather than on 'marginalized subjects' and their rights to difference). The festival coming community is in this respect a more inclusive form of affiliation, as it welcomes anyone regardless of their cultural background and their position within the existing structures of power. It is also a form of collective political practice which favours critical interventions to radical movements — in short, '"the project of constructing a form of knowledge that respects the other without absorbing it into the same", or... [into] the different' (Young, cited in Grossberg, 1996: 103).

To advance my argument one step further, I need to refer once again to Grossberg's interpretation of Agamben's *coming community*. He asserts that 'in specific contexts, identity can become a marker of people's abiding in such a singular community, where community defines an abode marking people's way of belonging within the structured mobilities of contemporary life' (Grossberg, 1996: 105). I suggest that music festivals can be understood as one such context — as that 'abode marking people's ways of belonging' and defining their singular becoming of the festival community as a trademark of their collective identity. What makes music festival places especially suited for a singular belonging is arguably a pronounced sense of *throwntogetherness*, a quality of coming together into a now-and-here (itself constituted by 'a history and a geography of thens and theres'), which confronts festival participants with an immediate challenge of negotiating multiplicity (cf. Massey, 2005: 140). This renders festivals a fertile ground for becoming of a community, a meeting place where engagement in a variety of cultural practices can foreground the coevalness of the different trajectories (different spatialities and temporalities) that create particular places and identities, but also point to the workings of power and exclusion in the social relations that construct those places and identities. Because of this truly democratic potential of festival spaces, an infinite number of possibilities for political action may mobilize and organize festival microcitizens into a coming community. And just as the festival coming community is always in a state of becoming, constantly changeable, unfinished, undetermined, and dependent on historically contingent processes and social practices, so is the scope of its political engagement, emerging on 'a continually receding horizon of the open-minded-space-to-come, which will not ever be reached but must constantly be worked towards' (Massey, 2005: 153).

References

Agamben, G. 1993. *The Coming Community*, translated by M. Hardt. Minneapolis: University of Minnesota Press.

Anderton, C. 2006. *(Re)Constructing Music Festival Places*. PhD thesis, University of Wales, Swansea.

Azara, I., and D. Crouch. 2006. 'La Cavalcata Sarda: Performing Identities in a Contemporary Sardinian Festival'. In *Festivals, Tourism and Social Change: Remaking Worlds*, edited by D. Picard, and M. Robinson, 32–45. Clevedon: Channel View Publications.

Bartleet, B. 2014. '"Pride in Self, Pride in Community, Pride in Culture": The Role of Stylin' Up in Fostering Indigenous Community and Identity'. In *The Festivalization of Culture*, edited by A. Bennett, J. Taylor, and I. Woodward, 69–85. Farnham: Ashgate.

Bennett, A., ed. 2004. *Remembering Woodstock*. Burlington, VT: Ashgate.

Bennett, A., and I. Woodward. 2014. 'Festival Spaces, Identity, Experience and Belonging'. In *The Festivalization of Culture*, edited by A. Bennett, J. Taylor, and I. Woodward, 11–25. Farnham: Ashgate.

Blackett, T. 2003. 'What Is a Brand?'. In *Brands and Branding*, edited by R. Clifton, and J. Simmons, 13–25. London: Profile Books.

Brah, A. 1996. *Cartographies of Diaspora: Contesting Identities*. London: Routledge.

Carah, N. 2010. *Pop Brands: Branding, Popular Music, and Young People*. New York: Peter Lang.

Carnegie, E., and M. Smith 2006. 'Mobility, Diaspora and the Hybridisation of Festivity: The Case of the Edinburgh Mela'. In *Festivals, Tourism and Social Change: Remaking Worlds*, edited by D. Picard, and M. Robinson, 255–268. Clevedon: Channel View Publications.

Caton, K., C. Pastoor, Y. Belhassen, B. Collins, and M. Wallin. 2013. 'Christian Music Festival Tourism and Positive Peace', *The Journal of Tourism and Peace Research* 3(2): 21–42.

Chalcraft, J., and P. Magaudda. 2011. '"Space is the Place". The Global Localities of the Sònar and Womad Music Festivals'. In *Festivals and the Cultural Public Sphere*, edited by G. Delanty, L. Giorgi, and M. Sassatelli, 173–189. London: Routledge.

Donald, J. 1996. 'The Citizen and the Man about Town'. In *Questions of Cultural Identity*, edited by S. Hall, and P. Du Gay, 170–190. London: Sage.

Dowd, T. 2014. 'Music Festivals as Trans-National Scenes: The Case of Progressive Rock in the Late Twentieth and Early Twenty-First Centuries'. In *The Festivalization of Culture*, edited by A. Bennett, J. Taylor, and I. Woodward, 147–168. Farnham: Ashgate.

Fabiani, J. 2011. 'Festivals: Local and Global. Critical Interventions and the Cultural Public Sphere'. In *Festivals and the Cultural Public Sphere*, edited by G. Delanty, L. Giorgi, and M. Sassatelli, 92–107. London: Routledge.

Gligorijević, J. 2019. *Contemporary Music Festivals as Micronational Spaces: Articulations of National Identity in Serbia's Exit and Guča Trumpet Festivals in the Post-Milošević Era*. PhD thesis, University of Turku.

_____. Forthcoming, 2021. 'Practices and Ideological Effects of Music Festival Branding: The Case Study of Serbia's Two Major Music Festivals, Exit and Guča'. In *Musik & Marken/Music & Brands*, edited by M. Ahlers, L. Grünewald-Schukalla, A. Jóri, and H. Schwetter. *Yearbook of the German Association for Music Business and Music Culture Research* (GMM).

Grossberg, L. 1996. 'Identity and Cultural Studies: Is That All There Is?' In *Questions of Cultural Identity*, edited by S. Hall, and P. du Gay, 88–107. London: Sage.

Harvey, D. 1989. *The Condition of Postmodernity: An Enquiry into the Origins of Cultural Change*. Cambridge, MA: Blackwell.

Hesmondhalgh, D. 2014. *Why Music Matters*. Chichester: John Wiley and Sons.

Hilton, S. 2003. 'The Social Value of Brands'. In *Brands and Branding*, edited by R. Clifton, and J. Simmons, 47–64. London: Profile Books.

Hughes, H. 2006. 'Gay and Lesbian Festivals: Tourism in the Change from Politics to Party'. In *Festivals, Tourism and Social Change: Remaking Worlds*, edited by D. Picard, and M. Robinson, 238–254. Clevedon: Channel View Publications.

Klein, N. 2000. *No Logo: Taking Aim at the Brand Bullies*. London: Flamingo.

Lash, S., and C. Lury. 2007. *Global Culture Industry*. Cambridge: Polity Press.

Lefebvre, H. [1974] 2009. *The Production of Space*, translated by D. Nicholson-Smith. Malden, MA: Blackwell Publishing.

Luckman, S. 2014. 'Location, Spatiality and Liminality at Outdoor Music Festivals: Doofs as Journey'. In *The Festivalization of Culture*, edited by A. Bennett, J. Taylor, and I. Woodward, 189–205. Farnham: Ashgate.

Lury, C. 2011. *Consumer Culture*. Second Edition. Cambridge: Polity Press.

Maffesoli, M. 1996. *The Time of the Tribes: The Decline of Individualism in Mass Society*. London: Sage.

Malpas, J. 2009. 'Cosmopolitanism, Branding, and the Public Realm'. In *Branding Cities: Cosmopolitanism, Parochialism, and Social Change*, edited by S. Hemelryk Donald, E. Kofman, and C. Kevin (ebook). London: Routledge.

Marschall, S. 2006. 'Creating the "Rainbow Nation": The National Women's Art Festival in Durban, South Africa'. In *Festivals, Tourism and Social Change: Remaking Worlds*, edited by D. Picard, and M. Robinson, 152–171. Clevedon: Channel View Publications.

Massey, D. 2005. *For Space*. London: Sage.

McKay, G., ed. 2015. *The Pop Festival: History, Music, Media, Culture*. London: Bloomsbury.

Milanović, B. 2016. *Global Inequality: A New Approach for the Age of Globalization*. Cambridge, MA: Harvard University Press.

Pisters, P. 2001. 'Introduction'. In *Micropolitics of Media Culture: Reading the Rhizomes of Deleuze and Guattari*, edited by P. Pisters, 7–26. Amsterdam: Amsterdam University Press.

Robinson, R. 2015. *Music Festivals and the Politics of Participation* (ebook). Farnham: Ashgate.

Sharpe, E. 2008. 'Festivals and Social Change: Intersections of Pleasure and Politics at a Community Music Festival', *Leisure Sciences* 30(3): 217–234.

Simić, M. 2009. *'Exit to Europe': State, Travel, Popular Music and 'Normal Life' in a Serbian Town*. PhD thesis, University of Manchester.

Soja, E. 1996. *Thirdspace: Journeys to Los Angeles and Other Real-and-Imagined Places*. Cambridge, MA: Blackwell.

St John, G. 2008. 'Protestival: Global Days of Action and Carnivalized Politics in the Present', *Social Movement Studies* 7(2): 167–190.

Stone, C. 2009. 'The British Pop Festival Phenomenon'. In *International Perspectives of Festivals and Events: Paradigms of Analysis*, edited by J. Ali-Knight, M. Robertson, A. Fyall, and A. Ladkin, 205–224. San Diego: Elsevier.

Taylor, J. 2014. 'Festivalizing Sexualities: Discourses of "Pride", Counter-Discourses of "Shame"'. In *The Festivalization of Culture*, edited by A. Bennett, J. Taylor, and I. Woodward, 27–48. Farnham: Ashgate.

Turner, V. W. 1969. *The Ritual Process: Structure and Anti-Structure*. London: Routledge.

Žižek, S. 2004. 'Passion in the Era of Decaffeinated Belief', *The Symptom: Online Journal for Lacan.com* 5. Available online at https://www.lacan.com/passionf.htm

Žižek, S. 2008. *In Defense of Lost Causes*. London. Verso.

Other Media

Belinski, I. 20 March 2017. 'Slavoj Žižek — Beyond Mandela Without Becoming Mugabe (aka How to Rebel)' (online video of public lecture). Available online at https://www.youtube.com/watch?v=a5DiZBb8f6A

Delistraty, C. 24 July 2014. 'Commercializing the Counterculture: How the Summer Music Festival Went Mainstream' (online article, *Pacific Standard*). Available online at https://psmag.com/social-justice/commercializing-counterculture-summer-music-festival-went-mainstream-86334

Fabiani, J. 3 April 2014. 'Changes in the Public Sphere (1983–2013)' (online article, *Eurozine*). Available online at https://www.eurozine.com/changes-in-the-public-sphere-1983-2013/

Farbman, J. 23 March 2017. 'Workers Hold the Keys: An Interview with Vivek Chibber' (online article, *Jacobin*). Available online at https://www.jacobinmag.com/2017/03/working-class-unions-labor-precarious-higher-education/

Freemuse. 6 February 2013. 'Freemuse Award Winner 2013: "Festival in the Desert" in Mali' (online press release, *Freemuse*). Available online at https://freemuse.org/def_art_freedom/freemuse-award-winner-2013-festival-in-the-desert-in-mali/

Harris, J. 6 September 2016. 'Does the Left Have a Future?' (news article, The *Guardian*). Available online at https://www.theguardian.com/politics/2016/sep/06/does-the-left-have-a-future

Kozlov, V. 13 April 2015. 'Russian Music Festival Kubana Feels Heat from Right-Wing Critics, Falling Ruble', (online article, *Billboard*). Available online at http://www.billboard.com/articles/columns/music-festivals/6531633/russia-kubana-festival-conservative-criticism-ruble

Pasbani, R. 28 November 2016. 'Montreal Fest Cancelled after Protests Against Black Metal Band with Neo-Nazi Ties Erupts' (online article, *Metal Injection*). Available online at http://www.metalinjection.net/politics/montreal-fest-cancelled-after-protests-against-black-metal-band-with-neo-nazi-ties-erupts

Simonsen, E. 15 June 2015. 'Music Festivals: The Rise of Commercialism and Its Effects' (online article, *Odyssey*) Available online at https://www.theodysseyonline.com/music-festivals-commercialization-discrimination-rape-death (last accessed 7 January 2017)

Spencer, L. n.d. 'The Modern Commercialised "Music" Festival', (blog, *Daily Dischord*). Available online at http://www.dailydischord.com/the-modern-commercialised-music-festival/ (last accessed 7 January 2017)

The Local. 18 October 2016. 'Outcry After "Neo-Nazi" Music Festival Held on Swiss Soil' (online article, The *Local*). Available online at https://www.thelocal.ch/20161018/outcry-after-neo-nazi-concert-held-on-swiss-soil

Žižek, S. 24 April 2012. 'Occupy Wall Street: What Is to Be Done Next?' (newspaper article, The *Guardian*). Available online at https://www.theguardian.com/commentisfree/cifamerica/2012/apr/24/occupy-wall-street-what-is-to-be-done-next

Chapter 19

Construction of Protest Space Through Chanting in the Egyptian Revolution (2011): Musical Dimensions of a Political Subject

Oscar Galeev
Yenching Academy, Peking University,
Beijing, P. R. China 100871
oscargaleev@gmail.com

19.1 Introduction: Politics of Protest Chanting

The waves of protests and, in some cases, mass uprisings in the early 2010s resonated from Zuccotti Park in New York City to Change Square in Sana'a, Yemen, and from Syntagma in Athens to Maidan in Ukraine. Despite their varying outcomes, the Arab Spring, the Occupy movements, and, in general, the revolutions of 2011 shared similar decentralized forms of organization and positioning, each with a particular urban space as its focal point (Mitchell, 2012: 8). Notably, the common protest dynamics and the very vocabulary

Musical Spaces: Place, Performance, and Power
Edited by James Williams and Samuel Horlor

ISBN 978-981-4877-85-5 (Hardcover), 978-1-003-18041-8 (eBook)
www.jennystanford.com

of these movements were centred on specific sites such as parks, streets, boulevards, and squares. In this context, Tahrir Square deserves special attention not only as the most iconic site of the Arab Spring, but also as a vivid example of popular politics emerging from a range of new uses and practices of space in the 2010s (Gregory, 2013: 235; Sassen, 2011: 573). Occupation and encampment in this small square — small compared to the scope of its impact — exposed the fragility of the Egyptian political order and shaped everyday practices of the Arab Spring protests (Bayat, 2012: 119). In order to understand how new street politics could become a successful tool in formulating a demand for political change, we must look into some key protesters' experiences of encampment, of the distinctive sensuous space constructed in Tahrir Square in 2011.

Slogans and chants such as 'People want the fall of the regime'[1] became the most transferrable and universal cultural elements of the Arab Spring all across the Middle East and North Africa. Accessible to all protesters, catchy chants and tunes arose as the most democratic form of expressing a demand for social and political change. Unlike any written information distributed on paper or on the internet, chants were equally comprehensible for all classes, including the urban poor, who were often illiterate and deprived of basic social and economic rights as a result of vast rural to urban migration in the region in the last decades. And it was chanting and the sounds of protesting crowds that often attracted more and more people to the streets and squares (Sanders and Visonà, 2012: 216). However, so far there has been very little research on how singing and chanting might be constitutive of spatial practices of protests. My study started with personal observations and conversations with some participants of the 2011 and 2013 protests on Tahrir, many of them emphasizing the role of musical uses of the human voice in their collective experience of the Egyptian Revolution. Trying to understand why soundscape and the human voice had such a power in shaping the aesthetics of the protests, I continued researching individual experiences of Tahrir protesters through their reports, interviews, videos, personal published stories, and tweets sent from Tahrir during the January 2011 Revolution and later collected by researchers into anthologies. In addition to these sources, I

[1] الشعب يريد إسقاط النظام

conducted an online field study, designed as an anthropological questionnaire and consisted of fifteen open questions in English and Arabic. Extensive responses from thirteen respondents, whose ages range from 17 to 40, describe both protesters' memories of Tahrir encampment events and their accompanying subjective feelings and observations. It might appear surprising that I use the data from tweets alongside later recollections of the events, such as memoires and stories. Nevertheless, the imposed limit of 140 characters, together with the immediacy and urgency of a message sent from the protesting crowd, creates an almost unmediated report, and this can be helpful in understanding the here-and-now experience of a protest (Gregory, 2013: 240). Personal accounts demonstrate how the sounds of the human voice are perceived by protesters and how they interplay with other noises accompanying the protest. However, it should be noted that this methodology, and the usage of a broad variety of sources, bears a risk of selectively picking the data relating to the protester's sonic experiences and excluding other complex facets of a comprehensive collective experience of the protest.

Overall, I will argue that the musicality of Tahrir lies at the core of its political aesthetics and of the sense of 'vivid present' in the days of the revolution. The soundscape of protest chants is instrumental in the formation of what Bakhtin (1984: 223) considers a 'terrain of sensuous embodied existence' (Gardiner, 2002: 66). This is what makes the soundscape of the street and the square into a politicized space in the Egyptian Revolution. Here, I am using a broader interpretation of the term 'soundscape' than simply meaning a sonic environment and the interactions of people, in this case, of city residents with this environment (Bijsterveld *et al.*, 2013: 35). In Schafer's (1994) original conceptualization, a soundscape is a relationship between individual experiences and the physical and cultural totality of space. But in a soundscape of a political protest, two additional levels of complexity arise. First, almost all people present in such a sounding space are the primary soundmakers; a site-specific soundscape, therefore, cannot be regarded as an ambient sound environment. It is rather a part of a coordinated and shared experience where sounds from all sources become a narrative medium (Kalinak, 2015: 99). Second, as protesters consciously construct a certain atmospheric effect in a sounding space, they all take part in a collective sound design (*ibid.*: 2). Using this expanded definition, I view a soundscape here as a designed acoustic backdrop without a clear demarcation between sound producers and an audience. This allows us to focus instead on a new collective political subject emerging in the revolution, a subject reflected in their embodied music cognition in Tahrir. As a result, a soundscape governed by the human voice and the chants of the protesters creates a form of musical communication that allows the message of the protest to be spread far beyond the physical limits of Tahrir Square.

19.2 Musical Rhythms of Protesting in Tahrir Square

Designed in the nineteenth century under Khedive Ismail, Tahrir remains a unique place in Cairo for its well-determined architectural planning (El-Menawy, 2012: 223). Nine government ministries, the Presidential Palace, the administrative Mogamma building, which notably was shut down during both 2011 and 2013 protests, and the headquarters of the National Democratic Party and of the Arab League surround the busy traffic circle that became the heart of the 2011 encampment. In such a conglomeration of symbols of state power around the square, the regime (النظام) opposed by the revolutionary crowd was not only the authoritarian Egyptian state in a narrow sense, but all that 'constituted the politically and socially familiar' in the Egypt of that time (Sabea, 2014: 72). A distinctive immersive auditory experience of Tahrir, or its soundscape under Mubarak's regime, also constituted a part of the *familiar* and of the normalized in the everyday life of Cairo. And from the very beginning of the protest, the collective resistance aimed to subvert all language of power in Tahrir, including through its soundscape (Kraidy, 2016: 155).

Schafer coins the term 'soundmarks' to describe sounds specific to a geographical area and easily recognizable by a community (1994: 78). How the soundmarks of a revolutionary Tahrir were identified by the dwellers of Cairo can be seen in the account of a participant of the eighteen days of the Egyptian Revolution who describes his experience of protesting as a repetitive movement through the city.

For this Egyptian young man, the entire revolution consisted of his daily walks from his grandmother's house to Tahrir Square and back; as he approached downtown Cairo and the streets leading to the square, he would hear the intensifying sounds of the crowd, of people '... chanting, discussing, planning, hoping...' (Al-Zubaidi and Cassel, 2013: 64). The most popular chants such as 'People want the fall of the regime' or 'Revolution until victory', for this protester, became familiar and constantly reproduced soundmarks with their own unchanging musical rhythm:

Rhythm: Ash-shàb yurìd isqàt àn-nizàm[2]

Translation: People want the fall of the regime!

Original: ماظنلا طاقسإ ديري بعشلا

Rhythm: Thàwra thàwra hàti nàsr

Thàwra fì kul shuàri màsr

Translation Revolution, revolution until victory!

Revolution in all streets of Egypt!

Original: رصم عراوش لك يف ةروث..رصنلا يتح ةروث ةروث

This illustrates the first obvious function of a protest soundscape: it allows the protesters to manifest their presence in the square and to attract more people. But hearing noises of the protest in this context is not merely a passive act; the senses mediate between the built environment and the self and, thus, become geographical as they create awareness of spatial relationships (Rodaway, 1994: 37). As a result, the soundscape of Tahrir is not static but rather evolving together with people's interpretation of the protests. Looking into the tweets sent during the January 2011 Revolution, we see that inhabitants of houses and hotels around the square regularly write about the slogans and chants that they could hear from their rooms (Idle and Nunns, 2011: 52, 70, 85, 90). On many occasions, the crowd would gather under residential apartment blocks and chant 'Come down to Tahrir!' with an accelerating tempo, waiting for people to come out of their apartments and into the streets.

---

[2]The rhythmic transcriptions in this chapter are not the standard Arabic transliteration; they only serve to represent accents and rhythm. Accented and underlined vowels in each line are the ones stressed in chants.

Rhythm:	Ìnzìl Tàhrìr
Translation:	Come down to Tahrir!
Original:	انزل تحرير

As for the protesters directly immersed in the soundscape of Tahrir, the sounds inside and around the square reflect both the fear and the excitement of the political protest; in such a way, the participants tweet about the sounds of police helicopters, army fighter jets, and bullets, contrasting them to the joy of hearing loud chants, songs, and poems in the streets (*ibid.*: 73, 86, 88, 127). A grave risk stemming both from a lack of authority and from participating openly in the demonstrations seems to amplify the emotional response to all forms of sensory stimuli. Consequently, feelings of vulnerability on the one hand and invincibility on the other hand, as well as the utopian visions of the revolution, become directly linked to the soundscape of the square. For instance, user @alla tweets on 31 January: '... massive crowds singing dancing we cannot be defeated' (*ibid.*: 146). Another protester answering the questionnaire for this study compares the feeling of unity and overwhelming collective excitement created by 'chanting together' to traditional chanting during football games. The soundscape of revolutionary Tahrir, therefore, exists in a constant interaction with the city as a whole; even years after the January Revolution, people of Cairo report how the atmospheric and sensuous aspects of the encampment produced a memorable and unmistakable experience of the protesting square.

Of course, sensuous geographies of Tahrir were formed by visual as well as auditory perception. However, as Rodaway argues, visual perception inevitably locates and abstracts each object in the surrounding space by focusing on it (1994: 114); the ear, on the other hand, 'favours sound from any direction' and constitutes a much more comprehensive social experience. Consequently, a soundscape is much harder to manipulate, but strategic control of soundscapes can be employed by those in power to achieve a desired atmospheric effect and social order; sound may act as an audio control tool, intensifying and demarcating space and fostering or hindering certain social interactions (Yang and Guaralda, 2013: 2). Such a difference between normal and revolutionary modes of the Tahrir soundscape is eloquently described by a protester who had heard the call to go to the street on the morning of 25 January. She says that she '... checked out the streets and the reaction of the people, and everything was quiet and there was no evidence of anything abnormal...' (Al-Saleh, 2015: 94). Later on, however, she notices a group of policemen heading to the city centre and finds her way to the protest by following the street chants: '... the voice of the people was our guide, and I melted in its melody' (*ibid.*).

Such a 'voice of the people' bears the connotations of both sounds uttered and of a capacity to express oneself freely; this duality can be traced throughout Tahrir protests as Egyptians grew more and more aware that the spaces of public protest enabled them to be heard, both metaphorically and literally (Al-Saleh, 2015: 12). For instance, an Egyptian immigrant in the US witnessing the January 2011 Revolution writes about an '... intangible sense of oppression and the insignificance of my own voice that I felt in my country...' (*ibid.*: 100). Hence, revolutionary Tahrir as a distinct soundscape involves active *sounding* to the same extent as *hearing*, but both are related to everyday experiences of auditory freedom and control. One of the protesters participating in the questionnaire describes the sounds of Tahrir during January 2011 as 'a storm, volcano, thunderstorms...' This imagery of eruptive and repetitive sounds establishes a rhythm for the protest that is defined by public musical practices different from those before the revolution. Thus, while a normal soundscape manifests acceptance of political control, with its disorganized shouts, chattering, and the passing of cars, or its quietness and the absence of sound, a loud and organized chanting in the contested space embodies the distinct aesthetics of revolt. For instance, a typical performance of a chant such as 'Change, freedom, and social justice' often includes multiple repetitions with a gradually accelerating tempo. After a number of repetitions and consecutive cheering and whistling, the crowd catches breath and in a few moments another protester initiates a new chant with a different rhythm:

Rhythm:	Taghìr, hurìya, eadìl ijtìma-ìya
Translation:	Change, freedom, and social justice!
Original:	ةيعامتجإ ةلادع ةيرح رييغت

Schutz explores voice as a means of musical or acoustic rather than linguistic communication and focuses on interpretation by re-creation (1976: 239); whether music is performed or merely listened to does not change the common scheme of interpretation, the mechanism by which each participant creates musical meaning. Moreover, the element of repetition creates a 'virtual unity' both within the musical piece itself and in a sense of accord between the performers and the audience. He claims that the musical province of meaning possesses its own cognitive style and leads to the emergence of new forms of spontaneity, self-experience, and sociality (*ibid.*: 232). Schutz concludes that the emerging musical community unifies individual experiences into the vivid present; this emphasis on the uniqueness of musical experiences has larger implications for collective performance and potential functionalities of singing (Skarda, 1979: 88). For instance, Shklovsky focuses on the rhythms of work songs and claims that collective activities become automatic through the musical rhythm (2004: 20). Therefore, virtual unity is based on unconscious automatization, for example, in marching accompanied by music, in chant, or in animated conversation. Such musical rhythmicity is often mentioned among the primary features of Tahrir's soundscape. One of the participants of the early days of the protests describes a pattern of the crowd repeatedly chanting a slogan and finishing in roaring applause (Al-Zubaidi and Cassel, 2013: 62). As the protest gained more and more supporters, the musical rhythms of Tahrir only multiplied; user @ashrafkhalil tweeted: '1:30 am and the crowds in Tahrir are still large and LOUD...' (Idle and Nunns, 2011: 148). The loudness of many voices arguing as well as aligning in this soundscape is usually mentioned by the participants in comparison to traditional Egyptian street festivals such as Mawlid.[3] Interestingly, user @beleidy employs this metaphor in a tweet, writing that 'Tahrir Square is a large festival' (*ibid.*: 90). Instead of saying that Tahrir is a place hosting a festival, this protester seems to suggest that Tahrir Square itself is celebrated.

A single slogan, a collective chant, a poem, and a song can serve various functions in a protest; they can all be used as a tool for both inciting an action against the authority and for restraining violence (Butler, 2011: 5). This is made possible by the combination

[3]The birthday of the Prophet celebrated as a national holiday in most Muslim countries.

of sounds and musical rhythms in the production of a 'vivid present' using Schutz' term. This vivid present has an agitating and galvanizing effect by combining the soundscape of the square with individual experiences of the revolutionaries. The loudness and pace of the chants only take effect at the moment of collective performance, but they are still built upon pre-revolutionary musical practices. Therefore, chanting allows for a creative rethinking of the soundscape that transforms protesters as well as the space in which the protest unfolds. Gregory argues that Tahrir Square prior to the revolution was reduced to a space of circulation (of people, vehicles, ideas) but not of communication (2013: 241). It should not be surprising, therefore, that chants during the protest were employed as a form of disruptive musical communication, aligning voices of the protesters in opposition to the controlled soundscape of everyday Egyptian political life.

19.3 Soundscape of the 'Carnivalesque'

The spontaneity of the musical creates a new relationship between the sensing body of a chanting protester and the immense number of sounds, voices, and noises filling Tahrir Square in the days of the revolution. Tahrir had '... no one sound', says one respondent to the questionnaire, a 29-year-old public health specialist: '... it was like a festival with different stages and groups organizing activities everywhere'. Hardt and Negri speak of a protest being carnivalesque not only in its atmosphere but also in its organization (2004: 198). The carnivalesque, a term they borrow from Bahktin, is rooted in the freedom of singularities that 'converge in the production of the common' (*ibid.*). Such a socio-political reality, expressed in the soundscape of Tahrir, is very different from the familiar political sphere in Egypt. Characterized by the absence of a centralized leadership, the protest of January and February 2011 in Tahrir is constantly referred to as a joyous celebration, what Bakhtin considers 'a feast for the whole world' (1984: 223). On Sunday 6 February, after hundreds were killed that very week through Egypt, a user by the name of @tarekshalaby tweets: 'Tahrir sq. Has been very festive/entertaining with songs n poetry [sic]', and twelve minutes later adds: 'At Tahrir sq. you can find popcorn, couscous, sweet potatoes, sandwiches, tea & drinks! Egyptians know how to revolt!' (Idle and Nunns, 2011: 157). The archetypical feast and celebratory spirit amidst the danger are made possible by breaking with the aesthetic limits of everyday formalism (Hardt and Negri, 2004: 209). As opposed to a monologic such as that of official state ideology, polyphony and heteroglossia — a multiplicity of voices and performances — create a dialogue between diverse singular subjects according to Bakhtin (1984: 159; Hardt and Negri, 2004: 211). Such a heteroglossia, of course, is projected into the soundscape of the protest, submerging the protesters into the polyphony of human voices that fills the streets of Cairo.

As a result of a constant dialogue involving multiple subjects, the carnivalesque celebration breaks with all conventional social divisions and turns Tahrir into a collection of *mawlids* or festivals (Keraitim and Mehrez, 2012: 44). What Bakhtin calls 'the carnivalistic misalliances' allows for the bringing together of the profane and the sacred, the young and the old (1984: 180). Keraitim and Mehrez note that, in such a manner, the *mawlids* of Tahrir '... undo established social, gender, and class boundaries' (2012: 44); poor and rich, villagers and urban dwellers, younger and older generations, and men and women share the public sphere here, and this enables them 'to speak, to sing and to interact' (*ibid.*). One of the respondents to the questionnaire, a middle-aged public health specialist, notes how unusual it was to see 'different types of people', in terms of social class, religion, and political affiliation, in one place and performing the same acts together in the streets. Notably, the festival-like episodes do not necessarily invent new modes of celebration but rather incorporate the traditional culture of festivities into the protest. In such a manner, many of the ceremonies and rites familiar to the people of Cairo and to all Egyptians are creatively used in the revolution. Mixing everyday experiences with the music of religious rituals such as Sufi *dhikr* circles or weddings became the defining practice of Tahrir (*ibid.*: 45). Two users, @sandmonkey and @tarekshlaby, tweet from Tahrir, respectively, 'Today, a Christian mass was held in Tahrir, two people got married, and a couple is spending their honeymoon there. Awesome', and 'There's a couple that's about to have their katb el ketab [wedding ceremony] at Tahrir sq. in front of the revolutionary crowd!' (Idle and Nunns, 2011:

161, 163). What emerges in the square in such moments is not an undifferentiated unity of the people demanding a political change, but quite the opposite, 'the plural singularities of the multitude', finding new commonalities despite habitual social distinctions (Hardt and Negri, 2004: 99). The chants, songs, poetry, and religious recitations accessible to almost everyone in everyday community festivals, prayers, weddings, and other celebratory occasions become absorbed into the protest as a result of its carnivalesque organization. Tahrir itself is celebrated and, as a result, music of all that is celebratory in everyday life gravitates to the square.

In addition to a capacity for creativity that is forged by the paradoxical organization of the carnivalesque, a protest is characterized by the domination of affect and polarized emotionality (Hardt and Negri, 2004: 220). In an example of Bakhtin's loud carnivals, imagination and desire are interlinked through 'the power of human passions' (*ibid.*: 210). Reading through the stories of Tahrir protesters, it is surprising how exalted and amplified their feelings of joy and terror are in public spaces, and how akin they are to Bakhtin's descriptions of medieval carnivals as contained moments of rebellion (1984: 49; Werbner, Webb, and Spellman-Poots, 2014: 11). The protesters from Cairo who responded to the questionnaire tell of extreme excitement, jocundity, and pride about their 'constant desire to scream loudly' in the square. On the other side of the emotional spectrum, a fear of police brutality and harassment, as well as anger and sheer terror of being killed or hurt in the protest, counterweighted the celebratory joy, and in the responses to questions about their personal experiences in Tahrir, the protesters speak in extreme terms of 'feeling alive' in one moment and fearing death in another. One protester associates the sounds and chants coming from the streets with the unimaginable, with a 'dream that just became a reality',[4] while another teenage girl tells how she would cry from joy and feel an urge to 'join the revolutionaries'[5] when hearing the crowd from her home. I specifically mention the emotional power of sounds heard from the streets not only to emphasize the enlargement of the soundscape of Tahrir into private

---

[4] لعفلاب كرحت عيمجلا نأ رعشت ثيح ةلاحلا هذه يف ةجهبلا ةديدش رعاشملا تناك و
ةقيقح ملحلا حبصأ و

[5] ىلإ يرجأ نأ ول دوأ تنك ، لوزنلا نم يل يلهأ عنمو ةحرفلا نم امنيح يكبأ تنك
راوثلاب قحلأو هحتفأل بابلا

spaces, but also to picture Tahrir as an audible and loud carnival projecting its emotional power back into the city.

The multitude of people engaged in a carnival undermined the ideological production of meaning in the square, and musical self-expression is instrumental to this process (Gardiner, 2002: 66). Ideology that rules the architecture and soundscape of the modern city, for Hardt and Negri, erodes any space for the common (2004: 202). But musical communication of the carnival allows for the bringing to light of a culture 'of the loud word spoken in the open in the street' (*ibid*.: 182). If we approach the political subject of Tahrir as a carnivalesque one, it becomes clear how chanting serves to resist the 'monolithic seriousness' of officialdom, leading to a new relationship between the protesters and the public space around them (Sanders, 2012: 144). The distinctive soundscape of a protest encampment in Tahrir featured a variety of musical genres, sounds of laughter, cries, and loud screams; as the protesters employed sounds and music to feel invincible in some moments and united in others, the musical dimension of the protest not only reflected the carnivalesque of its political organization, but also shaped and guided the protest subjects.

19.4 Conclusion

This chapter has argued than an encompassing sensory experience of the participants of the Egyptian Revolution was crucial in the making of its political subject, a multitude of equal voices fighting the regime power with the 'carnivalesque overtone in everyday life' (Bakhtin, 1984: 154). For Werbner, Webb, and Spellman-Poots, the lived moments of mass demonstrations unveil the centrality of bodily, earthy experience to the constitution of political subjects (2014: 12). This short study has looked further into the role that the musical qualities of loud speech might play in such a political act. If chanting helps the protesters in subverting the controlled space by embodied musical practices, the soundscape of a revolutionary street is continuously politicized and contested in the protest. As a result, the soundscape of a public protest and formation of a protest subject enable one another.

For this reason, silencing the soundscape, or annihilating the voices of people in the square, was an essential part in reinstalling the regime's ideological power, of bringing the revolution to an end. Literally silencing the protest allowed it to put an end to the distinctive emotional experience that drew the protesters into the streets and extended the protest further into the public spaces of Egyptian cities.

While this research has briefly outlined a connection between musical experiences of the protest and construction of the protest space and its political subject, there are potential directions to note for future research into this topic, both in the case of the Egyptian Revolution 2011 and more broadly concerning the nature of street protests in the 2010s. Firstly, does the human voice hold a privileged position in collective revolutionary aesthetics compared to other forms of sound production? How does it relate to the historical culture of loud speaking and music production in city squares and streets? Secondly, the radical carnivalesque in Tahrir managed to disrupt the political performance of the Egyptian state by creating a space that, in Bakhtin's words, 'could be touched, that is filled with aroma and sound' (1984: 185). Then how can the same sensuous experiences be used in everyday life as a political tool to counter state oppression? For instance, what are the soundmarks of political resistance and discontent in quotidian human interactions in public spaces? Hopefully, future research can pay attention to these questions to further explore the connection between musical cognition and contemporary politics.

References

Al-Saleh, A. 2015. *Voices of the Arab Spring: Personal Stories from the Arab Revolutions*. New York: Columbia University Press.

Al-Zubaidi, L., and M. Cassel. 2013. *Diaries of an Unfinished Revolution: Voices from Tunis to Damascus*. London: Penguin.

Bakhtin, M. 1984. *Rabelais and His World*, translated by H. Iswolsky. Bloomington: Indiana University Press.

Bayat, A. 2012. 'Politics in the City-Inside-Out', *City & Society* 24(2): 110–128.

Bijsterveld, K., A. Jacobs, J. Aalbers, and A. Fickers. 2013. 'Shifting Sounds: Textualization and Dramatization of Urban Soundscapes'. In

Soundscapes of the Urban Past: Staged Sound as Mediated Cultural Heritage, edited by K. Bijsterveld, 31–66. Bielefeld: Transcript Verlag.

Butler, J. 2011. 'Bodies in Alliance and the Politics of the Street', *Transversal* 10/2011. Available online at http://transversal.at/transversal/1011/butler/en

El-Menawy, A. 2012. *Tahrir: The Last 18 Days of Mubarak: An Insider's Account of the Uprising in Tahrir.* London: Gilgamesh.

Gardiner, M. 2002. *Critiques of Everyday Life: An Introduction*. London: Routledge.

Gregory, D. 2013. 'Tahrir: Politics, Publics and Performances of Space', *Middle East Critique* 22(3): 235–246.

Hardt, M., and A. Negri. 2004. *Multitude: War and Democracy in the Age of Empire*. London: Penguin.

Idle, N., and A. Nunns, eds. 2011. *Tweets from Tahrir: Egypt's Revolution as it Unfolded, in the Words of the People Who Made it.* New York: OR Books.

Kalinak, K. 2015. *Behind the Silver Screen: Sound*. London: Tauris.

Keraitim, S., and S. Mehrez. 2012. '*Mulid al-Tahrir*: Semiotics of a Revolution'. In *Translating Egypt's Revolution: The Language of Tahrir*, edited by S. Mehrez, 25–68. Cairo: American University in Cairo Press.

Kraidy, M. 2016. *The Naked Blogger of Cairo: Creative Insurgency in the Arab World*. Cambridge, MA: Harvard University Press.

Mitchell, W. 2012. 'Image, Space, Revolution: The Arts of Occupation', *Critical Inquiry* 39(1): 8–32.

Rodaway, P. 1994. *Sensuous Geographies: Body, Sense and Place*. London: Routledge.

Sabea, H. 2014. '"I Dreamed of Being a People": Egypt's Revolution, the People in Critical Imagination'. In *The Political Aesthetics of Global Protest: The Arab Spring and Beyond*, edited by P. Werbner, M. Webb, and K. Spellman-Poots, 67–92. Edinburgh: Edinburgh University Press.

Sanders, L. 2012. 'Reclaiming the City: Street Art of the Revolution'. In *Translating Egypt's Revolution: The Language of Tahrir*, edited by S. Mehrez, 143–182. Cairo: American University in Cairo Press.

Sanders, L., and M. Visonà. 2012. 'The Soul of Tahrir: Poetics of a Revolution'. In *Translating Egypt's Revolution: The Language of Tahrir*, edited by S. Mehrez, 213–248. Cairo: American University in Cairo Press.

Sassen, S. 2011. 'The Global Street: *Making* the Political', *Globalizations* 8(5): 573–579.

Schafer, R. M. 1994. *The Soundscape: Our Sonic Environment and the Tuning of the World*. Rochester, VT: Destiny Books.

Schutz, A. 1976. 'Fragments on the Phenomenology of Music', *Music and Man* 2(1–2): 5–71.

Shklovsky, V. 2004. 'Art as Technique'. In *Literary Theory: An Anthology*, edited by J. Rivkin, and M. Ryan, 15–21. Second Edition. Oxford: Blackwell.

Skarda, C. 1979. 'Alfred Schutz's Phenomenology of Music', *Journal of Musicological Research* 3(1–2): 75–132.

Werbner, P., M. Webb, and K. Spellman-Poots. 2014. 'Introduction'. In *The Political Aesthetics of Global Protest: The Arab Spring and Beyond*, edited by P. Werbner, M. Webb, and K. Spellman-Poots, 1–27. Edinburgh: Edinburgh University Press.

Yang, K., and M. Guaralda. 2013. 'Sensuous Geography: The Role of Sensuous Experience and their Contemporary Implications in Public Squares, a Lefebvrian Approach', *Ricerche e Progetti per il Territorio, la Cittá e l'Architettura* 4(1): 174–186.

Chapter 20

Bethlem, Music, and Sound as *Biopower* in Seventeenth-Century London

Joseph Nelson
Musicology/Ethnomusicology, University of Minnesota,
100 Ferguson Hall, 2106 – 4th Street South, Minneapolis,
MN 55455, United States
nels5698@umn.edu

20.1 Introduction

The streets of seventeenth-century London teemed with sound. Cries of street vendors, the braying of pack animals, and the rattling of wagon wheels echoed along corridors framed by multi-storey houses and shops. Areas of industry and marketplaces created sonic territories within that panoply of sound. Spots around the old Roman walls also bustled with life. Northwest of the city centre lay the open trading and festival areas of Smithfield, while to the east were alms houses, tenements, and dockyards. Among those buildings found outside the city walls were hospitals, places of rest and recuperation founded in medieval times and increasingly

Musical Spaces: Place, Performance, and Power
Edited by James Williams and Samuel Horlor

ISBN 978-981-4877-85-5 (Hardcover), 978-1-003-18041-8 (eBook)
www.jennystanford.com

serving the sick and impoverished. Along the northeastern portion of the old Roman wall sat the infamous Bethlem Hospital, first at the priory of St. Mary of Bethlehem and then at Moorfields. A tapestry of sounds and noises emerged from inside the asylum's buildings and from the city around it. The sonic environment inside doubtlessly contributed to its reputation for chaos as it echoed with the voices of the mad intermingled with the sounds of their treatment.

This chapter explores the intersection of sound and *biopower*, a term used by Michel Foucault (1994) to describe the sovereign state's power over life and death. That power put the ruling class and the poor, who suffered most under the elite's hegemonic control, into dialectical tension. In this context, the sovereign state — composed of the aristocracy and gentry — signalled implicit acceptance of institutions and practices such as confinement to Bethlem Hospital, thus exercising their biopower over the residents. This seventeenth-century use of biopower fell short of the later *biopolitics* that emerged through widespread institutional enforcement of biopower by the administrative state.

The Bethlem of early modern London was a place where sound phenomena and practices of listening and spectating intersected with structures of government and administrative control. However, it also provides an example of resistance to the biopower of the nascent state. By acting out their mental illness, residents of both Bethlem Hospital at the priory of St. Mary of Bethlehem and, later, the asylum's second incarnation at Moorfields contributed to the sonic landscape, or soundscape, inside the asylum. As they suffered under forms of medical intervention typical of the time — such as bleeding, purging, and cold-water baths — resistance to such medical treatment meant resistance to the regulatory power of the hospital staff. Residents' cries, howling, and rattling of chains emerged from and comprised part of that resistance. Visitors to the asylum noted this as part of the soundscape, and a record of it survived in the madsongs that circulated as broadside ballads on the streets and in taverns.

Perhaps the most famous representation of Bethlem's residents and of its interior sonic environment appeared in the broadside ballad 'A New Mad Tom' (Anonymous, 1658–1664[?]). Mad Tom, also called Poor Tom o' Bedlam, appeared in the literature dating back to the mid-sixteenth century and most famously in William

Shakespeare's *King Lear* (1606). Tom later featured in broadside ballads and political pamphlets, including with the references to him in songs such as 'Mad Maudlin', 'Loves Lunacie, or Mad Besses Fegary', and 'A Mad-mans Morrice'. In the political pamphlets of the late-seventeenth century, he served as a pseudonym for writers arguing for the reestablishment of the monarchy during the Civil War and in the early-eighteenth century for those expressing anxiety about politics. The archetypal nature of his madness and the variety of situations in which he appeared, be they comical or tragic, made him an ideal character through which writers and composers of songs could express their thoughts on politics, morality, or the absurdities of life (Cross, 2012: 29–30).

Some caution must be taken with the term 'madness', a word loaded with much historical baggage. It had no absolute, fixed definition in the early modern period, incorporating symptoms of the modern term 'mental illness' ranging from depression to mania and originating in conditions from religious fervour to love-melancholy. As scholars have discussed, such symptoms and underlying conditions often had class or race implications, but especially involved gender (Ingram *et al.*, 2011: 25–45). It remains important, however, to avoid reducing the collection of conditions suffered by asylum residents to 'animality' or any other term that implies them to be 'sub-human'. Although physicians at the time included such bestial language in descriptions of mad people (Scull, 1989: 58–75), the methods of treatment could prove just as brutal as any patient's outburst.

Mad Tom's importance to a study of the soundscape of Bethlem lies, in part, in the circulation of his ballad and in the specifics of male madness versus female madness. Although madwomen often appear in confinement, such as in 'Loves Lunacie' or Henry Purcell's 'From Silent Shades and Elysian Groves', madmen often wander or beg on the street. Indeed, perceptions of Mad Maudlin's disorderly behaviour came from bawdy lyrics being sung in public, with both the story of the song mentioning public spaces, and taverns and streets being the typical context for performing ballads. Maudlin thus exemplified the disruptive and taboo behaviour of a 'roaring girl', much as did Thomas Middleton's and Thomas Dekker's play *The Roaring Girl* from 1611 (Gates, 2013; Peery, 1948a, 1948b). William Hogarth would later depict Tom Rakewell's disorderly behaviour as happening in public or semi-private spaces, before the character reached Bedlam in the final print of the series, *A Rake's Progress* (1733/1735). In keeping with the fuzzy definition of madness given above, this chapter does not aim to draw a hard boundary between men's and women's madness. However, the image of the wandering vagabond begging for food was almost always attached to male characters, and men's madness posed a different danger to the body politic from that of women. Furthermore, the ways in which Mad Tom and his ballads circulated and occupied public spaces via street literature and popular entertainments, his connection to other practices, such as the Morris dance, and his penetration of various classes of ballad audiences make him an important lens through which to view the sonic landscape of early modern London.

20.2 Bethlem and Sonic Disorder

In his satirical account of London entitled *The London Spy*, Ned Ward described Bethlem Hospital through a caricature of the resident mad poor. He wrote that he thought it mad that the 'Magnificent Edifice' was built for 'Mad-Folks,' calling it 'so costly a *Colledge* for such a *Crack-brain'd Society*' (Ward, [1699] 1924: 63). Ward's account mirrors that of his contemporary Thomas Brown, who wrote that 'The Outside is a perfect Mockery to the Inside, and Admits of two amusing Queries, Whether the persons that Ordered the building of it or those that inhabit it, were the maddest?' (cited in Andrews *et al.*, 1997: 29). A similar sentiment appeared in the anonymous poem 'Bethlehems Beauty' (1676): 'So Brave, so Neat, so Sweet it does appear, Makes one Half-Madd to be a Lodger there...' (Aubin, 1943: 245–248). These writers likely visited the asylum, as did many wealthy Londoners in the seventeenth and eighteenth centuries.

The passages above referred to the second incarnation of the asylum, Bethlem at Moorfields, which Robert Hooke designed and that opened in 1676 after the Great Fire of 1666. The previous Bethlem building was relatively small, holding only about 30 residents (Andrews *et al.*, 1997: 59–65). This second incarnation of the asylum greatly expanded its physical dimensions as part of reforms instituted by its Board of Governors, who intended to improve the conditions and treatment of patients (Stevenson,

1996: 262–263). Despite the intentions behind those expansions, it remains unclear how much they improved the day-to-day life of residents. Administration of the old Bethlem Hospital moved to the City of London after the Dissolution of the Monasteries by Henry VIII and, later, to the Board of Governors who also oversaw Bridewell Prison. The Moorfields building was also built as a grand edifice to serve as a monument to the city as it emerged from the aftermath of the Great Fire. Its façade had the appearance of a palace, with two wings branching from a central atrium structure, one wing holding rooms for male patients and one for female patients. The northern-facing façade looked over the public park of Moorfields and had three decorated pavilions, two on either end of the building, and a central pavilion serving as the main entrance. Statues by the artists Caius Gabriel Cibber entitled *Melancholy* and *Raving Madness* loomed at the top of the columns flanking the gate. Iron grates pierced the stone fence around the north face to allow the public to view the asylum and its patients, who could be seen through large windows on the first and second floors. Passers-by could thus view Bethlem from outside its fence, in addition to the spectators admitted into the asylum itself.

Ward's account provided some clues as to the composition of Bethlem's sonic environment, writing that it included rattling chains and howling residents and reminded him of the vision of hell by the Spanish Baroque writer Don Quevedo. Robert Hooke, the architect of Bethlem's Moorfields building featured in Ward's account, wrote that 'Lunaticks cannot obtain that, which should, and in all Probability would, cure them, and that is a profound and quiet Sleep' (cited in Hunter and Macalpine, 1963: 220). Thomas Fitzgerald's 1733 poem goes further, describing the soundscape thus:

> Far other Views than these within appear,
> And Woe and Horror dwell for ever here.
> For ever from the echoing Roofs rebounds,
> A dreadful Din of heterogenous Sounds
> From this, from that, from ev'ry Quarter rise,
> Loud Shouts, and sullen Groans, and doleful Cries;
> Heart-soft'ning Plaints demand the pitying Tear,
> And Peals of hideous Laughter shock the Ear
>
> (Scull, 1993: 50)

This melange of intermingling sounds doubtlessly contributed to the perception of Bethlem as a place of chaos and disorder.

Exposure to mad people of the asylum and the sounds of their madness did not only occur during visitations. Patients were routinely released from Bethlem for the day to beg on the streets. Family members of residents visited, and in the priory hospital they helped care for relatives confined there. Bethlem Precinct, the land of the original priory surrounding the hospital, underwent much development during the sixteenth and seventeenth centuries. This development packed the precinct with residential and commercial buildings and exposed increasing numbers of neighbours to hospital residents. Sound formed an essential element of this interaction. Pollution from structures such as the nearby distillery impacted hospital residents, and neighbours commented on the haunting sounds emerging from the hospital (Andrews *et al.*, 1997: 51). Such sounds surely had a more profound impact on people in those times than on a modern auditor. Bruce R. Smith (1999) demonstrated the importance of aurality in the daily lives of early modern people, both in structuring daily routines and in shaping the cultural regimes through which they experienced the world. J. Martin Daughtry (2015) also demonstrated the impact sound could have on both mind and body.

Jacque Attali wrote that '*noise is violence*: it disturbs', and that '*music is a channelization of noise*' (1985: 26). He described music's relationship to power as one of ritual sacrifice. Music, in his estimation, became instrumentalized and commodified in the modern era as its potential was recognized by capitalism. He wrote that when those with power want to silence others, 'music is reproduced, normalized, *repetition*' (*ibid.*: 20). Much like music exists in a system of meaning, noise, too, is part of a network of signifiers attached to hearing and practices of listening. To Attali, noise acts as a rupture. It has limited value on its own but can be utilized in tactical ways if needed. Noise, then, lies at the heart of a discussion of Bethlem's soundscape and its representation in the madsongs of Mad Tom.

'Noise' had a nebulous definition in early modern England, with people in the seventeenth century applying the term 'noise' to everything from out-of-tune music to domestic arguments. However, Emily Cockayne has suggested that 'noisiness' was more narrowly defined. She wrote that 'Noisy sounds irritated the hearer because they were loud, clamorous, importune, irregular, intrusive, disturbing, distracting, inexplicable or shocking' (2002: 36). As Cockayne continued:

> Music was part of everyday life in early modern England, performed by skilled professionals and practicing amateurs alike. Exposure to musicians who were barely competent, or completely incompetent, would have been common, and attempts to perform music were not always appreciated by those in earshot.
>
> (2002: 35)

She reported that Roger North, a writer and noted amateur musician, described incidental sounds such as the slamming of doors as annoying due to their unequal rhythm (*ibid.*).

Penelope Gouk cited the work of Robert Hooke in describing music and hearing as an action of the ear tuning itself to harmonies, with harmonies understood in the 1680s as both a relationship between mathematical ratios and the relationship between pitch frequencies (Gouk, 1980: 598). In Hooke's theory, the ear hears combinations of pitches as pleasing when they conform to particular intervals (thirds, fourths, fifths, and sixths), and other intervals not conforming to standard harmonies of the time were non-musical and displeasing. Thus, the ear preferred musical sounds, a preference for consonance over dissonance.

The division of music from noisy sound often came down to class and geography, with rural music performed by rural musicians seen as less musical, and noises of the street, particularly at night and in poorer neighbourhoods and around taverns, seen as deleterious to health. Cockayne highlighted, for example, the laws that imposed fines on coaches with creaky wheels and joints, and she explained that the presence of livestock, especially pigs, in London's streets was considered both dirty and noisy (2007: 107). Indeed, Thomas Dekker wrote that 'for every street, carts and coaches make such a thundring as if the world ran upon wheels', and 'hammers are beating in one place, Tubs hooping in another, Pots clincking in a third, water-tankards running at tilt in a fourth' ([1606] 1879: 31). References to noise, street music, poor people, and animals indicate much overlap between all four in the minds of early modern Londoners. Comparing these to discourses on madness and its bestial nature also provides insight.

Urban and suburban landscapes were often demarcated by, among other things, sound. Eighteenth-century London gradually developed, sprawling and swallowing neighbouring towns and providing affluent residents means to escape to the areas outside the dense urban core. There was a softening of the prevailing disdain for suburbs, in part due to economics (Dyos, 1954), as plagues, fires, and pollution gradually caused the wealthy to move west. Phenomena such as pleasure gardens and public parks signalled a retreat from London's coal smoke-clogged streets in the late-seventeenth and early-eighteenth centuries, as evidenced in John Evelyn's *Fugifugium* (1661) and various diaries (Emmons, 2014; Jenner, 1995; Travis, 2014). The rise of other 'walking' literature by diarists such as Samuel Pepys and John Gay also attested to increasing discomfort with the pollution and poverty in the city.

Bethlem's sonic disorder stood at odds with the order embodied by musical conventions at the time. While early modern writers would call some music 'noise', especially if they found it displeasing, musical order and harmony followed the ideals of Neoplatonism. Such idealized music, as opposed to musical practice, appeared in philosophical writings from Robert Fludd's image of the *Templum Musicae* (1617) (Hauge, 2008) to Marin Mersenne's recalculation of the Pythagorean ratios in *Harmonie Universelle* (1630) (Gouk, 2008: 143–146) to the co-ordinance of musical order and the passions of the soul (Hanning, 2008: 115–117). Those same musical conventions appeared in other areas of musical production such as courtly dance (Berghaus, 1992).

20.3 'A New Mad Tom' and Bodily Disorder

Discussing ballads associated with Mad Tom requires some housekeeping. The ballad often identified as 'Mad Tom o' Bedlam' (Cross, 2012: 28–29) is not the text that appears in *Wit and Drollery* (Phillips, 1656: 126), a collection of licentious poems written by John Milton's nephew John Phillips (Campbell, 2004). This differs from the one under consideration here, 'A New Mad Tom' (1658–1664[?]). Instead, Cross's song, with the chorus about the 'bonny mad boys', appears under the title 'Mad Maudlin' with printed music in Thomas d'Urfey's *Songs Compleat, Pleasant and Divertive* (1719:

189). While 'Mad Maudlin' shares some text with that earlier song 'Mad Tom', they have different narrative voices and subtexts. The two have often been mixed up and it would seem foolish to claim any definitive rendition or original text of a ballad about Mad Tom. William Chappell, for example, comments on the part of 'A New Mad Tom' entitled 'The Man in the Moon Drinks Claret', the ballad printed on the second half of the 1658 broadsheet, as having been sung since the days of the Curtain Theatre. That theatre fell out of use by the reign of King Charles I in 1625 (Chappell, 1855: 328).

Claude Simpson (1966), in his re-examination of many of the songs in Chappell's song collections, did not refute this. Furthermore, Simpson stated that a copy of the tune used for 'A New Mad Tom', entitled 'Gray's Inn Maske', appeared in the form of lute tablature in Drexel MS 5612 with 'Tom of Bedlam' pencilled into the margin. The addition of Tom o' Bedlam as a title may have come later. However, the tune for 'A New Mad Tom' originated as a Morris dance or, as it appears in the stage direction, 'music extremely well fitted, having such a spirit of country jollity', in the antimasque of Francis Beaumont's *The Masque of the Inner-Temple and Gray's Inn* of 1613 (Fletcher, 1883: 688). The same music was heard later in Act III, scene 5 of another Shakespeare and Fletcher work *The Two Noble Kinsmen* of 1614 (1895: 100–101). The tune 'Gray's Inn Maske' remained popular and was included in John Playford's *An English Dancing Master* (1651).

These two ballads, 'Mad Maudlin' and 'A New Mad Tom', contain stark differences. The ballad that Cross referred to, 'Mad Maudlin', has a simple strophic form with multiple verses and a repeated refrain text. The repetitive nature of its chorus would lend it well to group singing, especially in a pub, much like modern pop songs. Although not harmonically complex, the ballad 'A New Mad Tom' shows the signs of its origin in a Stuart court masque, especially in its multi-section structure with alternating duple and triple metres. It lacks repeated text, which would seem to make it better suited to a single performer than group singing. Especially noteworthy is a slow section occurring midway through the song. Its text alludes to the famous interjection of Shakespeare's Tom, 'Poor Toms a-cold' (Shakespeare, 2005: 91), with the song text reading 'Cold and comfortlesse I lye'. This comes after two sections in which, respectively, eight bars and four bars are repeated. The disruptive

'Cold and comfortlesse' section injects a melancholic, laconic, and mournful quality to the song. It is followed by a jaunty compound duple metre and a return to the gregarious character of the first two sections. This disruption seems similar to the 5/8 section of George Frideric Handel's accompanied recitative 'Ah! stigie larvae' from *Orlando* (1733).

Another important aspect of 'A New Mad Tom', one shared with 'Mad Maudlin' and 'Loves Lunacie', is that of grotesquery. In Tom's ballad, grotesquery takes multiple forms. It includes text references to Hell and howling furies, bodily organs such as the brain and guts, and sexual allusions such as that to Vulcan's 'tackle'. The use of 'Gray's Inn Maske' makes sense, then, as it ties Mad Tom to the bodily labour involved in Morris dancing, a form of dance famously denounced by Philip Stubbs' *Anatomie of Abuses* (1583). John Forrest (1999) has made a persuasive case that this dance tradition emerged as an amalgamation of various others, including *charivari*, mumming plays, skimmingtons, and May Day celebrations, which ties it to the ritualized shaming of the broadside ballad and libel.

The origins of 'A New Mad Tom' in a rural or Morris dance, the dance tune 'Gray's Inn Maske', places the ballad in the category of 'rough music'. E. P. Thompson described the sonic qualities of 'rough music' as 'raucous, ear-shattering noise, unpitying laughter, and the mimicking of obscenities' (1992: 3). Violet Alford identified 'rough music', specifically *charivari*, as the 'beginning of popular justice' (1959: 505). An overlapping literary and musical practice of the time was libelling, which Andrew McRae has described as part of a network of news, rumour, and opinion by which early modern people would critique and hold accountable those in power through ritual shaming (2000: 381). Libels were satirical poems that would use graphic sexual language to roast their targets.

On the street, such political satire would often appear in the form of broadside ballads that ranged in their literary quality but provided fodder for ribald songs (Bellany, 2006; Fox, 1994). Eric Nebeker (2011) has written that broadside ballads and other forms of popular street literature circulated across class boundaries, allowing ballads such as 'A New Mad Tom' to penetrate literary cohorts of the landed gentry and aristocracy as well as among street labourers and pub goers. Balladeers and hawkers sold broadside ballads on the street for a penny and so participated in defining public spaces (Gordon,

2002), ensuring the presence of Mad Tom and the association between him and beggars in the markets and on street corners. Thus, while sexual content and 'rough music' might have acted as a Bakhtinian carnivalesque form of social resistance or catharsis for the working classes and poor, it also had a place in the entertainment and music consumption of the ruling classes. As Richard Leppert has written, music served in the capacity of 'demarcating sound as one means by which society constructs itself', authoring the distinction between upper- and lower-class people through performance and public ritual (1989: 27).

Mad Tom's character also requires some unpacking. He appeared in sixteenth-century texts such as those by John Awdeley in 1565, and Thomas Harman's work of 1566 cast Tom as a suspicious character who led bands of beggars. Seventeenth-century depictions of him in broadside ballads, however, tended to be more comical. William Carroll (2002) has written that Poor Tom o' Bedlam in *Lear* gave voice to the suffering poor with pathos and stood at a fulcrum point between the sixteenth-century rogue Tom o' Bedlam and the clownish figure of the broadside ballads of the latter part of the seventeenth century. Despite the humour in 'A New Mad Tom', that humour must be read in the context of his appearance in political broadsheets and pamphlets such as 'Mad Tom a Bedlams Desires of Peace' (Wortley, 1648), 'News from Bedlam, or, Tom of Bedlams Observations' (Anonymous, 1674), 'Bess o' Bedlam's Love to her Brother Tom' (Anonymous, 1709a), and 'Tom of Bedlam's Answer to his Brother Ben Hoadly' (Anonymous, 1709b). These texts voiced serious political and social commentaries, with Tom as interlocutor. Mad Tom should also be taken in the context of paratextual references via other ballads using tunes identified with him. One such ballad is 'The Cunning Northern Begger' (Anonymous, 1634[?]), which includes in its subtitle 'To the tune of Tom of Bedlam'. The continued reference to Mad Tom in vagabond and beggar ballads then ties him back into a network of meanings rooted in a class of outcasts and suspicious characters, an association that continued into the next century.

The madness associated with Mad Tom exhibited stereotypical symptoms of mania and melancholy, including delusions and a disjointed train of thought with unrelated or oddly juxtaposed ideas strung together. Bess of Bedlam exhibits similar behaviour in her ballads, particularly 'Loves Lunacie OR Mad Besses Fegary' (Crimsal, 1634–1658). Her most enduring appearance occurred in Henry Purcell's 'From Silent Shades and the Elysian Groves' (1683), a song that resembled Italian lament arias of the earlier seventeenth century (Eubanks Winkler, 2012; MacKinnon, 2001). Bess signalled the survival of those tropes of madness already appearing in Ophelia's text from *Hamlet* in 1609 (Bruster, 1995; Neely, 1991) and Monteverdi's *Arianna* (Cusick, 1994; Rosand, 1979). The songs for Mad Tom, Mad Maud, and Bess of Bedlam included descriptions of a range of unsavoury or disorderly behaviour, from excessive drinking and sexual language to hallucinations (see Achilleos, 2014; Fox, 1994; MacKinnon, 2001; McShane, 2011; Poulton, 1981; Shaw, 1996; Timberlake, 1996; Waage, 1977; Wiltenburg, 1988).

The musical rupture in 'A New Mad Tom', and Mad Maud's roaring and Bess's hallucinations, embodied disorder and placed the ballads at odds with the prevailing conventions of courtly music and Classical principles of *harmonia* that pervaded political discourse. These characters gave voice to a Hobbesian state of nature that mirrored the turbulent years of the seventeenth century. Sounds from the asylum echoed the cacophony of plague years, wars, and riots. It signalled social unrest and a class hierarchy thrown out of balance as apprentices revolted against their masters. Much like Hans Holbein's *The Dance of Death* (1538) did a century earlier, Mad Tom and the songs about beggars, vagrants, and madcaps exemplified the anxieties of early modern Londoners towards disorder. Furthermore, these songs offered a mirror that reflected the social ills of the seventeenth-century city.

20.4 Foucault and Aurality

Foucault wrote very little about music or sound. But that does not prohibit the use of his methods and theories in this study. As many subsequent theorists have noted, Foucault situates the ear as subsidiary to the eye in the formation of knowledge. However, two particular writings by Foucault offer insights into how aurality factors into a theory of power. Lauri Siisiäinen (2008) has recounted Foucault's explication of administrative power through the example of the clinic. In *The Birth of the Clinic*, the auditory faculty of the physician acts as a secondary means of perception to the eye. Yet this observation came from a passage on autopsies, the ultimate expression of medico-juridical control over a docile subject. Siisiäinen argued that another text by Foucault entitled '*Message ou Bruit*?' described how the physician listens to the noises of the body as a starting point of clinical practice. This positions aurality as a principal means of perception, and noise as a component in the composition of the living subject, giving the ear a primacy in the formation of knowledge.

'*Message ou Bruit*?' brought practices of listening into focus as part of the disciplinary apparatus of surveillance and the deployment of biopower (Foucault, 1994: 557–560). Indeed, defining aurality through practices of listening deconstructed the audio-visual litany in which sound lacks permanence and historical specificity or conditioning. Instead, identifying listening with clinical practice gave it a disciplinary framework. It illustrated its tactical use in separating different points of articulation in the sounding body. 'Noise' does not occur in an unmarked sense, instead it has qualities or characteristics demarcating points of origin, and it assigns meaning to the diagnostic ear. As mentioned above, aurality played a critical role in the lives of early modern people and especially in medical practice. The use of musical rhythm as a framework for taking a person's pulse in European medical practice, for example, dates back to at least Pietro d'Abano (c.1257–1315) (Lagerlund, 2008; Siraisi, 1975). Furthermore, numerous writers and philosophers have discussed music and sound in the development of science in the pre-modern and early modern periods. As Penelope Gouk wrote, 'Acoustics, the science of sound, properly took shape as a recognizably independent branch of natural philosophy in the course of the seventeenth century' (1999: 158). 'Acoustics' was often studied via music or by using musical ideas such as *harmony* and *resonance* to express scientific ideas (Erlmann, 2010; Gouk, 2008). Practices of listening and theories of sound were also associated with the erotic nature of the senses, pleasure, and music's effects on the body (Austern, 1993, 1998).

Listening as a process emerging from disciplinary regimes of medical or musical education and socialization, or learned from the experience of everyday life, also intersects with the idea of striated spaces, which are areas marked by sights and sounds (Deleuze and Guattari, 1987: 474–500). The meaning behind sound marks comprising the striated spaces of the soundscape came from earlier systems of meaning and disciplinary regimes. As Jean-Luc Nancy put it, sound is made of 'a totality of referrals: from a sign to a thing, from a state of things to a quality, from a subject to another subject or to itself, all simultaneously' (2007: 7). That process of marking was rooted in ideology (Attali, 1985: 19). In seventeenth-century London, demarcation of space emerged from religious and political theories about the role of different classes in society, the top-down distribution of authority, and the right to rule derived from divine or natural law. And, as Foucault wrote in an essay on Boulez:

> One is apt to think that a culture is more attached to its values than its forms, that these can easily be modified, abandoned, taken up again; that only meaning is deeply rooted... It is to ignore the fact that people cling to ways of seeing, saying, doing, and thinking, more than to what is seen, to what is thought, said, or done.
>
> (Foucault, 1998: 242)

20.5 Biopower and Immunopolitics

As mentioned above, the deployment of biopower in Bethlem comes from a nascent sovereign state, or the level of social hierarchy composed of the ruling classes. While Henry VIII had control of Bethlem after the Dissolution, he then turned it over to the City of London, who in turn granted its administration to the Board of Governors for Bridewell. The board consisted of various members of the gentry and ruling classes, and it doubtlessly included those who neither met in sessions held by the Court of Bridewell, which sent people into confinement in Bethlem, nor visited the hospital with any regularity. The court heard the cases of petitioners attempting to have relatives committed, of people brought before them by magistrates, and of patients whose family members had failed to pay for their upkeep. This fits with Foucault's concept of biopower, summarized by Eduardo Mendieta as being about the regulation of life, rather than the power of *sovereignty* that took life or let live (2014: 45). The difference might be thought of as a more diffuse expression of power in the form of actions by a body politic (*biopower*) versus a more specific power localized in a single sovereign to order or cancel an execution (*sovereignty*).

Yet, sovereignty in a broader sense was also at play. Foucault described this power as emerging through multiple points of articulation in the body politic and noted that it could not be acquired, seized, or shared. Furthermore, power existed in a network of intersecting spheres and apparatuses, one standing in dialectical tension with those without power or who lived under the rule of those with power. Foucault wrote, 'Where there is power, there is resistance' (1990: 95). This resistance did not lie in a 'position of exteriority' but as a required component of the whole field of power, as power depended on the resistance. Those in power required adversaries and targets against whom they could rally supporters. Such targets and adversaries did not form a single unified body or 'soul of resistance' (*ibid.*). Instead, they could come from anywhere in the power network, whether aristocrats, allied or adversarial states, or peasant farmers. For 'just as the network of power relations ends by forming a dense web that passes through apparatuses and institutions, without being exactly localized in them, so too the swarm of points of resistance traverses social stratification and individual unities' (*ibid.*). The circulation of street literature such as the ballad 'A New Mad Tom' exemplifies this diffuse form of resistance through its ability to touch multiple classes of people and to traverse geographic boundaries and through the variety of authors coming from across the spectrum of class.

Roberto Esposito (2006) provided an even more appropriate theory of biopower, one he called the *immunization paradigm*. This was part of an *immunopolitics* that sought to regulate life and maximize the production of economic and military power through actions like confinement. Esposito ruminated on the relationship between the biomedical definition of 'immunity' and the use of 'immunity' in juridical language. This latter definition 'alludes to a temporary or definitive exemption on the part of the subject with regard to concrete obligations or responsibilities that under normal circumstances would bind one to others' (*ibid.*: 24). The nature of immunity ties directly to the rights and obligations of membership in a community and the formation of an individual identity versus bending to the will or values of the community. In this paradigm of immunization, the politics of life are 'Not simply the relation that joins life to power, immunity is the power to preserve life' (*ibid.*). The goal, then, of immunization as a form of biopower is not the exercise of killing or making life, but the preservation of life as the only field on which power can exist.

Yet, preserving life as a field on which power can exist, and on which community exists, requires identification with the community — the negation of self in favour of *communitas* — or identifying and excising that which negatively impacts the community. In other words, if '*communitas* is that relation, which is binding its members to an obligation of reciprocal gift-giving... *immunitas* is the condition of dispensation from such an obligation and therefore the defence against the expropriating features of *communitas*' (*ibid.*: 28). That 'defence' comes at a cost, however, for 'the most incisive meaning of *immunitas* is inscribed in the reverse logic of *communitas*: immune is the "nonbeing" or the "not-having" anything in common' (*ibid.*). The immunity works both ways, with the subject of immunitas separated from the homogenizing effect of communitas and the community free of the individual subject and the traits or behaviours that made them individual. Frédéric Neyrat wrote that 'for Esposito, the Self of community never took place and will never take place' (2010: 33). The subject, to participate in the community, must give up selfhood to benefit from some degree of freedom from hunger and external dangers. However, as Neyrat went on to say, 'politics plays a dirty trick on life: while wanting to protect it, it can end up destroying it' (*ibid.*: 32). This 'strange reversal, this inversion or perversion, is at the heart of Esposito's questions and of our societies, in which we undergo the effects of *highly dubious protections*' (*ibid.*). Severing ties between the subject and the community equates to a symbolic death. This separation was evident in the purpose of confinement (Foucault, 1988: 38–64), both to provide a space for recovery from madness and to prevent the spread of madness to the surrounding population.

Residents of Bethlem were excised from the community, but they often recovered from their temporary madness or otherwise left Bethlem for some other form of care. Medical intervention did not impart a life sentence. However, the methods were grim and the hospital routinely fell into disrepair. By the end of the eighteenth century, the Moorfields building designed to house 150 residents held close to 300. Perhaps more disturbing, visitors were allowed to roam the asylum, as Ned Ward recounts and as appears in the 'Tom in Bedlam' print of Hogarth's *A Rake's Progress*. Their presence eroded the boundary provided by the walls of the building and the regulatory action of confinement. That act of spectating might have played its own part in the exercise of biopower. By creating the spectacle of the fantastical building at Moorfields, the placement of grates in the outer stone wall, and admitting visiting spectators, perhaps part of the point was to place those poor residents into a cage for viewing. Most of the residents moved freely about the asylum, but the walls of the building and Cibber's *Melancholy* and *Raving Madness* provided a boundary that confined residents could not cross.

20.6 Conclusion

Bedlam, as exemplified by Mad Tom's ballads, played an important role in the soundscape of urban London, both in the imagination of playwrights and writers and in the experiences of visitors to the asylum. Images such as Hogarth's 'Tom in Bedlam' and descriptions of chains and lice-ridden beds gave people a heuristic emblem of the consequences of madness. It also provided an object of spectacle and a vehicle through which they could enforce social order. Musical and unmusical sound, especially disorderly and noisome sound, was inextricably linked to those iconic characters of poor bedlamites and vagabonds, and agents of misrule, characters who threatened the safety of the streets or disturbed the nightly rituals of domestic life. They roared and hollered. They frightened young women and accosted gentlemen. Their carnivalesque nature came from their disorderly appearance, matched by their broken speech and disjunct melodies.

Just as the masques and airs of court conformed to the rules of social order and power, the songs of madmen and madwomen, particularly those that circulated in street literature and popular songs performed on street corners and in taverns, blazed as the embers of a distant catastrophe always on the horizon. In the minds of early modern Londoners, those madmen and madwomen represented the inverse of a rationally ordered society and were the ultimate expression of the passions driving the unwashed lower classes. While plague, wars, and fires swept through London every few years, the mad beggars and the attendant disease and poverty they exhibited posed an ever-present danger. Their disruptive utterances amidst a city of noise, and their inability to conform, acted as resistance to attempts by monarchs to consolidate and enforce sovereign power, to a nascent state of aristocrats and gentry forming their own power bases and civic institutions outside of court, and to a society lumbering towards a hegemonic order.

References

Achilleos, S. 2014. '"Drinking and Good Fellowship": Alehouse Communities and the Anxiety of Social Dislocation in Broadside Ballads of the 1620s and 1630s', *Early Modern Literary Studies*, Special Issue 22: 1–32.

Alford, V. 1959. 'Rough Music or Charivari', *Folklore* 70(4): 505–518.

Andrews, J., A. Briggs, R. Porter, P. Tucker, and K. Waddington. 1997. *The History of Bethlem*. London and New York: Routledge.

Anonymous. 1634(?). 'The Cunning Notherne Begger Who All the By-Standers Doth Earnestly Pray, To Bestow a Penny Upon Him Today'. London: F. Coules. University of Glasgow Library Euing Ballads 55, EBBA ID 31726.

Anonymous. 1658–1664(?). 'A New Mad Tom of Bedlam, Or, The Man in the Moon Drinks Claret'. University of Glasgow Library Euing Ballads 248, EBBA ID 31797.

Anonymous. 1674. *News from Bedlam, or, Tom of Bedlams Observations, Upon Every Month and Feastival Time in This Present Year, 1674*. London. ProQuest document ID 2264210805.

Anonymous. 1709a. *Bess o' Bedlam's Love to Her Brother Tom: With a Word in Behalf of Poor Brother Ben Hoadly*. London: The Booksellers of London and Westminster.

Anonymous. 1709b. *Tom of Bedlam's Answer to His Brother Ben Hoadly, St. Peter's-Poor Parson, Near the Exchange of Principles*. London: H. Hills.

Attali, J. 1985. *Noise: The Political Economy of Music*, translated by B. Massumi. Minneapolis: University of Minnesota Press.

Aubin, R. A. 1943. *London in Flames, London in Glory: Poems on the Fire and Rebuilding of London 1666–1709*. New Brunswick, NJ: Rutgers University Press.

Austern, L. P. 1993. '"Alluring the Auditorie to Effeminacie": Music and the Idea of the Feminine in Early Modern England', *Music & Letters* 74(3): 343–354.

_____. 1998. '"For, Love's a Good Musician": Performance, Audition, and Erotic Disorders in Early Modern Europe', *The Musical Quarterly* 82(3/4): 614–653.

Bellany, A. 2006. 'Singing Libel in Early Stuart England: The Case of the Staines Fidlers, 1627', *The Huntington Quarterly* 69(1): 177-193.

Berghaus, G. 1992. 'Neoplatonic and Pythagorean Notions of World Harmony and Unity and Their Influence on Renaissance Dance Theory', *Dance Research: The Journal of the Society for Dance Research* 10(2): 43–70.

Bruster, D. 1995. 'The Jailer's Daughter and the Politics of Madwomen's Language', *Shakespeare Quarterly* 46(3): 277–300.

Campbell, G. 2004. 'Phillips, John'. *Oxford Dictionary of National Biography.* Available online at https://doi.org/10.1093/ref:odnb/22161

Carroll, W. C. 2002. 'Songs of Madness: The Lyric Afterlife of Shakespeare's Poor Tom'. In *King Lear and Its Afterlife,* Shakespeare Survey: An Annual Survey of Shakespeare Studies and Production 55, edited by W. Holland, 82–95. Cambridge: Cambridge University Press.

Chappell, W. 1855. *Popular Music of the Olden Time.* Vol. 1. London: Cramer, Beale, & Chappell.

Cockayne, E. 2002. 'Cacophony, or Vile Scrapers on Vile Instruments: Bad Music in Early Modern English Towns', *Urban History* 29(1): 35–47.

_____. 2007. *Hubbub: Filth, Noise & Stench in England 1600–1770.* New Haven and London: Yale University Press.

Crimsal, R. 1634–1658(?). *Loves Lunacie. Or, Mad Besses Fegary,* British Library Roxburghe C.20.f.7.206-207, EBBA ID 30147.

Cross, S. 2012. 'Bedlam in Mind: Seeing and Reading Historical Images of Madness', *European Journal of Cultural Studies* 15(1): 19–34.

Cusick, S. G. 1994. '"There Was Not One Lady Who Failed to Shed a Tear": Arianna's Lament and the Construction of Modern Womanhood', *Early Music* 22(2): 21–43.

Daughtry, J. M. 2015. *Listening to War: Sound, Music, Trauma, and Survival in Wartime Iraq.* Oxford: Oxford University Press.

Dekker, T. [1606] 1879. *The Seven Deadly Sins of London Drawn in Seven Several Coaches, Through the Seven Several Gates of the City, Bringing the Plague with Them,* The English Scholar's Library of Old and Modern Works, No. 7, edited by E. Arbor. London: The English Scholar's Library.

Deleuze, G., and F. Guattari. 1987. *A Thousand Plateaus: Capitalism and Schizophrenia*, translated by B. Massumi. Minneapolis: University of Minneapolis Press.

d'Urfey, T. 1719. *Songs Compleat, Pleasant and Divertive; Set to Musick*. Vol. 4. London: William Pearson.

Dyos, H. J. 1954. 'The Growth of a Pre-Victorian Suburb: South London, 1580–1836', *The Town Planning Review* 25(1): 59–78.

Emmons, P. 2014. 'The Place of Odour in Modern Aerial Urbanism', *The Journal of Architecture* 19(2): 202–215.

Erlmann, V. 2010. *Reason and Resonance: A History of Modern Aurality*. New York: Verso Press.

Esposito, R. 2006. 'The Immunization Paradigm', translated by T. Campbell, *Diacritics* 36(2): 23–48.

Eubanks Winkler, A. 2012. 'Society and Disorder'. In *The Ashgate Research Companion to Henry Purcell*, edited by R. Herissone, 269–302. Abingdon: Ashgate.

Fletcher, J. 1883. *The Works of Beaumont and Fletcher*. Vol. 2, edited by G. Darley. London: George Routledge and Sons.

Forrest, J. 1999. *The History of Morris Dancing, 1458-1750*. Toronto: University of Toronto Press.

Foucault, M. 1988. *Madness and Civilization: A History of Insanity in the Age of Reason*, translated by R. Howard. New York: Vintage Books.

_____. 1990. *The History of Sexuality*. Vol. 1, translated by R. Hurley. New York: Vintage Books.

_____. 1994. *Dits et Écrits 1954–1998*. Vol. 1: 1954–1969, edited by D. Defert, F. Ewald, and J. Legrange. Paris: Éditions Gallimard.

_____. 1998. 'Pierre Boulez, Passing Through the Screen'. In *Aesthetics, Method, and Epistemology: Essential Works of Foucault 1954–1984*. Vol. 2, edited by J. Faubion, 241–244. New York: The New Press.

Fox, A. 1994. 'Ballads, Libels and Popular Ridicule in Jacobean England', *Past & Present* 145: 47–83.

Gates, D. 2013. 'The Roaring Boy: Contested Masculinity on the Early Modern Stage', *The Journal of the Midwest Modern Language Association* 46(1): 43–54.

Gordon, A. 2002. 'The Act of Libel: Conscripting Civic Space in Early Modern England', *Journal of Medieval and Early Modern Studies* 32(2): 375–397.

Gouk, P. 1980. 'The Role of Acoustics and Music Theory in the Scientific Work of Robert Hooke', *Annals of Science* 37(5): 573–605.

_____. 1999. *Music, Science and Natural Magic in Seventeenth-Century England*. New Haven and London: Yale University Press.

_____. 2008. 'Music and the Sciences'. In *The Cambridge History of Seventeenth-Century Music*, edited by T. Carter, and J. Butt, 132–157. Cambridge: Cambridge University Press.

Hanning, B. R. 2008. 'Music and the Arts'. In *The Cambridge History of Seventeenth-Century Music*, edited by T. Carter, and J. Butt, 111–131. Cambridge: Cambridge University Press.

Hauge, P. 2008. 'Robert Fludd (1574–1637): A Musical Charlatan? A Contextual Study of His "Temple of Music" (1617/18)', *International Review of the Aesthetics and Sociology of Music* 39(1): 3–29.

Hunter, R., and I. Macalpine. 1963. *Three Hundred Years of Psychiatry 1535–1860: A History Presented in Selected English Texts*. London: Oxford University Press.

Ingram, A., S. Sim, C. Lawlor, R. Terry, J. Baker, and L. Wetherall-Dickson. 2011. *Melancholy Experience in Literature of the Long Eighteenth Century: Before Depression, 1660–1800*. New York and Basingstoke: Palgrave Macmillan.

Jenner, M. 1995. 'The Politics of London Air: John Evelyn's *Fumifugium* and the Restoration', *The Historical Journal* 38(3): 535–551.

Lagerlund, H. 2008. 'Pietro d'Abano and the Anatomy of Perception'. In *Theories of Perception in Medieval and Early Modern Philosophy*, edited by S. Knuuttila and P. Kärkkäinen, 117–130. Dordrecht: Springer.

Leppert, R. 1989. 'Music, Representation, and Social Order in Early-Modern Europe', *Cultural Critique* 12: 25–55.

MacKinnon, D. 2001. '"Poor Senseless Bess, Clothed in her Rags and Folly": Early Modern Women, Madness, and Song in Seventeenth-Century England', *Parergon* 18(3): 119–151.

McRae, A. 2000. 'The Literary Culture of Early Stuart Libeling', *Modern Philology* 97(3): 364–392.

McShane, A. 2011. 'Ballads and Broadsides from the Beginning of Print to 1660'. In *The Oxford History of Popular Print Culture*. Vol. 1, edited by J. Raymond, 339–362. Oxford: Oxford University Press.

Mendieta, E. 2014. 'Biopower'. In *The Cambridge Foucault Lexicon*, edited by L. Lawlor, and J. Nale, 44–51. Cambridge: Cambridge University Press.

Nancy, J. L. 2007. *Listening*, translated by C. Mandell. New York: Fordham University Press.

Nebeker, E. 2011. 'The Broadside Ballad and Textual Publics', *SEL Studies in English Literature, 1500–1900* 51(1): 1–19.

Neely, C. 1991. '"Documents in Madness": Reading Madness and Gender in Shakespeare's Tragedies and Early Modern Culture', *Shakespeare Quarterly* 42(3): 315–338.

Neyrat. F. 2010. 'The Birth of Immunopolitics', translated by A. De Boever, *Parrhesia* 10: 31–38.

Peery, W. 1948a. 'The Roaring Boy Again (Part I)', *The Shakespeare Association Bulletin* 23(1): 12–16.

_____. 1948b. 'The Roaring Boy Again (Part II)', *The Shakespeare Association Bulletin* 23(2): 78–86.

Phillips, J. 1656. *Wit and Drollery, Joviall Poems. Never Before Printed.* London: Printed for Nath. Brook, at the Angel in Cornhil.

Poulton, D. 1981. 'The Black-Letter Broadside Ballad and Its Music', *Early Music* 9(4): 427–437.

Rosand, R. 1979. 'The Descending Tetrachord: An Emblem of Lament,' *The Musical Quarterly* 65(3): 346–359.

Scull, A. 1989. *Social Order/Mental Disorder: Anglo-American Psychiatry in Historical Perspective.* Berkeley and Los Angeles: University of California Press.

_____. 1993. *The Most Solitary Afflictions: Madness and Society in Britain, 1700–1900.* New Haven and London: Yale University Press.

Shakespeare, W. 2005. *The Tragedy of King Lear,* edited by J. Halio. Cambridge: Cambridge University Press.

Shakespeare, W. and J. Fletcher. 1895. *The Two Noble Kinsmen,* edited by W. J. Rolfe. New York: Harper & Brothers Publishers.

Shaw, P. 1996. 'Mad Moll and Merry Meg: The Roaring Girl as Popular Heroine in Elizabethan and Jacobean Writings', *Sederi: Yearbook of the Spanish and Portuguese Society for English Renaissance Studies* 7: 129–139.

Siisiäinen, L. 2008. 'From the Empire of the Gaze to Noisy Bodies: Foucault, Audition and Medical Power', *Theory & Event* 11(1). Available online at https://muse.jhu.edu/article/233871.

Simpson, C. F. 1966. *The British Broadside Ballad and Its Music.* New Brunswick: Rutgers University Press.

Siraisi, N. G. 1975. 'The Music of Pulse in the Writings of Italian Academic Physicians (Fourteenth and Fifteenth Centuries)', *Speculum* 50(4): 689–710.

Smith, B. R. 1999. *The Acoustic World of Early Modern England: Attending to the O-Factor.* Chicago and London: The University of Chicago Press.

Stevenson, C. 1996. 'Robert Hooke's Bethlem', *Journal of the Society of Architectural Historians* 55(3): 254–275.

Timberlake, C. 1996. 'Practica Musicae: Mad Songs and Englishmen', *Journal of Singing* 52(3): 53–58.

Thompson, E. P. 1992. 'Rough Music Reconsidered', *Folklore* 103(1): 3–26.

Travis, T. 2014. '"Belching it Forth their Sooty Jaws": John Evelyn's Vision of a "Volcanic" City', *The London Journal: A Review of Metropolitan Society Past and Present* 39(1): 1–20.

Waage, F. 1977. 'Social Themes in Urban Broadsides of Renaissance England', *The Journal of Popular Music* 11(3): 731–742.

Ward, N. [1699] 1924. *London Spy Compleat, in Eighteen Parts*. London: The Casanova Society.

Wiltenburg, J. 1988. 'Madness and Society in the Street Ballads of Early Modern England', *Journal of Popular Culture* 21(4): 101–127.

Wortley, F. 1648. 'Mad Tom a Bedlams Desires of Peace: Or His Benedicities for Diestracted Englands Restauration to Her Wits Again'. National Library of Scotland Crawford.EB.1003, EBBA ID 33617.

Epilogue: Towards More Geographic Musicologies

James Williams
Department of Therapeutic Arts, University of Derby,
Britannia Mill, Derby, DE22 3BL, UK
j.williams@derby.ac.uk

As it is with so many publishing processes, deciding on the title of this book took almost as long as the collating of the chapters. Sam and I sent numerous emails back and forth to one another before reaching an agreed title — Sam called this *scholarly sparring*. Part of the issue of titling this volume was related to the discipline of geography, and the *use* of the word 'geography' (or '*geographic*') itself — mainly because whilst so many of the chapters in this volume pertain to a sense (or senses) of geography, the research does not always rest on concepts from scholarship in that discipline or explore music explicitly through its theoretical lenses. All of the chapters do, however, engage closely with relevant principles; on the most basic level, they find significance in location, in the places and spaces of music-making. After suggestions such as *Geographic Musicologies*, *Geographic Ethnomusicologies*, and *Musical Geographies*, we ultimately decided to drop the reference to 'geography' from the title altogether, after which we also decided to drop any reference to musicology or ethnomusicology — after all, not all contributing authors to this volume would identify themselves as musicologists or ethnomusicologists or similarly refer to their work as musicology or ethnomusicology. The much clearer and more accurate title *Musical Spaces: Place, Performance, and Power* was agreed.

But in and amongst the plethora of (in many cases, shared) angles from which the contributing authors look at their subjects — including power, environment, space, learning, (trans)locality, and creativity — there sits a diverse range of themes also important in the discipline of geography and in other closely related humanities and social sciences. This book makes reference to ideas of ancestry, boundaries, cartography, citizenship, climate, community, ecology, environment, folklore, global- and glocalization, indigenization, landscape, migration, nature, race, religion, regionalism, sovereignty, terrain and territory, and universality, to name a few. But, significantly, whilst the chapters draw reference to and explore many of these key issues, we, as musicologists, ethnomusicologists, music educationalists, musicians, music sociologists, music psychologists, and so on, are only really starting to unpick these ideas gradually in our scholarship. This book calls for an increased focus on combining the geographical with the musical, and whilst George Revill's 'Geographies of Music, Sound, and Auditory Culture' (2017) picks out some key texts already in circulation, especially on the study of *music in geography*, there is much more to learn about many of the themes identified above and throughout this book through the study of *geography in music*.

As music researchers, grasping and grappling with concepts of geography may seem a giant leap from what we typically know and do. But an important interdisciplinary territory lies at the point where our interests, theories, and research methods overlap — a territory requiring growth. An alliance between all those interested in geography and music, much as the delegates of the original 'Geography, Music, Space' conference at Durham University witnessed, promotes dialogue on many of the themes and issues raised by the authors in this book, and this is something to harness and grow to help us better understand a vast range of current issues. Thank you once again to so many of you who were involved in this project and pulled this volume together. As we move forward, I hope that much of this interdisciplinarity will continue to grow.

Reference

Revill, G. 2017. 'Geographies of Music, Sound, and Auditory Culture', *Oxford Bibliographies*. Available online at https://doi.org/10.1093/OBO/9780199874002-0130

Index